EAA Series

EAA series is successor of the EAA Lecture Notes and supported by the European Actuarial Academy (EAA GmbH), founded on the 29 August, 2005 in Cologne (Germany) by the Actuarial Associations of Austria, Germany, the Netherlands and Switzerland. EAA offers actuarial education including examination, permanent education for certified actuaries and consulting on actuarial education.

actuarial-academy.com

For further titles published in this series, please go to
http://www.springer.com/series/7879

Łukasz Delong

Backward Stochastic Differential Equations with Jumps and Their Actuarial and Financial Applications

BSDEs with Jumps

Łukasz Delong
Institute of Econometrics, Division
of Probabilistic Methods
Warsaw School of Economics
Warsaw, Poland

ISSN 1869-6929 ISSN 1869-6937 (electronic)
EAA Series
ISBN 978-1-4471-5330-6 ISBN 978-1-4471-5331-3 (eBook)
DOI 10.1007/978-1-4471-5331-3
Springer London Heidelberg New York Dordrecht

Library of Congress Control Number: 2013942002

Mathematics Subject Classification: 60G51, 60G57, 60H05, 60H30

Printed on acid-free paper

Springer is part of Springer Science+Business Media (www.springer.com)

The farther backward you can look, the farther forward you are likely to see.
Winston Churchill

Preface

A linear backward stochastic differential equation was introduced by Bismut (1973) in an attempt to solve an optimal stochastic control problem by the maximum principle. The general theory of nonlinear backward stochastic differential equations with Lipschitz generators was first presented by Pardoux and Peng (1990). Since then, BSDEs have been thoroughly studied and found numerous applications. Backward stochastic differential equations can be used to solve stochastic optimal control problems, establish probabilistic representations of solutions to partial differential equations and define nonlinear expectations. Since many financial problems can be related to stochastic optimization problems and nonlinear expectations, it is not surprising that BSDEs have become a very important tool in financial mathematics. Nowadays, backward stochastic differential equations are an active field of research which is stimulated by new financial and actuarial applications.

The first motivation for this book is to provide a self-contained overview of the theory of backward stochastic differential equations with jumps and their applications to insurance and finance. Two classical books on BSDEs: "*Backward Stochastic Differential Equations*" by El Karoui and Mazliak (1997) and "*Forward-Backward Stochastic Differential Equations and Their Applications*" by Ma and Yong (2000) target theory-oriented readers and miss some important applications which were developed in financial mathematics in recent years. Possible insurance applications are not mentioned at all in these books. The recent monograph "*Some Advances on Quadratic BSDE: Theory–Numerics–Applications*" by Dos Reis (2011) points out an actuarial and financial application but the author focuses on advanced theory of quadratic BSDEs, which definitely is not the first step in the study of BSDEs. All three books deal with BSDEs driven by Brownian motions and omit BSDEs with jumps which are very important for actuarial and financial modelling. There exists a considerable volume of mathematical papers on BSDEs and BSDEs with jumps. However, these papers are quite difficult to access by a beginner in the field of BSDEs and stochastic processes. Our goal is to present a book on BSDEs with jumps which covers key theoretical results and focuses on applications and which can be followed by nonspecialists in stochastic methods.

The second motivation for this book is to promote backward stochastic differential equations in the actuarial community. BSDEs seem not to be well-known in insurance mathematics, despite their recognized advantages in financial mathematics and optimal control theory. This state should be changed as many actuarial problems are closely related to financial problems, hence they can be approached with BSDEs. Since optimization problems are gaining importance in actuarial mathematics, efficient and modern solution methods for stochastic control problems should be presented. While the monograph "*Stochastic Control in Insurance*" by Schmidli (2007) deals with Hamilton-Jacobi-Bellman equations, our goal is to show how to apply BSDEs to solve optimization problems.

Jump processes play a leading role in actuarial modelling. Following Mikosch (2009), we can say that modelling of claim numbers by point processes is bread and butter for the actuary. Jump processes are also used in financial mathematics. Let us remark that Lévy processes have been introduced with success to financial models, see the monograph by Cont and Tankov (2004) and the textbook by Øksendal and Sulem (2004) where HJB equations are applied to solve financial optimization problems for Lévy-driven processes. Due to the importance of jump processes in actuarial and financial applications, we investigate BSDEs driven by a Brownian motion and a compensated random measure (called BSDEs with jumps). Since BSDEs can be used in a general stochastic framework, we consider general (quasi-left continuous) jump processes. Consequently, we also extend the actuary's toolbox for stochastic modelling.

We hope that this book will make BSDEs more accessible to those who are interested in applying these equations to actuarial and financial problems. Our book should be beneficial to students and researchers in applied probability and practitioners. Students and researchers in applied probability should get a strong mathematical introduction to the theory and applications of BSDEs. Practitioners should learn how to derive asset-liability strategies in sophisticated internal models (advocated by Solvency II Directive), set up hedging strategies and price complex insurance products with financial guarantees. This book may also be useful in actuarial education since it covers applied stochastic calculus and stochastic optimal control theory, which are included in the educational syllabuses of the Groupe Consultatif and the International Actuarial Association.

Warsaw, Poland
April 2013

Łukasz Delong

Contents

Chapter 1
Introduction

Abstract We discuss advantages of solving optimal control problems and defining nonlinear expectations by backward stochastic differential equations. We comment on applications of backward stochastic differential equations to pricing and hedging of liabilities and modelling of dynamic risk measures.

A backward stochastic differential equation (BSDE) with jumps is an equation of the form

$$\begin{aligned} Y(t) = \xi + \int_t^T f\big(s, Y(s), Z(s), U(s,.)\big)ds \\ - \int_t^T Z(s)dW(s) - \int_t^T \int_{\mathbb{R}} U(s,z)\tilde{N}(ds,dz), \quad 0 \le t \le T, \end{aligned} \tag{1.1}$$

where W is a Brownian motion and $\tilde{N}$ is a compensated random measure. Given a terminal condition ξ and a generator f, we are interested in finding a triple (Y, Z, U) which solves (1.1). More precisely, we aim to find an adapted process $Y := (Y(t), 0 \le t \le T)$ which is modelled by the dynamics

$$dY(t) = -f\big(t, Y(t), Z(t), U(t,.)\big)dt + Z(t)dW(t) + \int_{\mathbb{R}} U(t,z)\tilde{N}(dt,dz), \tag{1.2}$$

and satisfies $Y(T) = \xi$ where ξ is an $\mathscr{F}_T$-measurable random variable. At first sight it seems to be a hopeless task to construct such a process. However, the dynamics (1.2) is driven by two predictable processes $Z := (Z(t), 0 \le t \le T)$ and $U := (U(t,z), 0 \le t \le T, z \in \mathbb{R})$ which are allowed to be chosen as the part of the solution to the BSDE (1.1). The processes Z and U are called control processes. They control the process Y so that Y satisfies the terminal condition.

It should be pointed out that we would not be able to find an adapted solution to an equation with a random terminal condition if we did not introduce control processes. Let us consider the equation

$$dY(t) = 0, \quad Y(T) = \xi, \tag{1.3}$$

Ł. Delong, *Backward Stochastic Differential Equations with Jumps and Their Actuarial and Financial Applications*, EAA Series, DOI 10.1007/978-1-4471-5331-3_1,

where ξ is an $\mathscr{F}_T$-measurable random variable. It is straightforward to conclude that $Y(t) = \xi$, $0 \leq t \leq T$, is the only solution to (1.3). Unfortunately, this solution is not adapted to the underlying filtration $\mathscr{F}$ and it would not be useful for applications. The situation changes if we modify (1.3). Recalling the property of predictable representation, we deduce that for a square integrable random variable ξ and the square integrable martingale $M(t) = \mathbb{E}[\xi|\mathscr{F}_t]$, $0 \leq t \leq T$, we can find predictable processes $(\mathscr{Z}, \mathscr{U})$ such that

$$M(t) = \mathbb{E}[\xi|\mathscr{F}_t] = \mathbb{E}[\xi] + \int_0^t \mathscr{Z}(s)dW(s) + \int_0^t \int_{\mathbb{R}} \mathscr{U}(s,z)\tilde{N}(ds,dz),$$
$$0 \leq t \leq T.$$

Instead of (1.3) let us deal with the equation

$$dY(t) = Z(t)dW(t) + \int_{\mathbb{R}} U(t,z)\tilde{N}(dt,dz), \quad Y(T) = \xi. \tag{1.4}$$

We can now conclude that there exists an adapted solution to (1.4) which is of the form $Y(t) = M(t) = \mathbb{E}[\xi|\mathscr{F}_t]$, $0 \leq t \leq T$. The processes (Z, U), which are needed for the complete characterization of the dynamics (1.4), coincide with the processes $(\mathscr{Z}, \mathscr{U})$ which are derived from the predictable representation of ξ. The property of predictable representation plays the crucial role in the theory of BSDEs since it allows us to find an adapted solution to an equation with a random terminal condition.

Equation (1.4) is called a backward stochastic differential equation with zero generator. A backward stochastic differential equation with zero generator is the simplest example of a BSDE. The second key example of a BSDE is a linear backward stochastic differential equation which has the dynamics

$$dY(t) = \left(\alpha(t)Y(t) + \beta(t)Z(t) + \int_{\mathbb{R}} \gamma(t,z)U(t,z)Q(t,dz)\eta(t)\right)dt$$
$$+ Z(t)dW(t) + \int_{\mathbb{R}} U(t,z)\tilde{N}(dt,dz),$$
$$Y(T) = \xi,$$

where $\vartheta(dt,dz) = Q(t,dz)\eta(t)dt$ is the compensator of the random measure N. Clearly, a BSDE with zero generator is the special case of a linear BSDE. Linear BSDEs arise in many financial and actuarial applications. In this book we investigate both linear and nonlinear BSDEs.

Backward stochastic differential equations have become a central method of stochastic control theory. BSDEs have proved to be a useful and powerful alternative to Hamilton-Jacobi-Bellman (HJB) equations. Let us recall that an HJB equation is a partial differential equation which characterizes the optimal value function and the optimal control strategy of an optimization problem. We now point out advantages of characterizing optimal value functions and optimal control strategies by BSDEs.

Firstly, BSDEs are applicable in the case of non-Markovian dynamics. Let us recall that the Bellman programming principle, which leads to HJB equations, can only be applied to Markovian state processes. Markovian BSDEs may arise as a special type of BSDEs and they are called forward-backward stochastic differential equations.

The second advantage of BSDEs concerns differentiability issues. We point out that a classical Hamilton-Jacobi-Bellman equation for an optimal value function can be derived provided that the value function is differentiable (sufficiently smooth) with respect to state variables. Consequently, the main effort lies in proving differentiability of the optimal value function. Let us remark that in sophisticated models, including models with jumps, differentiability may not hold. In order to weaken differentiability requirements in stochastic control problems, viscosity solutions can be used. The notion of a viscosity solution allows us to characterize the optimal value function as a viscosity solution of a HJB equation. However, in applications of stochastic control models we are interested in the optimal control strategy which is defined by derivatives of the optimal value function with respect to the state variables. In a viscosity setting we cannot use such strategies. Therefore, differentiability of the value function has to be proved and strong (and cumbersome) assumptions are introduced to succeed in the proof. At the same time, the existence of a solution (Y, Z, U) to the BSDE (1.1) is not determined by differentiability of ξ and f. We will see that a BSDE has a unique solution under square integrability assumptions. It is mathematically convenient and beneficial to characterize the optimal value function of a stochastic control problem as a solution to a BSDE. If the optimal value function is characterized by a BSDE, then the optimal control strategy is characterized by the control processes of the same BSDE. Hence, we can define optimal value functions and optimal strategies by BSDEs without imposing smoothness assumptions.

Thirdly, BSDEs can be applied to solve optimization problems in models with multiple state variables. Under the Bellman programming principle the optimal value function is characterized as a function of the state variables and the time variable. The reader who is familiar with HJB equations should recall that it is difficult to derive value functions with more than two state variables. Mixed partial derivatives complicate HJB equations and optimal strategies. Numerical schemes for HJB equations use finite difference methods which are not efficient in high dimensions. Consequently, independence of the state variables is often assumed in order to separate the variables in the optimal value function and reduce dimension of the HJB equation. In our applications we investigate BSDEs driven by a two-dimensional Brownian motion and a random measure on $\mathscr{B}(\mathbb{R})$. The extension to cover an n-dimensional Brownian motion and a random measure on $\mathscr{B}(\mathbb{R}^m)$ is straightforward in most cases. We would have to add more control processes into our equations. In some cases the generators would also change to reflect more control process. Fortunately, we do not have to assume independence of the state variables. The BSDE is always driven by orthogonal martingale terms. We point out that numerical schemes for BSDEs use Monte Carlo methods which are efficient in high dimensions. Moreover, under the schemes proposed for solving BSDEs we do not have to estimate all control processes, we only need to estimate these control processes which appear in the generator and are used in the optimal control strategy.

Optimization is often used in finance and insurance. Consequently, BSDEs are a useful tool in financial and insurance mathematics. In this book we focus on optimization problems which are related to perfect replication, partial hedging and asset portfolio selection. In some of these applications, BSDEs arise very naturally. Let us start by studying the classical problem of replicating a claim with a traded asset. In this case the goal is to find an investment strategy under which the investment portfolio process matches the terminal liability. Notice that the logic behind solving BSDEs (1.1) is exactly the same: we aim to find a control process under which the solution matches the terminal condition.

Example 1.1 We consider the Black-Scholes financial model with a bank account S_0 and a stock S. The prices of the assets are modelled by the equations

$$\begin{aligned}\frac{dS_0(t)}{S_0(t)} &= r\big(S(t)\big)dt,\\ \frac{dS(t)}{S(t)} &= \mu\big(S(t)\big)dt + \sigma\big(S(t)\big)dW(t).\end{aligned} \tag{1.5}$$

If r, μ, σ are constant, then we deal with the classical Black-Scholes model. We are interested in finding a replicating strategy and a replicating portfolio for a claim $F(S(T))$ contingent on the stock. The classical results from mathematical finance yield that the price of the claim can be characterized as a unique solution to the partial differential equation

$$\begin{aligned}&u_t(t,s) + r(s)su_s(t,s) + \frac{1}{2}\sigma^2(s)s^2u_{ss}(t,s) - r(s)u(t,s) = 0,\\ &\quad (t,s) \in [0,T) \times (0,\infty),\\ &u(T,s) = F(s), \quad s \in (0,\infty),\end{aligned}$$

and the replicating strategy for the claim is determined by the derivative $u_s(t,s)$. The solution to the perfect replication problem can be characterized with a partial differential equation only if we deal with Markovian asset price processes modelled by forward stochastic differential equations, the pay-off is contingent on the terminal value of the stock and the price of the claim is sufficiently smooth in all state variables. In general financial models these assumptions may not hold. First of all, we may deal with non-Markovian price processes and path-dependent pay-offs. In some cases we can introduce another state variable modelled by an auxiliary forward stochastic differential equation and we can recover the Markovian structure. However, this is not doable for all path-dependent pay-offs, and lookback options are the key example when this technique fails. Secondly, existence of a smooth solution to a partial differential equation is not guaranteed. It is a delicate matter to impose conditions on the coefficients and the pay-off which lead to a sufficiently smooth price of the claim. Those conditions usually exclude many interesting and practically relevant cases from considerations, such as binary options.

The difficulties, which we have mentioned, arise since the solution is derived by exploiting the Markovian structure of the problem and applying the Itô's formula. However, we may follow a different approach. Let us consider a general non-Markovian financial model with the price processes

$$\begin{aligned}\frac{dS_0(t)}{S_0(t)} &= r(t)dt,\\ \frac{dS(t)}{S(t)} &= \mu(t)dt + \sigma(t)dW(t),\end{aligned} \tag{1.6}$$

where r, μ, σ are predictable processes. We are interested in finding a replicating strategy and a replicating portfolio for a path-dependent claim $F = F(S(t), 0 \leq t \leq T)$. It is easy to notice that finding a solution to our replication problem is equivalent to finding a solution (X, π) to the equation

$$\begin{aligned}&dX(t) = \pi(t)\big(\mu(t)dt + \sigma(t)dW(t)\big) + \big(X(t) - \pi(t)\big)r(t)dt,\\ &X(T) = F,\end{aligned} \tag{1.7}$$

which describes the dynamics of the replicating portfolio for F. We can observe that (1.7) is a linear BSDE. Hence, finding a solution to the replication problem is equivalent to finding a solution to a linear BSDE. Surprisingly, the derivation of the replicating strategy and the replicating portfolio by solving a BSDE turns out to be very intuitive. We will show that the linear BSDE (1.7) has a unique solution under mild assumptions. The replicating strategy is now derived from the predictable representation of the claim and in order to use the predictable representation property we do not need to impose any smoothness (continuity, differentiability) assumptions on the coefficients and the pay-off. Of course, the characterization of the replicating strategy in a non-Markovian model as a unique solution to a linear BSDE is less explicit than in a Markovian model where we characterize the strategy as a unique solution to a partial differential equation. Fortunately, the Malliavin calculus for BSDEs allows us to derive more explicit results even in general models. Perfect replication of claims is discussed in Sects. 9.2 and 9.4.

Since perfect replication of liabilities with traded assets is not always possible, investors are also interested in finding investment portfolios which hedge claims with a minimal replication error. More generally, the investor's goal could be to optimize the investment performance of the assets while limiting the risk of not covering the liabilities. Such investment strategies can be derived by solving stochastic optimal control problems. We would like to point out that optimization problems have become an important part of asset-liability management and solvency capital modelling. Nowadays, actuaries and risk managers model assets and liabilities and try to identify actions which lead to optimal results.

Example 1.2 We consider the Black-Scholes model (1.5) and an insurer who issues a unit-linked endowment policy with a claim $F(S(T))$. The claim is paid if a policyholder is alive at time T. We assume that the future life-time τ of the policyholder is exponentially distributed with parameter λ. We investigate the Markowitz portfolio selection problem for the insurer. Let us remark that many financial institutions based their asset-liability management programmes on the Markowitz risk-return objective. We are interested in finding an investment strategy π which minimizes the mean-square error

$$\mathbb{E}\big[\big|X^{\pi}(T)-F\big(S(T)\big)\mathbf{1}\{\tau>T\}-K_1\mathbf{1}\{\tau>T\}-K_2\mathbf{1}\{\tau\le T\}\big|^2\big],\tag{1.8}$$

where K_1, K_2 are profit targets set by the insurer, and the insurer's wealth process X^{π} is described by the dynamics

$$\begin{aligned}
&dX(t)=\pi(t)\big(\mu\big(S(t)\big)dt+\sigma\big(S(t)\big)dW(t)\big)+\big(X(t)-\pi(t)\big)r\big(S(t)\big)dt,\\
&X(0)=x.
\end{aligned}$$

Recalling classical results from the stochastic control theory, we have to solve the system of HJB equations

$$\begin{aligned}
&u_t(t,x,s)+xr(s)u_x(t,x,s)\\
&\quad+\inf_{\pi}\Big\{\big(\mu(s)-r(s)\big)\pi u_x(t,x,s)+\frac{1}{2}\sigma^2(s)\pi^2u_{xx}(t,x,s)+\sigma^2(s)\pi s u_{xs}(t,x,s)\Big\}\\
&\quad+\mu(s)su_s(t,x,s)+\frac{1}{2}\sigma^2(s)s^2u_{ss}(t,x,s)\\
&\quad+\big(v(t,x,s)-u(t,x,s)\big)\lambda=0,\quad (t,x,s)\in[0,T)\times\mathbb{R}\times(0,\infty),\\
&u(T,x,s)=\big(x-F(s)-K_1\big)^2,\quad (x,s)\in\mathbb{R}\times(0,\infty),
\end{aligned}\tag{1.9}$$

and

$$\begin{aligned}
&v_t(t,x,s)+xr(s)v_x(t,x,s)\\
&\quad+\inf_{\pi}\Big\{\big(\mu(s)-r(s)\big)\pi v_x(t,x,s)+\frac{1}{2}\sigma^2(s)\pi^2v_{xx}(t,x,s)+\sigma^2(s)\pi s v_{xs}(t,x,s)\Big\}\\
&\quad+\mu(s)sv_s(t,x,s)+\frac{1}{2}\sigma^2(s)s^2v_{ss}(t,x,s)=0,\\
&\qquad(t,x,s)\in[0,T)\times\mathbb{R}\times(0,\infty),\\
&v(T,x,s)=(x-K_2)^2,\quad (x,s)\in\mathbb{R}\times(0,\infty).
\end{aligned}\tag{1.10}$$

If there exist smooth solutions to the HJB equations, then the optimal investment strategy is defined with first and second derivatives of the optimal value function u or v. However, the existence of differentiable solutions to the HJB equations (1.9)–(1.10) is not guaranteed and, as already commented, (unnecessary) restrictions on the coefficients and the pay-off have to be introduced to conclude that the HJB equations have smooth solutions. In real applications we would assume that the interest rate and the volatility are not perfectly correlated with the traded stock and that the claim is contingent on a financial asset that is only partially correlated with the traded stock. These assumptions would further complicate the HJB equations and make the proof of the regularity of the optimal value function harder. Even if we succeed in establishing the existence of smooth solutions to the HJB equations under restrictive conditions, then solving the partial differential equations with multiple state variables and mixed partial derivatives would be numerically difficult. We should also keep in mind that HJB equations cannot be applied if we deal with path-dependent pay-offs, which are often embedded in insurance contracts called variable annuities.

BSDEs can be helpful since they can efficiently handle the difficulties which we have discussed. In Sect. 10.2 we prove that in the non-Markovian financial model (1.6) the optimal investment strategy for the quadratic hedging problem (1.8) with a path-dependent claim $F = F(S(t), 0 \leq t \leq T)$ can be characterized with the unique solutions to the linear BSDEs

$$
\begin{aligned}
d\hat{Y}(t) &= -\left(\left|\frac{\mu(t)-r(t)}{\sigma(t)}\right|^2 - 2r(t)\right)\hat{Y}(t)dt + 2\frac{\mu(t)-r(t)}{\sigma(t)}\hat{Z}(t)dt + \hat{Z}(t)dW(t), \\
\hat{Y}(T) &= 1,
\end{aligned}
\tag{1.11}
$$

and

$$
\begin{aligned}
d\mathscr{Y}(t) &= r(t)\mathscr{Y}(t)dt + \frac{\mu(t)-r(t)}{\sigma(t)}\mathscr{Z}(t)dt + \mathscr{Z}(t)dW(t) + \mathscr{U}(t)\tilde{N}(dt), \\
\mathscr{Y}(T) &= F\mathbf{1}\{\tau > T\} + K_1\mathbf{1}\{\tau > T\} + K_2\mathbf{1}\{\tau \leq T\},
\end{aligned}
\tag{1.12}
$$

where $\tilde{N}$ is the compensated random measure generated by the point process $\mathbf{1}\{t \geq \tau\}$. We will show that the BSDEs (1.11)–(1.12) have solutions under mild assumptions and smoothness (continuity, differentiability) assumptions are not relevant. To some extent, we can introduce correlated risk factors into the model and the solution can still be efficiently derived by solving linear BSDEs and applying Monte Carlo methods. We would like to point out that in the case when we deal with many risk factors solving a BSDE with Monte Carlo methods is much more efficient than solving an HJB equation with finite difference methods, see Chap. 5. We remark that a quadratic hedging problem in a general combined financial and insurance model with correlated risk factors leads to a stochastic Riccati equation which is a nonlinear BSDE.

Apart from applications in the field of optimal control, backward stochastic differential equations are also used to define nonlinear expectations called g-expectations. A nonlinear expectation is an operator which preserves all essential properties of the standard expected value operator except linearity. The original motivation for studying nonlinear expectations comes from the theory of decision making. It was shown that decisions made in the real world contradicted optimal decisions based on additive probabilities and the expected utility theory. Consequently, economists and mathematicians begun to look for a new notion of expectation. The g-expectation, which is defined by a BSDE with a nonlinear generator g, is the fundamental example of a nonlinear expectation. The g-expectation has become an important concept in probability since it gave rise to g-martingales, g-supermartingales, g-submartingales and nonlinear versions of classical results such as a nonlinear Doob-Meyer decomposition.

In financial and insurance applications we use g-expectations to define dynamic risk measures. Static risk measures such as Value-at-Risk or Tail-Value-at-Risk over 5-day or 1-year horizon are well-understood. However, it is still challenging to model dynamic risk measures which quantify the riskiness of financial positions continuously during a specified period of time. It is clear that financial positions should be consistently valued over time until they are liquidated. Properties of BSDEs indicate that g-expectations can be useful for modelling dynamic risk measures.

Example 1.3 We consider an aggregate insurance claims process modelled by a step process J. We are interested in valuating the risk of a contract $F(J(T))$ contingent on the claim process. The insurer may face a so-called model ambiguity and he may not know the true claim intensity and the true claim distribution. This may be the case if the historical data are scarce, if the intensity and the severity fluctuate due to seasonal effect or if the intensity and the severity change in the case of a random event (contagion risk). It is reasonable to assume that we measure the risk of the liability F by considering all possible scenarios for the evolution of the claim intensity and the claim distribution and taking the maximum loss from the liability under all scenarios. We end up with the expectation

$$Y(t) = \sup_{\mathbb{Q}\in\mathscr{Q}} \mathbb{E}^{\mathbb{Q}}\big[F\big(J(T)\big)|\mathscr{F}_t\big], \quad 0 \le t \le T, \tag{1.13}$$

where $\mathscr{Q}$ denotes the set of all possible (and equivalent) scenarios for the claim intensity and the claim distribution of the claim process J. The risk measure (1.13) is called a generalized-scenario-based risk measure, see Chap. 2.2.1 in McNeil et al. (2005), and it is applied in practice. The set $\mathscr{Q}$ contains possible claim distributions and claim intensities of J and it consists of the claim distributions and the claim intensities which are of the form

$$q^{\mathbb{Q}}(t,dz) = \frac{1+\kappa(t,z)}{\int_{\mathbb{R}}(1+\kappa(t,z))q(dz)} q(dz), \quad 0 \le t \le T, \; z \in \mathbb{R},$$

$$\lambda^{\mathbb{Q}}(t) = \int_{\mathbb{R}} \big(1+\kappa(t,z)\big)q(dz)\lambda, \quad 0 \le t \le T,$$

where q is a probability distribution function, $\lambda > 0$ is a constant and κ is a process which determines the set $\mathscr{Q}$. Under the basic scenario, $\kappa = 0$, the aggregate claims process J is the compound Poisson process with the jump size distribution q and the intensity λ, and under any other scenario, $\kappa \neq 0$, the aggregate claims process J

is a step process with the jump size distribution $q^{\mathbb{Q}}$ and the intensity $\lambda^{\mathbb{Q}}$. The probability measure $q^{\mathbb{Q}}$ is the Esscher transform of the probability measure q, and $\lambda^{\mathbb{Q}}$ is defined by scaling the intensity λ in the way which guarantees that the set $\mathscr{Q}$ contains equivalent scenarios for J. The set of equivalent scenarios is obtained by applying Girsanov's theorem, see Sect. 2.5. Let us assume that the distortion process κ is bounded, i.e. $|\kappa(t,z)| \leq \delta < 1$. In Sect. 3.4 we show that Y solves the nonlinear BSDE

$$\begin{aligned} dY(t) &= -\delta \int_{\mathbb{R}} \big|U(t,z)\big|\lambda q(dz)dt + \int_{\mathbb{R}} U(t,z)\tilde{N}(dt,dz), \\ Y(T) &= F\big(J(T)\big), \end{aligned}$$

where $\tilde{N}$ is the compensated random measure generated by the compound Poisson process with the intensity λ and the jump size distribution q. The theory of BSDEs yields that the operator (1.13) defines a time-consistent dynamic risk measure. Dynamic risk measures under model ambiguity play an important role in the theory of dynamic risk measures and we use them, and their BSDE representations, in Chap. 12 to price and hedge insurance contracts.

This book is divided into three parts. In Part I we present the theory of BSDEs with Lipschitz generators driven by a Brownian motion and a compensated random measure. We put an emphasis on random measures generated by step processes and Lévy processes. We present key results and techniques (including numerical algorithms) for BSDEs with jumps. We also study filtration-consistent nonlinear expectations and g-expectations. We remark that BSDEs with jumps are still at the stage of development and we are not able to present the most general statements of mathematical results in all cases. We sometimes resign from presenting the most general result since it would becloud the main idea. In Part I we focus on mathematical tools and proofs which are crucial for understanding the theory and are useful for applications. We try to explain advanced mathematics behind BSDEs in detail. In Part II we investigate actuarial and financial applications of BSDEs with jumps. We consider a general financial and insurance model and we deal with pricing and hedging of insurance equity-linked claims and asset-liability management problems. Different pricing and hedging objectives are studied. We investigate perfect hedging, superhedging, quadratic optimization, utility maximization, indifference pricing, ambiguity risk minimization and no-good-deal pricing. We also consider dynamic risk measures. In Part III we present some other useful classes of BSDEs and we comment on their applications. Biographical notes do not represent a complete survey on BSDEs and they only refer the reader to works which are closely related to the topics considered.

Part I
Backward Stochastic Differential Equations—The Theory

Chapter 2
Stochastic Calculus

Abstract We review important results of stochastic calculus. We introduce a Brownian motion, a random measure and a compensated random measure. Examples of Lévy processes, step processes and their jump measures are given. We investigate stochastic integrals with respect to Brownian motion and compensated random measures and we recall their properties. We discuss the weak property of predictable representation for local martingales. Equivalent probability measures are defined, and Girsanov's theorem for Brownian motion and random measures is stated. We give differentiation rules of the Malliavin calculus.

We review important results of stochastic calculus which we use in this book. This chapter is written in the spirit of a rèsume and we collect facts needed to investigate BSDEs driven by Brownian motions and compensated random measures.

2.1 Brownian Motion and Random Measures

Let us consider a probability space $(\Omega, \mathscr{F}, \mathbb{P})$ with a filtration $\mathscr{F} = (\mathscr{F}_t)_{0\le t\le T}$ and a finite time horizon $T < \infty$. We assume that the filtration $\mathscr{F}$ satisfies the usual hypotheses of completeness ($\mathscr{F}_0$ contains all sets of $\mathbb{P}$-measure zero) and right continuity ($\mathscr{F}_t = \mathscr{F}_{t+}$).

A stochastic process $V(\omega, t)$ is a real function defined on $\Omega \times [0, T]$ such that $\omega \mapsto V(\omega, t)$ is $\mathscr{F}$-measurable for any $t \in [0, T]$. A stochastic process V is called $\mathscr{F}$-adapted if $\omega \mapsto V(\omega, t)$ is $\mathscr{F}_t$-measurable for any $t \in [0, T]$. The natural filtration generated by a process V is denoted by $\mathscr{F}^V$. We always assume that the natural filtration is completed with sets of measure zero. By $\mathscr{B}(A)$ we denote the Borel sets of $A \subset \mathbb{R}$, by $\mathscr{P}$ we denote the σ-field on $\Omega \times [0, T]$ generated by all left-continuous and adapted processes. The field $\mathscr{P}$ is called the predictable σ-field. A process $V : \Omega \times [0, T] \to \mathbb{R}$, or $V : \Omega \times [0, T] \times \mathscr{E} \to \mathbb{R}$, is called $\mathscr{F}$-predictable if it is $\mathscr{F}$-adapted and $\mathscr{P}$-measurable, or $\mathscr{P} \otimes \mathscr{B}(\mathscr{E})$-measurable. Clearly, the limit of a converging sequence of predictable processes is a predictable process. If there is no confusion, the reference to the filtration $\mathscr{F}$ is omitted. A process is called càdlàg if its trajectories are right-continuous and have left limits. By K we denote

Ł. Delong, *Backward Stochastic Differential Equations with Jumps and Their Actuarial and Financial Applications*, EAA Series, DOI 10.1007/978-1-4471-5331-3_2,

constants, which are allowed to vary from line to line. The term *a.s.* means *almost surely* with respect to the probability measure, and, unless specified, the term *a.e.* means *almost everywhere* with respect to the Lebesgue measure. All statements for random variables and stochastic processes should be understood a.s.

We introduce a Brownian motion and a random measure. Brownian motion and random measures are used to develop financial and actuarial stochastic models.

Definition 2.1.1 An $\mathscr{F}$-adapted process $W := (W(t), 0 \leq t \leq T)$ with $W(0) = 0$ is called a Brownian motion if

(i) for $0 \leq s < t \leq T$, $W(t) - W(s)$ is independent of $\mathscr{F}_s$,
(ii) for $0 \leq s < t \leq T$, $W(t) - W(s)$ is a Gaussian random variable with mean zero and variance $t - s$.

There exists a modification of a Brownian motion which has continuous paths.

Definition 2.1.2 A function N defined on $\Omega \times [0, T] \times \mathbb{R}$ is called a random measure if

(i) for any $\omega \in \Omega$, $N(\omega, .)$ is a σ-finite measure on $\mathscr{B}([0, T]) \otimes \mathscr{B}(\mathbb{R})$,
(ii) for any $A \in \mathscr{B}([0, T]) \otimes \mathscr{B}(\mathbb{R})$, $N(., A)$ is a random variable on $(\Omega, \mathscr{F}, \mathbb{P})$.

We remark that $N(\omega, [0, t], A)$ may be equal to infinity (see Example 2.3 and the case of Lévy processes).

Example 2.1 Let $(T_n)_{n\geq 1}$ denote the sequence of jump times of a Poisson process. The function

$$N\big(\omega, [s, t]\big) = \sharp\big\{n \geq 1, T_n \in [s, t]\big\}, \quad 0 \leq s < t \leq T,$$

which counts the number of jumps of the Poisson process in the time interval $[s, t]$, defines a random measure. If we fix ω, then the sequence of jump times $(T_n)_{n\geq 1}$ of the Poisson process is given on the time axis, and N as a function of $[s, t]$ is a finite measure which counts the number of $(T_n)_{n\geq 1}$ which are in the interval $[s, t]$. If we fix $[s, t]$, then N is a Poisson distributed random variable which counts the number of random jump times $(T_n)_{n\geq 1}$ of the Poisson process which are in the interval $[s, t]$.

Next, we introduce a predictable compensator of a random measure.

Definition 2.1.3 A random measure N is called $\mathscr{F}$-predictable if for any $\mathscr{F}$-predictable process V such that the integral $\int_0^T \int_{\mathbb{R}} |V(s, z)| N(ds, dz)$ exists, the process $(\int_0^t \int_{\mathbb{R}} V(s, z) N(ds, dz), 0 \leq t \leq T)$ is $\mathscr{F}$-predictable.

Definition 2.1.4 For a random measure N we define

$$E_N(A) = \mathbb{E}\left[\int_{[0,T]\times\mathbb{R}} \mathbf{1}_A(\omega, t, z) N(\omega, dt, dz)\right], \quad A \in \mathscr{F} \otimes \mathscr{B}\big([0, T]\big) \otimes \mathscr{B}(\mathbb{R}).$$

If there exists an $\mathscr{F}$-predictable random measure ϑ such that

(i) E_ϑ is a σ-finite measure on $\mathscr{P} \otimes \mathscr{B}(\mathbb{R})$,
(ii) the measures E_N and E_ϑ are identical on $\mathscr{P} \otimes \mathscr{B}(\mathbb{R})$,

then we say that the random measure N has a compensator ϑ.

We remark that the compensator is uniquely determined, see Theorem 11.6 in He et al. (1992).

Given the compensator ϑ of a random measure N, we can define the compensated random measure

$$\tilde{N}(\omega, dt, dz) = N(\omega, dt, dz) - \vartheta(\omega, dt, dz).$$

Random measures are usually related to jumps of discontinuous processes. We state the following result, see Theorem 11.15 He et al. (1992).

Proposition 2.1.1 *Let $J := (J(t), 0 \le t \le T)$ be an $\mathscr{F}$-adapted, càdlàg process, and set $D = \{\Delta J \neq 0\}$. Then*

$$N(dt, dz) = \sum_{s \in (0,T]} \mathbf{1}_{(s, \Delta J(s))}(dt, dz) \mathbf{1}_{\{\Delta J(s) \neq 0\}}(s) \mathbf{1}_D\{s\}, \tag{2.1}$$

is an integer-valued random measure which has a unique $\mathscr{F}$-predictable compensator.

The measure N defined in Proposition 2.1.1 is called the jump measure of the process J. The measure $N([0, T], A)$ counts the number of jumps of the process J of size specified in the set A in the time interval $[0, T]$.

Two important families of discontinuous processes should be pointed out. In financial and actuarial applications we usually deal with Lévy processes and step processes.

Definition 2.1.5 An $\mathscr{F}$-adapted process $J := (J(t), 0 \le t \le T)$ with $J(0) = 0$ is called a Lévy process if

(i) for $0 \le s < t \le T$, $J(t) - J(s)$ is independent of $\mathscr{F}_s$,
(ii) for $0 \le s < t \le T$, $J(t) - J(s)$ has the same distribution as $J(t - s)$,
(iii) the process J is continuous in probability, for any $t \in [0, T]$ and $\varepsilon > 0$ we have $\lim_{s \to t} \mathbb{P}(|L(t) - L(s)| > \varepsilon) = 0$.

There exists a modification of a Lévy process which has càdlàg paths.

Example 2.2 The Poisson process and the compound Poisson process are the prime examples of discontinuous Lévy processes. It is easy to conclude that the jump measure of a compound Poisson process with intensity λ and jump distribution q has the compensator $\vartheta(dt, dz) = \lambda q(dz)dt$. The jump measure of a compound Poisson process is a finite random measure.

Example 2.3 The family of Lévy processes contains Variance Gamma, Normal Inverse Gaussian and stable processes, see Chap. 4 in Cont and Tankov (2004). In general, the jump measure of a Lévy process has the compensator $\vartheta(dt, dz) = \nu(dz)dt$ where ν is a σ-finite measure (called a Lévy measure) satisfying $\int_{|z|<1} z^2\nu(dz) < \infty$, see Proposition 3.7 in Cont and Tankov (2004). The measure ν determines properties of the Lévy process (we can have a finite variation or an infinite variation process with an infinite number of small jumps in every finite time interval), see Chaps. 3, 4 in Cont and Tankov (2004). For all Lévy processes except the compound Poisson process, the jump measure of a Lévy process is a σ-finite random measure with $N([0, T], \mathbb{R}) = +\infty$.

If the random measure (2.1) is generated by a Lévy process, then it is called a Poisson random measure.

Definition 2.1.6 A process J is called a step process if its trajectories are càdlàg step functions having a finite number of jumps in every finite time interval. An $\mathscr{F}$-adapted step process J with $J(0) = 0$ has the representation

$$J(t) = \sum_{n=1}^{\infty} \xi_n \mathbf{1}\{T_n \leq t\}, \tag{2.2}$$

where

(i) $(T_n)_{n\geq 1}$ is a sequence of $\mathscr{F}$-stopping times such that $0 \leq T_1 \leq T_2 \leq \cdots \leq T_n \uparrow \infty$, $n \to \infty$,
(ii) $\xi_n \in \mathscr{F}_{T_n}$, $n \geq 1$,
(iii) for each $n \geq 1$, $T_n < \infty \Rightarrow T_n < T_{n+1}$,
(iv) for each $n \geq 1$, $\xi_n \neq 0 \Leftrightarrow T_n < \infty$.

In representation (2.2), T_n denotes the nth jump time of J and ξ_n denotes the jump size of J at time T_n. The sequence $(T_n)_{n\geq 1}$ defines a non-explosive point process. The jump measure of a step process is a finite random measure.

Example 2.4 The compound Poisson process is a step process.

Example 2.5 The compound Cox process is a second example of a step process. The compound Cox process J can be defined by $J(t) = j(\int_0^t \lambda(s)ds)$ where λ is a stochastic intensity process and j is an independent compound Poisson process with intensity 1 and jump size distribution q, see Theorem 12.2.3 in Rolski et al. (1999). We can deduce that the compensator of the corresponding jump measure is of the form $\vartheta(dt, dz) = \lambda(t)q(dz)$.

Example 2.6 Take a continuous process $\lambda : \Omega \times [0, T] \to (0, \infty)$ and define the hazard process $\Psi(t) = \int_0^t \lambda(s)ds$. We introduce a random time τ which has the conditional distribution

$$\mathbb{P}(\tau > t | \mathscr{F}_t^{\lambda}) = e^{-\Psi(t)} = e^{-\int_0^t \lambda(s)ds}, \quad 0 \leq t \leq T,$$

and we define the step process

$$J(t) = \mathbf{1}\{t \geq \tau\}, \quad 0 \leq t \leq T.$$

The compensated process $J(t) - \Psi(t \wedge \tau)$ is an $\mathscr{F}^{\lambda} \vee \mathscr{F}^{J}$-martingale, see Proposition 2.13 in Jeanblanc and Rutkowski (2000), and the jump measure of J has the compensator $\vartheta(dt, \{1\}) = (1 - J(t-))\lambda(t)dt$.

We remark that given the conditional distribution of (T_{n+1}, ξ_{n+1}) with respect to $\mathscr{F}_{T_n}$, it is possible to derive the compensator of the step process, see Theorem 11.49 in He et al. (1992).

We need some assumptions concerning the random measure and its compensator. We always assume that

(RM) the random measure N is an integer-valued random measure with the compensator

$$\vartheta(dt, dz) = Q(t, dz)\eta(t)dt, \tag{2.3}$$

where $\eta : \Omega \times [0, T] \to [0, \infty)$ is a predictable process, and Q is a kernel from $(\Omega \times [0, T], \mathscr{P})$ into $(\mathbb{R}, \mathscr{B}(\mathbb{R}))$ satisfying

$$\int_0^T \int_{\mathbb{R}} z^2 Q(t, dz)\eta(t)dt < \infty. \tag{2.4}$$

We also set $N(\{0\}, \mathbb{R}) = N((0, T], \{0\}) = \vartheta((0, T], \{0\}) = 0$.

This is our standing assumption and any random measure N considered in this book satisfies (RM). From the definition of a kernel we recall that for $(\omega, t) \in \Omega \times [0, T]$, $Q(t, .)$ is a measure on $\mathscr{B}(\mathbb{R})$, and for $A \in \mathscr{B}(\mathbb{R})$, $Q(., A)$ is a predictable process. Notice that the compensators considered in our examples satisfy assumption (2.3). In fact, the representation of the compensator (2.3) holds in most practical cases, we refer to Theorem II.1.8 in Jacod and Shiryaev (2003) for a general representation of the compensator of a random measure. In (2.3) we assume that the compensator is absolutely continuous with respect to the Lebesgue measure dt. The absolute continuity of the compensator with respect to the Lebesgue measure dt can be motivated by financial and actuarial applications in which we investigate jump measures of quasi-left continuous, càdlàg, adapted processes. Let us recall that a càdlàg, adapted process is called quasi-left continuous if a sequence of totally inaccessible stopping times exhausts its jump times, see Proposition I.2.26 and Corollary II.1.19 in Jacod and Shiryaev (2003). In other words, quasi-left continuous processes and absolutely continuous compensators of jump measures arise if we model jumps that arrive in an unpredictable way. Indeed, this is the right probabilistic framework for discontinuous processes used in insurance and finance. Assumption (2.4) implies that the quasi-left continuous process related to the jump measure is locally square integrable, see Theorem 11.31 in He et al. (1992) (in applications we deal with square integrable processes). The measure zero of the set $\{0\}$ indicates that N is indeed a jump measure, see Theorem 11.25 in He et al. (1992). In many actuarial applications, in which we deal with step processes, η can be interpreted as a claim intensity and Q as a claim distribution. If we consider a Lévy process, then we simply set $\eta(t) = 1$ and $Q(t, dz) = \nu(dz)$ where ν is a σ-finite Lévy measure.

2.2 Classes of Functions, Random Variables and Stochastic Processes

We start with defining spaces of functions, random variables and stochastic processes which we use in this book.

- Let $L^2_\nu(\mathbb{R})$ denote the space of measurable functions $\varphi : \mathbb{R} \to \mathbb{R}$ satisfying

$$\int_{\mathbb{R}} |\varphi(z)|^2 \nu(dz) < \infty,$$

  where ν is a σ-finite measure,
- Let $\mathscr{C}^{1,2}([0,T], \mathbb{R})$ denote the space of continuous functions $\varphi : [0,T] \times \mathbb{R} \to \mathbb{R}$ which have continuous partial derivatives $\frac{\partial}{\partial t}\varphi(t,x)$, $\frac{\partial}{\partial x}\varphi(t,x)$ and $\frac{\partial^2}{\partial x^2}\varphi(t,x)$. Partial derivatives are denoted by $\phi_t, \phi_x, \phi_{xx}$. If there is no confusion, first derivative is denoted by ϕ'.
- Let $\mathbb{L}^2(\mathbb{R})$ denote the space of random variables $\xi : \Omega \to \mathbb{R}$ satisfying

$$\mathbb{E}\big[|\xi|^2\big] < \infty.$$

- Let $\mathbb{H}^2(\mathbb{R})$ denote the space of predictable processes $Z : \Omega \times [0,T] \to \mathbb{R}$ satisfying

$$\mathbb{E}\left[\int_0^T |Z(t)|^2 dt\right] < \infty.$$

- Let $\mathbb{H}^2_N(\mathbb{R})$ denote the space of predictable processes $U : \Omega \times [0,T] \times \mathbb{R} \to \mathbb{R}$ satisfying

$$\mathbb{E}\left[\int_0^T \int_{\mathbb{R}} |U(t,z)|^2 Q(t,dz)\eta(t)dt\right] < \infty,$$

  where we integrate with respect to the predictable compensator of the random measure N.
- Let $\mathbb{S}^2(\mathbb{R})$ denote the space of adapted, càdlàg processes $Y : \Omega \times [0,T] \to \mathbb{R}$ satisfying

$$\mathbb{E}\left[\sup_{t\in[0,T]} |Y(t)|^2\right] < \infty.$$

- Let $\mathbb{S}^2_{inc}(\mathbb{R})$ denote the subspace of $\mathbb{S}^2(\mathbb{R})$ which contains processes with non-decreasing trajectories, and let $\mathbb{S}^\infty(\mathbb{R})$ denote the subspace of $\mathbb{S}^2(\mathbb{R})$ which contains bounded processes.

The spaces $\mathbb{H}^2(\mathbb{R})$, $\mathbb{H}^2_N(\mathbb{R})$ and $\mathbb{S}^2(\mathbb{R})$ are endowed with the norms:

$$\|Z\|^2_{\mathbb{H}^2} = \mathbb{E}\left[\int_0^T e^{\rho t}\left|Z(t)\right|^2 dt\right],$$

$$\|U\|^2_{\mathbb{H}^2_N} = \mathbb{E}\left[\int_0^T \int_{\mathbb{R}} e^{\rho t}\left|U(t,z)\right|^2 Q(t,dz)\eta(t)dt\right],$$

$$\|Y\|^2_{\mathbb{S}^2} = \mathbb{E}\left[\sup_{t\in[0,T]} e^{\rho t}\left|Y(t)\right|^2\right],$$

with some $\rho \geq 0$.

We also define classes of processes which are differentiable in the Malliavin sense. First, we present the idea behind the Malliavin derivative.

If we investigate Malliavin differentiability, then we deal with the completed filtration generated by a Lévy process. We work with the product of two canonical spaces $(\Omega_W \times \Omega_N, \mathscr{F}_W \otimes \mathscr{F}_N, \mathbb{P}_W \otimes \mathbb{P}_N)$ completed with sets of measure zero. The space $(\Omega_W, \mathscr{F}_W, \mathbb{P}_W)$ is the usual canonical space for a one-dimensional Brownian motion (the space of continuous functions on $[0,T]$ with the σ-algebra generated by the topology of uniform convergence and Wiener measure). The space $(\Omega_N, \mathscr{F}_N, \mathbb{P}_N)$ is a canonical space for a pure jump Lévy process, and for its proper definition we refer to Solé et al. (2007). In the product space $(\Omega_W \times \Omega_N, \mathscr{F}_W \otimes \mathscr{F}_N, \mathbb{P}_W \otimes \mathbb{P}_N)$ we can study a two-parameter Malliavin derivative.

We follow the exposition from Solé et al. (2007). Let ν be a Lévy measure such that $\int_{\mathbb{R}} |z|^2\nu(dz) < \infty$. Consider the finite measure υ

$$\upsilon(A) = \int_{A(0)} \sigma^2 dt + \int_{A'} z^2\nu(dz)dt, \quad A \in \mathscr{B}\big([0,T]\big) \otimes \mathscr{B}(\mathbb{R}),$$

where $A(0) = \{t \in [0,T]; (t,0) \in A\}$ and $A' = A \setminus A(0)$. We define the martingale-valued random measure $\varUpsilon$

$$\varUpsilon(A) = \int_{A(0)} \sigma dW(t) + \int_{A'} z\tilde{N}(dt,dz), \quad A \in \mathscr{B}\big([0,T]\big) \otimes \mathscr{B}(\mathbb{R}),$$

and its continuous and discontinuous parts

$$\varUpsilon^c(t) = \int_0^t \sigma dW(s), \quad \varUpsilon^d(t,A) = \int_0^t \int_A z\tilde{N}(ds,dz), \quad A \in \mathscr{B}(\mathbb{R}).$$

We introduce the multiple two-parameter integral with respect to $\varUpsilon$

$$I_n(\varphi_n) = \int_{([0,T]\times\mathbb{R})^n} \varphi\big((t_1,z_1),\dots(t_n,z_n)\big)\varUpsilon(dt_1,dz_1)\cdot\ldots\cdot\varUpsilon(dt_n,dz_n),$$

for functions $\varphi \in L^2_\upsilon(([0,T]\times\mathbb{R})^n)$ satisfying

$$\|\varphi_n\|^2_{L^2_\upsilon} = \int_{([0,T]\times\mathbb{R})^n} \left|\varphi_n\big((t_1,z_1),\dots,(t_n,z_n)\big)\right|^2 \upsilon(dt_1,dz_1)\cdot\ldots\cdot\upsilon(dt_n,dz_n) < \infty.$$

We finally recall the chaotic decomposition property which states that any square integrable random variable ξ measurable with respect to the completed natural filtration generated by a Lévy process has the unique representation

$$\xi = \sum_{n=0}^{\infty} I_n(\varphi_n), \tag{2.5}$$

where $\varphi_n \in L_\upsilon^2(([0,T]\times\mathbb{R})^n)$ are symmetric in the n pairs (t_i, z_i), $1 \le i \le n$. The Malliavin derivative uses the chaotic decomposition property (2.5).

We consider the following spaces:

- Let $\mathbb{D}^{1,2}(\mathbb{R})$ denote the space of random variables $\xi \in \mathbb{L}^2(\mathbb{R})$ which are measurable with respect to the natural filtration generated by a Lévy process and have the representation $\xi = \sum_{n=0}^{\infty} I_n(\varphi_n)$ such that

$$\sum_{n=1}^{\infty} n n! \|\varphi_n\|_{L_\upsilon^2}^2 < \infty.$$

  For a random variable $\xi \in \mathbb{D}^{1,2}(\mathbb{R})$ we define the Malliavin derivative $D\xi : \Omega \times [0,T]\times\mathbb{R} \to \mathbb{R}$ to be a stochastic process of the form

$$D_{t,z}\xi = \sum_{n=1}^{\infty} n I_{n-1}\big(\varphi_n\big((t,z),\cdot\big)\big). \tag{2.6}$$

- Let $\mathbb{L}^{1,2}(\mathbb{R})$ denote the space of adapted and product measurable processes $V : \Omega \times [0,T]\times\mathbb{R} \to \mathbb{R}$ satisfying

$$\mathbb{E}\bigg[\int_{[0,T]\times\mathbb{R}} \big|V(s,y)\big|^2 \upsilon(ds,dy)\bigg] < \infty,$$

$$V(s,y) \in \mathbb{D}^{1,2}(\mathbb{R}), \quad \upsilon\text{-a.e. } (s,y) \in [0,T]\times\mathbb{R},$$

$$\mathbb{E}\bigg[\int_{([0,T]\times\mathbb{R})^2} \big|D_{t,z}V(s,y)\big|^2 \upsilon(ds,dy)\upsilon(dt,dz)\bigg] < \infty.$$

Let us define a stopping time, (local) martingale, quadratic variation and *BMO* martingale.

Definition 2.2.1 A random variable $\tau : \Omega \to [0,T]$ is called an $\mathscr{F}$-stopping time if $\{\tau \le t\} \in \mathscr{F}_t$ for every $t \in [0,T]$.

Definition 2.2.2 An $\mathscr{F}$-adapted process $M := (M(t), 0 \le t \le T)$ is called an $\mathscr{F}$-martingale (supermartingale/submartingale) if

(i) $\mathbb{E}[|M(t)|] < \infty$, $0 \le t \le T$,
(ii) $\mathbb{E}[M(t)|\mathscr{F}_s] = M(s)$, $0 \le s < t \le T$, $(\mathbb{E}[M(t)|\mathscr{F}_s] \le M(s) / \mathbb{E}[M(t)|\mathscr{F}_s] \ge M(s))$.

Definition 2.2.3 An $\mathscr{F}$-adapted process $M := (M(t), 0 \le t \le T)$ is called an $\mathscr{F}$-local martingale if there exists a sequence of $\mathscr{F}$-stopping times $(\tau_n, n \in \mathbb{N})$ such that $\tau_n \to T$, $n \to \infty$, and $(M(t \wedge \tau_n), 0 \le t \le T)$ is an $\mathscr{F}$-martingale.

Definition 2.2.4 The quadratic variation process of a càdlàg semimartingale V is defined by

$$[V, V](t) = \lim_{n\to\infty} \sum_{i=1}^{n} \big(V\big(t_{i+1}^n \wedge t\big) - V\big(t_i^n \wedge t\big)\big)^2, \quad 0 \le t \le T,$$

where $\lim_{n\to\infty} \sup_{i=1,\dots,n} |t_{i+1}^n - t_i^n| = 0$, and the convergence is uniform in probability.

Example 2.7 The quadratic variation of a Brownian motion is given by $[W, W](t) = t$, and the quadratic variation of a quadratic pure jump process J (a purely discontinuous Lévy process or a step process) is given by $[J, J](t) = \sum_{s\le t} |\Delta J(s)|^2$, see Theorems II.28 and II.39 in Protter (2004).

Definition 2.2.5 Let $M := (M(t), 0 \le t \le T)$ be an $\mathscr{F}$-local martingale. The process M is called a BMO (bounded mean oscillation) martingale if there exists a constant K such that

$$\mathbb{E}\big[[M, M](T) - [M, M](\tau)|\mathscr{F}_\tau\big] \le K,$$
$$\big|\Delta M(\tau)\big| \le K,$$

for any $\mathscr{F}$-stopping time $\tau \in [0, T]$.

We end this chapter with two important martingale inequalities, which are often applied in this book. We state the Burkholder-Davis-Gundy inequalities, see Theorem IV.48 in Protter (2004).

Theorem 2.2.1 *Let M be a local martingale. For any $p \ge 1$ there exist constants K_1, K_2, depending on p but independent from M, such that*

$$\mathbb{E}\Big[\sup_{0\le t\le T} \big|M(t)\big|^p\Big] \le K_1 \mathbb{E}\big[\big|[M, M](T)\big|^{p/2}\big] \le K_2 \mathbb{E}\Big[\sup_{0\le t\le T} \big|M(t)\big|^p\Big]. \quad (2.7)$$

We also recall the Doob's inequality, see Theorem I.20 in Protter (2004).

Theorem 2.2.2 *Let M be a positive submartingale. For any $p > 1$ we have*

$$\mathbb{E}\Big[\sup_{0\le t\le T} \big|M(t)\big|^p\Big] \le K \sup_{0\le t\le T} \mathbb{E}\big[\big|M(t)\big|^p\big]. \quad (2.8)$$

As a corollary, we can conclude that the martingale $M(t) = \mathbb{E}[\xi|\mathscr{F}_t]$, $0 \le t \le T$, $\xi \in \mathbb{L}^2(\mathbb{R})$, satisfies the inequality

$$\mathbb{E}\Big[\sup_{0\le t\le T}\big|M(t)\big|^2\Big] \le K\mathbb{E}\big[|\xi|^2\big].$$

2.3 Stochastic Integration

We state main properties of stochastic integrals with respect to Brownian motion and compensated random measures.

Theorem 2.3.1

(a) *Let $V : \Omega \times [0,T] \to \mathbb{R}$ be a predictable process satisfying*

$$\int_0^T \big|V(t)\big|^2 dt < \infty,$$

Then $(\int_0^t V(s)dW(s), 0 \le t \le T)$ is a continuous local martingale with the quadratic variation process

$$\Big[\int_0^{\cdot} V(s)dW(s), \int_0^{\cdot} V(s)dW(s)\Big](t) = \int_0^t \big|V(s)\big|^2 ds, \quad 0 \le t \le T.$$

(b) *Let $V : \Omega \times [0,T] \times \mathbb{R} \to \mathbb{R}$ be a predictable process satisfying*

$$\int_0^T \int_{\mathbb{R}} \big|V(t,z)\big|^2 Q(t,dz)\eta(t)dt < \infty,$$

where we integrate with respect to the compensator of a random measure N. Then $(\int_0^t \int_{\mathbb{R}} V(s,z)\tilde{N}(ds,dz), 0 \le t \le T)$ is a càdlàg local martingale with the quadratic variation process

$$\Big[\int_0^{\cdot}\int_{\mathbb{R}} V(s,z)\tilde{N}(ds,dz), \int_0^{\cdot}\int_{\mathbb{R}} V(s,z)\tilde{N}(ds,dz)\Big](t)$$
$$= \int_0^t \int_{\mathbb{R}} \big|V(s,z)\big|^2 N(ds,dz), \quad 0 \le t \le T.$$

Proof Case (a) follows from Theorems IV.22 and IV.28 in Protter (2004). Case (b) follows from Definition 11.16 and Theorem 11.21 in He et al. (1992). □

We also use the following result, see Theorem 11.21 in He et al. (1992).

Theorem 2.3.2 *Let $V : \Omega \times [0,T] \times \mathbb{R} \to \mathbb{R}$ be a predictable process satisfying*

$$\int_0^T \int_{\mathbb{R}} \big|V(t,z)\big| Q(t,dz)\eta(t)dt < \infty,$$

where we integrate with respect to the compensator of a random measure N. Then $(\int_0^t \int_{\mathbb{R}} V(s,z)\tilde{N}(ds,dz), 0 \le t \le T)$ is a càdlàg local martingale and $(\int_0^t \int_{\mathbb{R}} V(s,z)N(ds,dz), 0 \le t \le T)$ is a càdlàg process. Let N be the jump measure of a càdlàg process J. We also have the property

$$\int_0^t \int_{\mathbb{R}} V(s,z)N(ds,dz) = \sum_{s\in(0,t]} V\big(s, \Delta J(s)\big)\mathbf{1}_{\Delta J(s)\neq 0}(s), \quad 0 \le t \le T.$$

Notice that if V is a non-negative predictable process satisfying $\mathbb{E}[\int_0^T \int_{\mathbb{R}} V(t,z)Q(t,dz)\eta(t)dt] < \infty$, then

$$\mathbb{E}\left[\int_0^T \int_{\mathbb{R}} V(t,z)N(dt,dz)\right] = \mathbb{E}\left[\int_0^T \int_{\mathbb{R}} V(t,z)Q(t,dz)\eta(t)dt\right]. \tag{2.9}$$

To prove (2.9), from Theorem 2.3.2 we first deduce

$$\mathbb{E}\left[\int_0^{\tau_n} \int_{\mathbb{R}} V(t,z)N(dt,dz)\right] = \mathbb{E}\left[\int_0^{\tau_n} \int_{\mathbb{R}} V(t,z)Q(t,dz)\eta(t)dt\right],$$

where $(\tau_n)_{n\ge 1}$ is a sequence of stopping times, and we next apply the monotone convergence theorem.

We need a stronger version of Theorem 2.3.1.

Theorem 2.3.3

(a) *Let $V \in \mathbb{H}^2(\mathbb{R})$. Then $(\int_0^t V(s)dW(s), 0 \le t \le T)$ is a continuous, square integrable martingale which satisfies*

$$\mathbb{E}\left[\left|\int_0^T V(s)dW(s)\right|^2\right] = \mathbb{E}\left[\int_0^T |V(s)|^2 ds\right].$$

(b) *Let $V \in \mathbb{H}^2_N(\mathbb{R})$. Then $(\int_0^t \int_{\mathbb{R}} V(s,z)\tilde{N}(ds,dz), 0 \le t \le T)$ is a càdlàg, square integrable martingale which satisfies*

$$\mathbb{E}\left[\left|\int_0^T \int_{\mathbb{R}} V(s,z)\tilde{N}(ds,dz)\right|^2\right] = \mathbb{E}\left[\int_0^T \int_{\mathbb{R}} |V(s,z)|^2 Q(s,dz)\eta(s)ds\right].$$

Proof Case (a) follows from Lemma IV.27 and Theorem IV.22 in Protter (2004). We prove case (b). By Theorem 2.3.1 the process $\int_0^t \int_{\mathbb{R}} V(s,z)\tilde{N}(ds,dz)$ is a càdlàg local martingale. By Theorem 2.3.2 and property (2.9) we obtain

$$\mathbb{E}\left[\int_0^T \int_{\mathbb{R}} |V(s,z)|^2 N(ds,dz)\right] = \mathbb{E}\left[\int_0^T \int_{\mathbb{R}} |V(s,z)|^2 Q(s,dz)\eta(s)ds\right] < \infty.$$

Since $\int_0^t \int_{\mathbb{R}} V(s,z)\tilde{N}(ds,dz)$ is a local martingale with integrable quadratic variation, it is a square integrable martingale, see Corollary II.26.3 in Protter (2004). By Corollary II.26.3 in Protter (2004) we also derive

$$\begin{aligned}
&\mathbb{E}\left[\left|\int_0^T\int_{\mathbb{R}} V(s,z)\tilde{N}(ds,dz)\right|^2\right]\\
&=\mathbb{E}\left[\left[\int_0^\cdot\int_{\mathbb{R}} V(s,z)\tilde{N}(ds,dz), \int_0^\cdot\int_{\mathbb{R}} V(s,z)\tilde{N}(ds,dz)\right](T)\right]\\
&=\mathbb{E}\left[\int_0^T\int_{\mathbb{R}} \left|V(s,z)\right|^2 N(ds,dz)\right],
\end{aligned}$$

and the proof is complete. □

From Sect. II.6 in Protter (2004) we also recall that

$$\left[\int_0^\cdot V_1(s)dW(s), \int_0^\cdot\int_{\mathbb{R}} V_2(s,z)\tilde{N}(ds,dz)\right](T)=0.$$

Finally, let us present the Itô's formula, see Theorem II.32 in Protter (2004).

Theorem 2.3.4 *Consider a process $\mathscr{X} := (\mathscr{X}(t), 0 \le t \le T)$ which satisfies the dynamics*

$$\begin{aligned}
\mathscr{X}(t) = \mathscr{X}(0) &+ \int_0^t \mu(s)ds + \int_0^t \sigma(s)dW(s)\\
&+ \int_0^t\int_{\mathbb{R}} \gamma(s,z)\tilde{N}(ds,dz), \quad 0 \le t \le T,
\end{aligned}$$

where μ, σ and γ are predictable processes such that $\int_0^T |\mu(s)|ds < \infty$, $\int_0^T |\sigma(s)|^2 ds < \infty$, $\int_0^T \int_{\mathbb{R}} |\gamma(s,z)|^2 Q(s,dz)\eta(s)ds < \infty$. Let $\varphi \in \mathscr{C}^{1,2}([0,T]\times\mathbb{R})$. Then

$$\begin{aligned}
\varphi\big(\tau,\mathscr{X}(\tau)\big) = \varphi\big(0,\mathscr{X}(0)\big) &+ \int_0^\tau \varphi_t\big(s,\mathscr{X}(s-)\big)ds + \int_0^\tau \varphi_x\big(s,\mathscr{X}(s-)\big)d\mathscr{X}(s)\\
&+ \int_0^\tau \frac{1}{2}\varphi_{xx}\big(s,\mathscr{X}(s-)\big)\sigma^2(s)ds + \int_0^\tau\int_{\mathbb{R}} \Big(\varphi\big(s,\mathscr{X}(s-)+\gamma(s,z)\big)\\
&- \varphi\big(s,\mathscr{X}(s-)\big) - \varphi_x\big(s,\mathscr{X}(s-)\big)\gamma(s,z)\Big)N(ds,dz),
\end{aligned}$$

for any stopping time $0 \le \tau \le T$.

Example 2.8 Let $M(t) = e^{W(t)-\frac{1}{2}t}$, $0 \le t \le T$. Then

$$M(t) = 1 + \int_0^t M(s)dW(s), \quad 0 \le t \le T.$$

Let $M(t) = e^{\int_0^t\int_{\mathbb{R}} z\tilde{N}(ds,dz) - \int_0^t\int_{\mathbb{R}}(e^z - z - 1)Q(s,dz)\eta(s)ds}$, $0 \le t \le T$, where N is a random measure with a compensator satisfying $\int_0^T\int_{\mathbb{R}} z^2 Q(s,dz)\eta(s)ds < \infty$ and

$\int_0^T \int_{\mathbb{R}}(e^z-1)^2 Q(ds,dz)\eta(s)ds<\infty$. Then

$$M(t)=1+\int_0^t\int_{\mathbb{R}} M(s-)(e^z-1)\tilde{N}(ds,dz),\quad 0\le t\le T.$$

We also use the following result, which is a special case of the multidimensional Itô's formula, see Theorem II.33 in Protter (2004).

Proposition 2.3.1 *Consider the processes* $\mathscr{X}_i:=(\mathscr{X}_i(t),0\le t\le T)$, $i=1,2$, *which satisfy the dynamics*

$$\mathscr{X}_i(t)=\mathscr{X}_i(0)+\int_0^t\mu_i(s)ds$$
$$+\int_0^t\sigma_i(s)dW(s)+\int_0^t\int_{\mathbb{R}}\gamma_i(s,z)\tilde{N}(ds,dz),\quad 0\le t\le T,\ i=1,2,$$

where μ_i, σ_i *and* γ_i *are predictable processes such that* $\int_0^T|\mu_i(s)|ds<\infty$, $\int_0^T|\sigma_i(s)|^2ds<\infty$, $\int_0^T\int_{\mathbb{R}}|\gamma_i(s,z)|^2Q(s,dz)\eta(s)ds<\infty$, *for* $i=1,2$. *Then*

$$\mathscr{X}_1(\tau)\mathscr{X}_2(\tau)=\mathscr{X}_1(0)\mathscr{X}_2(0)+\int_0^\tau\mathscr{X}_1(s-)d\mathscr{X}_2(s)+\int_0^\tau\mathscr{X}_2(s-)d\mathscr{X}_1(s)$$
$$+\int_0^\tau\sigma_1(s)\sigma_2(s)ds+\int_0^\tau\int_{\mathbb{R}}\gamma_1(s,z)\gamma_2(s,z)N(ds,dz),$$

for any stopping time $0\le\tau\le T$.

2.4 The Property of Predictable Representation

We now introduce the property of predictable representation, see Sect. XIII.2 in He et al. (1992) and Sect. III.4 in Jacod and Shiryaev (2003). The predictable representation property is the key concept in the theory of BSDEs which allows us to construct a solution to a BSDE. From the practical point of view, the predictable representation yields hedging strategies for financial claims.

Let us consider a probability space $(\Omega,\mathscr{F},\mathbb{P})$ with a filtration $\mathscr{F}=(\mathscr{F}_t)_{0\le t\le T}$. In this book we always assume that the weak property of predictable representation holds, that is

(PR) any $\mathscr{F}$-local martingale M has the representation

$$M(t)=M(0)+\int_0^t Z(s)dW(s)+\int_0^t\int_{\mathbb{R}}U(s,z)\tilde{N}(ds,dz)\quad 0\le t\le T,\tag{2.10}$$

where Z and U are $\mathscr{F}$-predictable processes integrable with respect to W and $\tilde{N}$.

This is our second standing assumption, next to (RM) from Sect. 2.1. If M is a locally square integrable local martingale, then the processes Z and U are locally square integrable in the sense of the assumptions from Theorem 2.3.1, see Definition III.4.2 in Jacod and Shiryaev (2003) and Theorem 11.31 in He et al. (1992). By Theorems 2.3.1–2.3.2 we also get

$$\begin{aligned}&\mathbb{E}\big[[M,M](\tau_n)\big]\\&\quad = M^2(0)+\mathbb{E}\bigg[\int_0^{\tau_n}\big|Z(s)\big|^2 ds\bigg]+\mathbb{E}\bigg[\int_0^{\tau_n}\int_{\mathbb{R}}\big|U(s,z)\big|^2 Q(s,dz)\eta(s)ds\bigg],\end{aligned}\tag{2.11}$$

where $(\tau_n)_{n\geq 1}$ is a sequence of stopping times. If we now assume that M is a square integrable martingale, then $\mathbb{E}[[M,M](T)] < \infty$, see Corollary II.26.3 in Protter (2004), and applying the monotone convergence theorem and Fatou's lemma to (2.11) we can conclude that $Z \in \mathbb{H}^2(\mathbb{R})$ and $U \in \mathbb{H}^2_N(\mathbb{R})$. Moreover, we can easily deduce that the representation of a square integrable martingale M is unique in $\mathbb{H}^2(\mathbb{R}) \times \mathbb{H}^2_N(\mathbb{R})$. Consequently, in this book we assume that any square integrable $\mathscr{F}$-martingale M has the unique representation

$$M(t) = M(0) + \int_0^t Z(s)dW(s) + \int_0^t\int_{\mathbb{R}} U(s,z)\tilde{N}(ds,dz), \quad 0 \leq t \leq T, \tag{2.12}$$

where $(Z,U) \in \mathbb{H}^2(\mathbb{R}) \times \mathbb{H}^2_N(\mathbb{R})$. We can also assume that any square integrable $\mathscr{F}_T$-measurable random variable ξ has the unique representation

$$\xi = \mathbb{E}[\xi] + \int_0^T Z(s)dW(s) + \int_0^T\int_{\mathbb{R}} U(s,z)\tilde{N}(ds,dz), \tag{2.13}$$

where $(Z,U) \in \mathbb{H}^2(\mathbb{R}) \times \mathbb{H}^2_N(\mathbb{R})$. Representation (2.13) follows immediately from (2.12) by taking the martingale $M(t) = \mathbb{E}[\xi|\mathscr{F}_t]$, $0 \leq t \leq T$.

We point out that we introduce the predictable representation property (PR) as an assumption. In general, the predictable representation property does not have to hold. However, in our case it is possible to construct a probability space $(\Omega, \mathscr{F}, \mathbb{P})$ in such a way that any $\mathscr{F}$-local martingale has the predictable representation. It is known that the weak property of predictable representation holds for a Brownian motion, a Lévy process, a step process and the corresponding completed natural filtration, see Theorems 13.19 and 13.49 in He et al. (1992). Moreover, given a Brownian motion W and an independent jump process J (a Lévy process or a step process), the weak property of predictable representation holds for (W,J) and the product of their completed natural filtrations. Finally, the weak property of predictable representation holds for (W,J) under any equivalent probability measure, see Theorem 13.22 in He et al. (1992). Hence, by the change of measure we can establish the predictable representation for a Brownian motion and a jump process with a random compensator (depending on W and J), see Sect. 2.5. For such a construction we refer to Becherer (2006) and Chap. 7 in Crépey (2011). We comment on the predictable representation in our financial and insurance model in Sect. 7.2.

2.5 Equivalent Probability Measures

Let us recall that for a semimartingale V such that $V(0) = 0$ there exists a unique càdlàg solution $\mathscr{E}$ to the forward stochastic differential equation

$$d\mathscr{E}(t) = \mathscr{E}(t-)dV(t), \quad \mathscr{E}(0) = 1,$$

given by

$$\mathscr{E}(t) = e^{V(t)-\frac{1}{2}[V,V](t)} \prod_{0<u\leq t} \big(1 + \Delta V(u)\big)e^{-\Delta V(u)+\frac{1}{2}|\Delta V(u)|^2}, \quad 0 \leq t \leq T. \quad (2.14)$$

The process $\mathscr{E}$ is called the stochastic exponential of V, see Theorem II.37 in Protter (2004). If $\Delta V(t) > -1$, $0 \leq t \leq T$, then the stochastic exponential $\mathscr{E}$ is positive.

Let $\mathbb{P}$ and $\mathbb{Q}$ be two equivalent probability measures, $\mathbb{Q} \sim \mathbb{P}$. There exists a positive martingale $M := (M(t), 0 \leq t \leq T)$ such that

$$\frac{d\mathbb{Q}}{d\mathbb{P}}\Big|\mathscr{F}_t = M(t), \quad 0 \leq t \leq T, \quad (2.15)$$

see Definition III.8.1 in Protter (2004) and Theorem 12.4 in He et al. (1992). In the view of the predictable representation property, we define

$$\frac{dM(t)}{M(t-)} = \phi(t)dW(t) + \int_{\mathbb{R}} \kappa(t,z)\tilde{N}(dt,dz), \quad M(0) = 1, \quad (2.16)$$

where $\phi := (\phi(t), 0 \leq t \leq T)$ and $\kappa := (\kappa(t,z), 0 \leq t \leq T, z \in \mathbb{R})$ are $\mathscr{F}$-predictable processes satisfying

$$\int_0^T |\phi(t)|^2 dt < \infty, \quad \int_0^T \int_{\mathbb{R}} |\kappa(t,z)|^2 Q(t,dz)\eta(t)dt < \infty,$$
$$\kappa(t,z) > -1, \quad 0 \leq t \leq T, \ z \in \mathbb{R}. \quad (2.17)$$

The process M defined by (2.16) under assumptions (2.17) is only a local martingale, see Theorem 2.3.1. We have to impose stronger assumptions on (ϕ, κ) so that the local martingale M is a true martingale. In this book we use the following proposition.

Proposition 2.5.1 *Let $M := (M(t), 0 \leq t \leq T)$ be the stochastic exponential defined by*

$$\frac{dM(t)}{M(t-)} = \phi(t)dW(t) + \int_{\mathbb{R}} \kappa(t,z)\tilde{N}(dt,dz), \quad M(0) = 1,$$

where ϕ and κ are predictable processes such that

$$|\phi(t)| \leq K, \quad \int_{\mathbb{R}} |\kappa(t,z)|^2 Q(t,dz)\eta(t) \leq K, \quad 0 \leq t \leq T,$$
$$\kappa(t,z) > -1, \quad 0 \leq t \leq T, \ z \in \mathbb{R}.$$

The process M is a square integrable, positive martingale.

Proof From Theorem 2.3.1 and (2.14) we conclude that M is a positive local martingale. We define the sequence of stopping times $\tau_n = \inf\{t : |M(t)| \geq n\} \wedge T$. We can derive the inequality

$$\begin{aligned}
\mathbb{E}\big[|M(t)|^2 \mathbf{1}\{t \leq \tau_n\}\big] &\leq \mathbb{E}\big[|M(\tau_n \wedge t)|^2\big] \\
&\leq K\mathbb{E}\Bigg[1 + \bigg|\int_0^{\tau_n \wedge t} M(s-)\phi(s)dW(s)\bigg|^2 \\
&\quad + \bigg|\int_0^{\tau_n \wedge t}\int_{\mathbb{R}} M(s-)\kappa(s,z)\tilde{N}(ds,dz)\bigg|^2\Bigg] \\
&= K\Bigg(1 + \mathbb{E}\bigg[\int_0^{\tau_n \wedge t} |M(s-)\phi(s)|^2 ds\bigg] \\
&\quad + \mathbb{E}\bigg[\int_0^{\tau_n \wedge t}\int_{\mathbb{R}} |M(s-)\kappa(s,z)|^2 Q(s,dz)\eta(s)ds\bigg]\Bigg) \\
&\leq K\bigg(1 + \int_0^t \mathbb{E}\big[|M(s)|^2 \mathbf{1}\{s \leq \tau_n\}\big]ds\bigg), \quad 0 \leq t \leq T,
\end{aligned}$$

where we use Theorem 2.3.3. By the Gronwall's inequality, see Theorem V.68 in Protter (2004), we obtain

$$\mathbb{E}\big[|M(t)|^2 \mathbf{1}\{t \leq \tau_n\}\big] \leq K, \quad 0 \leq t \leq T.$$

We let $n \to \infty$, apply Fatous' lemma and we can deduce that M is uniformly square integrable. The uniform integrability yields that the local martingale M is a true martingale, see Theorem I.51 in Protter (2004). □

We state Girsanov's theorem which plays an important role in stochastic calculus and financial mathematics.

Theorem 2.5.1 *Let W and N be a $(\mathbb{P}, \mathscr{F})$-Brownian motion and a $(\mathbb{P}, \mathscr{F})$-random measure with compensator $\vartheta(ds,dz) = Q(s,dz)\eta(s)ds$. We define an equivalent probability measure $\mathbb{Q} \sim \mathbb{P}$ with a positive $\mathscr{F}$-martingale* (2.16). *The processes*

$$\begin{aligned}
W^{\mathbb{Q}}(t) &= W(t) - \int_0^t \phi(s)ds, \quad 0 \leq t \leq T, \\
\tilde{N}^{\mathbb{Q}}(t,A) &= N(t,A) \\
&\quad - \int_0^t \int_{\mathbb{R}} \big(1 + \kappa(s,z)\big) Q(s,dz)\eta(s)ds, \quad 0 \leq t \leq T, \ A \in \mathscr{B}(\mathbb{R}),
\end{aligned} \tag{2.18}$$

are a $(\mathbb{Q}, \mathscr{F})$-Brownian motion and a $(\mathbb{Q}, \mathscr{F})$-compensated random measure.

Proof Let M denote the martingale (2.16) which changes the measure. The result of our theorem follows from the Girsanov-Meyer theorem, see Theorem III.40 in Protter (2004), which states that if for a $\mathbb{P}$-local martingale V the sharp bracket process $\langle V, M\rangle$ exists under $\mathbb{P}$, then

$$V(t) - \int_0^t \frac{1}{M(s-)} d\langle V, M\rangle(s), \quad 0 \leq t \leq T,$$

is a $\mathbb{Q}$-local martingale. The first assertion for the Brownian motion can be deduced from Theorem III.46 in Protter (2004). We prove the second assertion for the compensated random measure. The measure $\vartheta(dt, dz) = (1 + \kappa(t,z))Q(t,dz)\eta(t)dt$ is an $\mathscr{F}$-predictable random measure, see Definition 2.1.3. We choose a nonnegative, predictable function V such that $\int_0^t \int_{\mathbb{R}} V(s,z)N(ds,dz)$ is locally integrable under $\mathbb{Q}$. We set $V^m(s,z) = V(s,z) \wedge (m|z|)$. We can now deal with the $\mathbb{P}$-local martingale $\int_0^t \int_{\mathbb{R}} V^m(s,z)\tilde{N}(ds,dz)$, see Theorem 2.3.1. We define the quadratic covariation process

$$\begin{aligned}
&\left[\int_0^\cdot \int_{\mathbb{R}} V^m(s,z)\tilde{N}(ds,dz), M\right](t) \\
&\quad = \int_0^t \int_{\mathbb{R}} M(s-)\kappa(s,z)V^m(s,z)N(ds,dz), \quad 0 \leq t \leq T. \qquad (2.19)
\end{aligned}$$

Since the martingale M is càdlàg, we get

$$\begin{aligned}
&\int_0^T \int_{\mathbb{R}} \left|M(s-)\kappa(s,z)V^m(s,z)\right| Q(s,dz)\eta(s)ds \\
&\leq K\sqrt{\int_0^T \int_{\mathbb{R}} \left|\kappa(s,z)\right|^2 Q(s,dz)\eta(s)ds \int_0^{\tau_n} \int_{\mathbb{R}} m|z|^2 Q(s,dz)\eta(s)ds} < \infty,
\end{aligned}$$

and from Theorem 2.3.2 we deduce that the process $\int_0^t \int_{\mathbb{R}} M(s-)\kappa(s,z)V^m(s,z) \times \tilde{N}(ds,dz)$ is a $\mathbb{P}$-local martingale and the quadratic covariation process (2.19) is locally integrable under $\mathbb{P}$. Hence, the compensator of the covariation process (2.19) (the sharp bracket) exists under $\mathbb{P}$, see Sect. III.5 in Protter (2004), and it takes the form

$$\left\langle \int_0^\cdot \int_{\mathbb{R}} V^m(s,z)\tilde{N}(ds,dz), M\right\rangle(t) = \int_0^t \int_{\mathbb{R}} M(s-)\kappa(s,z)V^m(s,z)Q(s,dz)\eta(s)ds.$$

The Girsanov-Meyer theorem now yields that

$$\int_0^t \int_{\mathbb{R}} V^m(s,z)\big(N(ds,dz) - \big(1 + \kappa(s,z)\big)Q(s,dz)\eta(s)ds\big), \quad 0 \leq t \leq T,$$

is a $\mathbb{Q}$-local martingale. Let $(\tau_k)_{k\geq 1}$ be a localizing sequence of stopping times for $\int_0^t \int_{\mathbb{R}} V^m(s,z)\tilde{N}^{\mathbb{Q}}(ds,dz)$, let $(\tau_n)_{n\geq 1}$ be a localizing sequence of stopping times for $\int_0^t \int_{\mathbb{R}} V(s,z)N(ds,dz)$, and let τ be a stopping time. We have

$$\mathbb{E}^{\mathbb{Q}}\Big[\int_0^{\tau_k\wedge\tau_n\wedge\tau}\int_{\mathbb{R}} V^m(s,z)N(ds,dz)\Big]$$
$$=\mathbb{E}^{\mathbb{Q}}\Big[\int_0^{\tau_k\wedge\tau_n\wedge\tau}\int_{\mathbb{R}} V^m(s,z)\big(1+\kappa(s,z)\big)Q(s,dz)\eta(s)ds\Big].$$

Taking the limit $k\to\infty$, $m\to\infty$ and applying the Lebesgue monotone convergence theorem, we show

$$\mathbb{E}^{\mathbb{Q}}\Big[\int_0^{\tau_n\wedge\tau}\int_{\mathbb{R}} V(s,z)N(ds,dz)\Big]$$
$$=\mathbb{E}^{\mathbb{Q}}\Big[\int_0^{\tau_n\wedge\tau}\int_{\mathbb{R}} V(s,z)\big(1+\kappa(s,z)\big)Q(s,dz)\eta(s)ds\Big].$$

Hence, by Lemma I.1.44 in Jacod and Shiryaev (2003) the process $\int_0^t\int_{\mathbb{R}} V(s,z)\times\tilde{N}^{\mathbb{Q}}(ds,dz)$ is a $\mathbb{Q}$-local martingale. We now choose a predictable function V such that $\int_0^t\int_{\mathbb{R}}|V(s,z)|N(ds,dz)$ is locally integrable under $\mathbb{Q}$. Following the same reasoning, we show that $\int_0^t\int_{\mathbb{R}} V^+(s,z)\tilde{N}^{\mathbb{Q}}(ds,dz)$ and $\int_0^t\int_{\mathbb{R}} V^-(s,z)\tilde{N}^{\mathbb{Q}}(ds,dz)$ are $\mathbb{Q}$-local martingales, and $\int_0^t\int_{\mathbb{R}} V(s,z)\tilde{N}^{\mathbb{Q}}(ds,dz)$ is a $\mathbb{Q}$-local martingale. The proof is complete by Theorem II.1.8 in Jacod and Shiryaev (2003) and Definition 2.1.4. □

We give two examples which illustrate the change of measure.

Example 2.9 Consider the dynamics

$$\frac{dS(t)}{S(t)}=\mu(t)dt+\sigma(t)dW(t),\qquad S(0)=s,$$

where μ,σ are predictable, bounded processes. Let r be a predictable, nonnegative, bounded process. Define the stochastic exponential

$$\frac{dM(t)}{M(t)}=-\frac{\mu(t)-r(t)}{\sigma(t)}dW(t),\qquad M(0)=1,\tag{2.20}$$

and assume that $t\mapsto\frac{\mu(t)-r(t)}{\sigma(t)}$ is a.s bounded. By Proposition 2.5.1 the stochastic exponential M is a square integrable martingale. Hence, we can define an equivalent probability measure $\mathbb{Q}$ by $\frac{d\mathbb{Q}}{d\mathbb{P}}|\mathscr{F}_T=M(T)$. From Theorem 2.5.1 we deduce that the dynamics of S under the new measure $\mathbb{Q}$ is given by

$$\frac{dS(t)}{S(t)}=r(t)dt+\sigma(t)dW^{\mathbb{Q}}(t).$$

The Itô's formula and Proposition 2.5.1 yield that $e^{-\int_0^t r(s)ds}S(t)$ is a $\mathbb{Q}$-martingale.

Example 2.10 Consider a compound Poisson process J with intensity λ and jump size distribution q. Let N denote the corresponding jump measure. Choose a predictable process κ such that $|\kappa(t,z)|<1$, $(t,z)\in[0,T]\times\mathbb{R}$. We define the stochastic exponential

$$\frac{dM(t)}{M(t-)}=\int_{\mathbb{R}}\kappa(t,z)\tilde{N}(dt,dz),\quad M(0)=1. \tag{2.21}$$

By Proposition 2.5.1 the stochastic exponential M is a square integrable martingale. Hence, we can define an equivalent probability measure $\mathbb{Q}$ by $\frac{d\mathbb{Q}}{d\mathbb{P}}|\mathscr{F}_T=M(T)$. From Theorem 2.5.1 we deduce that

$$N(dt,dz)-\big(1+\kappa(t,z)\big)\lambda q(dz)dt,$$

is the compensated random measure of the process J under the equivalent probability measure $\mathbb{Q}$. Consequently, under the equivalent probability measure $\mathbb{Q}$ the process J has the jump size distribution and the intensity

$$\begin{aligned}
q^{\mathbb{Q}}(t,dz)&=\frac{1+\kappa(t,z)}{\int_{\mathbb{R}}(1+\kappa(t,z))q(dz)}q(dz),\quad 0\le t\le T,\ z\in\mathbb{R},\\
\lambda^{\mathbb{Q}}(t)&=\int_{\mathbb{R}}\big(1+\kappa(t,z)\big)q(dz)\lambda,\quad 0\le t\le T.
\end{aligned}$$

In general, since κ is a stochastic process then the new distribution $q^{\mathbb{Q}}$ and the new intensity $\lambda^{\mathbb{Q}}$ are stochastic processes as well. The set of equivalent probability measures determined by the martingales (2.21) with processes κ such that $|\kappa(t,z)|<1$, $(t,z)\in[0,T]\times\mathbb{R}$ defines the set of equivalent scenarios for the compound Poisson process J, see Example 1.3.

2.6 The Malliavin Calculus

The Malliavin calculus plays an important role in the theory of BSDEs. It allows us to characterize a solution to a BSDE, prove path regularities of a solution and develop numerical schemes for finding a solution. Since Definition 2.6 of the Malliavin derivative is not very useful in calculations, we present some practical differentiation rules.

Consider the canonical Lévy space $(\Omega_W\times\Omega_N,\mathscr{F}_W\otimes\mathscr{F}_N,\mathbb{P}_W\otimes\mathbb{P}_N)$ and recall the Malliavin derivatives $D_{t,0}$, $D_{t,z}$ and the measures ν, υ, Υ from Sect. 2.2. The derivative $D_{t,0}$ is derivative with respect to the continuous component of a Lévy process (the Brownian motion) and we can apply the classical Malliavin calculus for Hilbert space-valued random variables, see Nualart (1995). By D_t we denote the classical Malliavin derivative on the Wiener space $(\Omega_W,\mathscr{F}_W,\mathbb{P}_W)$. We state the first result, see Proposition 3.5 in Solé et al. (2007).

Proposition 2.6.1 *If for $\mathbb{P}^N$-a.e. $\omega_N \in \Omega_N$ a random variable $\xi(.,\omega_N)$ on $(\Omega_W, \mathscr{F}_W, \mathbb{P}_W)$ is Malliavin differentiable, then*

$$D_{t,0}\xi(\omega_W, \omega_N) = \frac{1}{\sigma} D_t \xi(.,\omega_N)(\omega_W), \quad a.s., a.e., (\omega, t) \in \Omega \times [0, T], \tag{2.22}$$

where D_t denotes the Malliavin derivative on the Wiener space.

The derivative $D_{t,z}$, for $z \neq 0$, is derivative with respect to the pure jump component of a Lévy process. In order to calculate this derivative, we use the following increment quotient operator

$$\mathscr{I}_{t,z}\xi(\omega_W, \omega_N) = \frac{\xi(\omega_W, \omega_N^{t,z}) - \xi(\omega_W, \omega_N)}{z}, \tag{2.23}$$

where $\omega_N^{t,z}$ transforms a family $\omega_N = ((t_1, z_1), (t_2, z_2), \ldots) \in \Omega_N$ into a new family $\omega_N^{t,z} = ((t, z), (t_1, z_1), (t_2, z_2), \ldots) \in \Omega_N$ by adding a jump of size z at time t into the trajectory of the Lévy process. We can state the second result, see Propositions 5.4 and 5.5 in Solé et al. (2007).

Proposition 2.6.2 *Consider $\xi \in \mathbb{L}^2(\mathbb{R})$ which is measurable with respect to the natural filtration generated by a Lévy process. If $\mathbb{E}[\int_0^T \int_{\mathbb{R}\setminus\{0\}} |\mathscr{I}_{t,z}\xi|^2 z^2 \nu(dz)dt] < \infty$, then*

$$D_{t,z}\xi = \mathscr{I}_{t,z}\xi, \quad a.s., \upsilon\text{-}a.e.\ (\omega, t, z) \in \Omega \times [0, T] \times \big(\mathbb{R} \setminus \{0\}\big). \tag{2.24}$$

Let us now present some differentiation rules.

Proposition 2.6.3 *Consider the natural filtration $\mathscr{F}$ generated by a Lévy process and let $\xi \in \mathbb{D}^{1,2}(\mathbb{R})$. For $0 \leq s \leq T$ we have $\mathbb{E}[\xi|\mathscr{F}_s] \in \mathbb{D}^{1,2}(\mathbb{R})$, and*

$$D_{t,z}\mathbb{E}[\xi|\mathscr{F}_s] = \mathbb{E}[D_{t,z}\xi|\mathscr{F}_s]\mathbf{1}\{t \leq s\}, \quad a.s., \upsilon\text{-}a.e.\ (\omega, t, z) \in \Omega \times [0, T] \times \mathbb{R}.$$

Proof The result follows by adapting the proof of Proposition 1.2.8 from Nualart (1995) into our setting. □

It follows from Proposition 2.6.3 that if ξ is $\mathscr{F}_s$-measurable then $D_{t,z}\xi = 0$ a.s., υ-a.e. $(\omega, t, z) \in \Omega \times (s, T] \times \mathbb{R}$, see Corollary 1.2.1 in Nualart (1995).

We state the chain rule.

Proposition 2.6.4 *Let $\varphi : \mathbb{R} \to \mathbb{R}$ be a Lipschitz continuous function. Under the assumptions of Propositions 2.6.1 and 2.6.2 we have $\varphi(\xi) \in \mathbb{D}^{1,2}(\mathbb{R})$. Moreover:*

(a) *There exists an a.s. bounded random variable ζ such that*

$$D_{t,0}\varphi(\xi) = \zeta D_{t,0}\xi, \quad a.s., a.e. (\omega, t) \in \Omega \times [0, T].$$

If the law of ξ *is absolutely continuous with respect to the Lebesgue measure or* φ *is continuously differentiable, then* $\zeta = \varphi'(\xi)$.

(b) *We have the relation*

$$D_{t,z}\varphi(\xi) = \frac{\varphi(\xi + zD_{t,z}\xi) - \varphi(\xi)}{z},$$
$$a.s., \upsilon\text{-}a.e.\ (\omega, t, z) \in \Omega \times [0, T] \times \mathbb{R} \setminus \{0\}.$$

Proof Case (a) follows from Proposition 1.2.4 in Nualart (1995) and Proposition 2.6.1. Case (b) follows from Proposition 2.6.2 and the definition of the operator (2.23). □

The next two results are taken from Delong and Imkeller (2010b).

Proposition 2.6.5 *Consider a finite measure* q *on* $\mathbb{R}$. *Let* $\varphi : \Omega \times [0, T] \times \mathbb{R} \to \mathbb{R}$ *be a product measurable, adapted process which satisfies*

$$\begin{aligned}
&\mathbb{E}\Big[\int_{[0,T]\times\mathbb{R}} |\varphi(s, y)|^2 q(dy)ds\Big] < \infty,\\
&\varphi(s, y) \in \mathbb{D}^{1,2}(\mathbb{R}), \quad a.e.\ (s, y) \in [0, T] \times \mathbb{R}, \qquad (2.25)\\
&\mathbb{E}\Big[\int_{([0,T]\times\mathbb{R})^2} |D_{t,z}\varphi(s, y)|^2 q(dy)ds\upsilon(dt, dz)\Big] < \infty.
\end{aligned}$$

Then $\int_{[0,T]\times\mathbb{R}} \varphi(s, y)q(dy)ds \in \mathbb{D}^{1,2}(\mathbb{R})$ *and we have the differentiation rule*

$$D_{t,z}\int_0^T\int_{\mathbb{R}} \varphi(s, y)q(dy)ds = \int_t^T\int_{\mathbb{R}} D_{t,z}\varphi(s, y)q(dy)ds,$$

$a.s., \upsilon\text{-}a.e.\ (\omega, t, z) \in \Omega \times [0, T] \times \mathbb{R}$.

Proposition 2.6.6 *Let* $\varphi : \Omega \times [0, T] \times \mathbb{R} \to \mathbb{R}$ *be a predictable process which satisfies* $\mathbb{E}[\int_{[0,T]\times\mathbb{R}} |\varphi(s, y)|^2\upsilon(ds, dy)] < \infty$. *Then*

$$\varphi \in \mathbb{L}^{1,2}(\mathbb{R}) \quad \text{if and only if} \quad \int_{[0,T]\times\mathbb{R}} \varphi(s, y)\Upsilon(ds, dy) \in \mathbb{D}^{1,2}(\mathbb{R}).$$

Moreover, if $\int_{[0,T]\times\mathbb{R}} \varphi(s, y)\Upsilon(ds, dy) \in \mathbb{D}^{1,2}(\mathbb{R})$, *then*

$$D_{t,z}\int_0^T\int_{\mathbb{R}} \varphi(s, y)\Upsilon(ds, dy) = \varphi(t, z) + \int_t^T\int_{\mathbb{R}} D_{t,z}\varphi(s, y)\Upsilon(ds, dy),$$

$a.s., \upsilon\text{-}a.e.\ (\omega, t, z) \in \Omega \times [0, T] \times \mathbb{R}$, *and* $\int_{[0,T]\times\mathbb{R}} D_{t,z}\varphi(s, y)\Upsilon(ds, dy)$ *is a stochastic integral in the Itô sense.*

Notice that we can also establish the following relation

$$\begin{aligned}
&D_{t,z}\int_s^T\int_{\mathbb{R}}\varphi(r,y)\Upsilon(dr,dy)\\
&=D_{t,z}\bigg(\int_0^T\int_{\mathbb{R}}\varphi(r,y)\Upsilon(dr,dy)-\int_0^s\int_{\mathbb{R}}\varphi(r,y)\Upsilon(dr,dy)\bigg)\\
&=\int_s^T\int_{\mathbb{R}}D_{t,z}\varphi(r,y)\Upsilon(dr,dy),\quad 0\le t\le s\le T,\ s>0.
\end{aligned}$$

We now give examples which illustrate the differentiation rules.

Example 2.11 Consider a square integrable function $V:\mathbb{R}\to\mathbb{R}$ and a Lipschitz continuous function $\varphi:\mathbb{R}\to\mathbb{R}$. Let

$$\xi=\varphi\bigg(\int_0^T V(s)dW(s)\bigg).$$

We can write $\xi=\varphi(\int_0^T\frac{V(s)}{\sigma}d\Upsilon^c(s))$. It is know that the random variable $\int_0^T V(s)dW(s)$ is normally distributed, see Lemma 4.3.11 in Applebaum (2004). By Propositions 2.6.4 and 2.6.6 we obtain

$$D_{t,0}\xi=\varphi'\bigg(\int_0^T V(s)dW(s)\bigg)\frac{V(t)}{\sigma},$$

a.s., a.e. $(\omega,t)\in\Omega\times[0,T]$.

Example 2.12 Consider the put option $\xi=(K-V(T))^+$ where $V(T)=e^{\sigma W(T)-\frac{1}{2}\sigma^2T}$ models the terminal value of a stock. In applications we would like to use the Malliavin derivative of ξ. Unfortunately, we cannot use the result from Example 2.11 since the exponential function is not Lipschitz continuous. We follow a different approach. First, we find the Malliavin derivative of $V(T)$. Let us define the process $V(t)=e^{\sigma W(t)-\frac{1}{2}\sigma^2t}$, $0\le t\le T$, and by the Itô's formula we get

$$V(t)=1+\int_0^t V(s)\sigma dW(s),\quad 0\le t\le T.$$

In Sect. 4.1 we show that the process V, which solves a linear forward stochastic differential equation, is Malliavin differentiable, see Theorem 4.1.2. We can now apply Proposition 2.6.6 and we derive the equation

$$D_{u,0}V(t)=V(u)+\int_u^t D_{u,0}V(s)\sigma dW(s),\quad 0\le u\le t\le T.$$

Since $D_{u,0}V$ turns out to be a stochastic exponential of the Brownian motion W, we conclude that $D_{u,0}V(t)=V(u)e^{\int_u^t\sigma dW(s)-\frac{1}{2}\int_u^t\sigma^2ds}=V(t)$, $0\le u\le t\le T$. By Proposition 2.6.4 we now get the Malliavin derivative

$$D_{t,0}\xi = -e^{\sigma W(T)-\frac{1}{2}\sigma^2 T}\mathbf{1}\{e^{\sigma W(T)-\frac{1}{2}\sigma^2 T} < K\},$$

a.s., a.e. $(\omega, t) \in \Omega \times [0, T]$.

Example 2.13 Let N be the jump measure of a compound Poisson process with jump size distribution q. Consider a function $V : [0, T] \times \mathbb{R} \to \mathbb{R}$ such that $\int_0^T \int_{\mathbb{R}} |V(s, y)|^2 q(dy)ds < \infty$ and a Lipschitz continuous function $\varphi : \mathbb{R} \to \mathbb{R}$. Let

$$\xi = \varphi\left(\int_0^T \int_{\mathbb{R}} V(s, y)\tilde{N}(ds, dy)\right).$$

We can write $\xi = \varphi(\int_0^T \int_{\mathbb{R}} \frac{V(s,y)}{y} \Upsilon^d(ds, dy))$. By Propositions 2.6.6 and 2.6.4 we obtain

$$D_{t,z}\xi = \frac{\varphi(\int_0^T \int_{\mathbb{R}} V(s, y)\tilde{N}(ds, dy) + V(t, z)) - \varphi(\int_0^T \int_{\mathbb{R}} V(s, y)\tilde{N}(ds, dy))}{z},$$

a.s., υ-a.e. $(\omega, t, z) \in \Omega \times [0, T] \times \mathbb{R} \setminus \{0\}$.

Example 2.14 Let N be the jump measure of a compound Poisson process with jump size distribution q. Consider the stop-loss contract $\xi = (J(T) - K)^+$ where $J(t) = \int_0^t \int_0^\infty yN(ds, dy)$, $0 \le t \le T$, is the compound Poisson process used for modelling insurer's claims. We assume that the claim size distribution q is supported on $(0, \infty)$ and satisfies $\int_0^\infty y^2 q(dy) < \infty$. In applications we would like to use the Malliavin derivative of ξ. From Example 2.13 we immediately deduce that

$$D_{t,z} = \frac{(J(T) + z - K)^+ - (J(T) - K)^+}{z},$$

a.s., υ-a.e. $(\omega, t, z) \in \Omega \times [0, T] \times (0, \infty)$.

Bibliographical Notes Definitions are taken from He et al. (1992) and Protter (2004). Propositions and theorems are taken from the sources cited in the text. Stochastic integration and the theory of semimartingales are studied in Applebaum (2004) (see for Lévy processes), Brémaud (1981) (see for point processes), Karatzas and Shreve (1988) (see for Brownian motions), Nualart (1995) (see for the Malliavin calculus) and He et al. (1992), Jacod and Shiryaev (2003), Protter (2004) (see for the general theory). Financial and actuarial applications of Brownian motions, Lévy processes and step processes are investigated in Cont and Tankov (2004), Mikosch (2009), Øksendal and Sulem (2004), Pham (2009), Rolski et al. (1999), Schmidli (2007) and Shreve (2004).

Chapter 3
Backward Stochastic Differential Equations—The General Case

Abstract We investigate BSDEs driven by a Brownian motion and a compensated random measure. The case of Lipschitz continuous generators is considered. We derive so-called a priori estimates, which are crucial in the study of BSDEs. We prove existence and uniqueness of a solution. We state two versions of a comparison principle which allows us to compare the solutions to BSDEs based on the terminal conditions and the generators. Explicit solutions to some important types of BSDEs (including the linear BSDE) are derived. In the case of a Lévy process, we prove Malliavin differentiability of the solution and we characterize the solution by the Malliavin derivative. The Clark-Ocone formula is obtained.

We deal with backward stochastic differential equations driven by a Brownian motion and a compensated random measure and we consider the case of Lipschitz continuous generators. In this chapter we lay the foundations of BSDEs. Key results for BSDEs are presented, which are further developed in next chapters.

3.1 Existence and Uniqueness of Solution

Our goal is to investigate the backward stochastic differential equation

$$\begin{aligned} Y(t) = \xi &+ \int_t^T f\big(s, Y(s-), Z(s), U(s,.)\big)ds \\ &- \int_t^T Z(s)dW(s) - \int_t^T \int_{\mathbb{R}} U(s,z)\tilde{N}(ds,dz), \quad 0 \le t \le T. \quad (3.1) \end{aligned}$$

Given a terminal condition ξ and a generator f, we are interested in finding a triple $(Y, Z, U) \in \mathbb{S}^2(\mathbb{R}) \times \mathbb{H}^2(\mathbb{R}) \times \mathbb{H}^2_N(\mathbb{R})$ which satisfies (3.1). The processes Z and U are called control processes. They control an adapted process Y so that Y satisfies the terminal condition. Since for a càdlàg process Y the set $\{s \in [0,T], \Delta Y(s) \neq 0\}$ is countable, we may also deal with the equation

$$Y(t) = \xi + \int_t^T f\big(s, Y(s), Z(s), U(s,.)\big)ds$$

Ł. Delong, *Backward Stochastic Differential Equations with Jumps and Their Actuarial and Financial Applications*, EAA Series, DOI 10.1007/978-1-4471-5331-3_3,

$$-\int_t^T Z(s)dW(s) - \int_t^T \int_{\mathbb{R}} U(s,z)\tilde{N}(ds,dz), \quad 0 \leq t \leq T.$$

Before we study the BSDE (3.1), let us have a look at a BSDE with zero generator and a BSDE with generator independent of (Y, Z, U). Those equations are the simplest examples of BSDEs. First, we are interested in finding a triple (Y, Z, U) which satisfies

$$Y(t) = \xi - \int_t^T Z(s)dW(s) - \int_t^T \int_{\mathbb{R}} U(s,z)\tilde{N}(ds,dz), \quad 0 \leq t \leq T. \tag{3.2}$$

Equation (3.2) is called a BSDE with zero generator.

Proposition 3.1.1 *Assume that* $\xi \in \mathbb{L}^2(\mathbb{R})$. *There exists a unique solution* $(Y, Z, U) \in \mathbb{S}^2(\mathbb{R}) \times \mathbb{H}^2(\mathbb{R}) \times \mathbb{H}^2_N(\mathbb{R})$ *to the BSDE* (3.2). *The process* Y *has the representation*

$$Y(t) = \mathbb{E}[\xi|\mathscr{F}_t], \quad 0 \leq t \leq T,$$

and the control processes (Z, U) *are derived from the representation*

$$\xi = \mathbb{E}[\xi] + \int_0^T Z(s)dW(s) + \int_0^T \int_{\mathbb{R}} U(s,z)\tilde{N}(ds,dz).$$

Proof Notice that any solution (Y, Z, U) to the BSDE (3.2) must satisfy the equation

$$\xi = Y(0) + \int_0^T Z(s)dW(s) + \int_0^T \int_{\mathbb{R}} U(s,z)\tilde{N}(ds,dz).$$

By the predictable representation property the processes (Z, U) are determined by the predictable representation of ξ and $Y(0) = \mathbb{E}[\xi]$. Moreover, $(Z, U) \in \mathbb{H}^2(\mathbb{R}) \times \mathbb{H}^2_N(\mathbb{R})$ and the representation of ξ is unique. Taking the conditional expected value on both sides of the BSDE (3.2), we immediately get $Y(t) = \mathbb{E}[\xi|\mathscr{F}_t]$, $0 \leq t \leq T$. We take the càdlàg modification of the martingale Y, see Theorem I.9 in Protter (2004). By the Doob's inequality we can show that $Y \in \mathbb{S}^2(\mathbb{R})$. □

We point out that finding a solution to the BSDE (3.2) is equivalent to finding the predictable representation of the random variable ξ. The predictable representation property is the key concept in the theory of BSDEs since it allows us to find a solution to an equation with a random terminal condition.

Let us deal with the equation

$$\begin{aligned} Y(t) &= \xi + \int_t^T f(s)ds \\ &\quad - \int_t^T Z(s)dW(s) - \int_t^T \int_{\mathbb{R}} U(s,z)\tilde{N}(ds,dz), \quad 0 \leq t \leq T, \end{aligned} \tag{3.3}$$

which we call a BSDE with generator independent of (Y, Z, U). We can immediately prove the following result.

Proposition 3.1.2 *Assume that $\xi \in \mathbb{L}^2(\mathbb{R})$ and $f : \Omega \times [0, T] \to \mathbb{R}$ is a predictable process satisfying $\mathbb{E}[\int_0^T |f(s)|^2 ds] < \infty$. There exists a unique solution $(Y, Z, U) \in \mathbb{S}^2(\mathbb{R}) \times \mathbb{H}^2(\mathbb{R}) \times \mathbb{H}^2_N(\mathbb{R})$ to the BSDE* (3.2). *The process Y has the representation*

$$Y(t) = \mathbb{E}\left[\xi + \int_t^T f(s)ds|\mathscr{F}_t\right], \quad 0 \le t \le T,$$

and the control processes (Z, U) are derived from the representation

$$\xi + \int_0^T f(s)ds = \mathbb{E}\left[\xi + \int_0^T f(s)ds\right] + \int_0^T Z(s)dW(s) + \int_0^T \int_{\mathbb{R}} U(s, z)\tilde{N}(ds, dz).$$

Since $t \mapsto \int_0^t f(s)ds$ is a.s. continuous, see Sect. 4.3 in Applebaum (2004), the càdlàg modification of Y can be defined.

We point out that in many applications we are able to reduce a BSDE to a BSDE with zero generator or a BSDE with generator independent of (Y, Z, U), and use Propositions 3.1.1–3.1.2 and the predictable representation property to derive the solution.

Let us now investigate the BSDE (3.1). We recall that $\vartheta(dt, dz) = Q(t, dz)\eta(t)dt$ denote the compensator of the random measure N. We assume that

(A1) the terminal value $\xi \in \mathbb{L}^2(\mathbb{R})$,
(A2) the generator $f : \Omega \times [0, T] \times \mathbb{R} \times \mathbb{R} \times L^2_Q(\mathbb{R}) \to \mathbb{R}$ is predictable and Lipschitz continuous in the sense that

$$\begin{aligned} &\left|f(\omega, t, y, z, u) - f\left(\omega, t, y', z', u'\right)\right|^2 \\ &\le K\left(\left|y - y'\right|^2 + \left|z - z'\right|^2 + \int_{\mathbb{R}} \left|u(x) - u'(x)\right|^2 Q(t, dx)\eta(t)\right), \end{aligned}$$

a.s., a.e. $(\omega, t) \in \Omega \times [0, T]$, for all $(y, z, u), (y', z', u') \in \mathbb{R} \times \mathbb{R} \times L^2_Q(\mathbb{R})$,
(A3) $\mathbb{E}[\int_0^T |f(t, 0, 0, 0)|^2 dt] < \infty$.

Assumptions (A1)–(A3) are called standard in the theory of BSDEs. These assumptions hold in all our applications and they should hold in most actuarial and financial applications. Let us recall that the predictability of the generator f means that $f : \Omega \times [0, T] \times \mathbb{R} \times \mathbb{R} \times L^2_Q(\mathbb{R}) \to \mathbb{R}$ is $\mathscr{P} \otimes \mathscr{B}(\mathbb{R}) \otimes \mathscr{B}(\mathbb{R}) \otimes \mathscr{B}(L^2_Q(\mathbb{R}))$ measurable.

We derive so-called a priori estimates for a solution to (3.1).

Lemma 3.1.1 *Assume that (ξ, f) and (ξ', f') satisfy* (A1)–(A3). *Let $(Y, Z, U) \in \mathbb{S}^2(\mathbb{R}) \times \mathbb{H}^2(\mathbb{R}) \times \mathbb{H}^2_N(\mathbb{R})$ and $(Y', Z', U') \in \mathbb{S}^2(\mathbb{R}) \times \mathbb{H}^2(\mathbb{R}) \times \mathbb{H}^2_N(\mathbb{R})$ be solutions* *to the BSDE* (3.1) *with* (ξ, f) *and* (ξ', f'). *We have the estimates*

$$\begin{aligned}
&\|Z-Z'\|^2_{\mathbb{H}^2}+\|U-U'\|^2_{\mathbb{H}^2_N}\\
&\quad\le \hat{K}\Big(\mathbb{E}\big[e^{\rho T}|\xi-\xi'|^2\big]\\
&\qquad+\mathbb{E}\Big[\int_0^T e^{\rho t}\big|f\big(t,Y(t),Z(t),U(t)\big)-f'\big(t,Y'(t),Z'(t),U'(t)\big)\big|^2dt\Big]\Big),
\end{aligned} \tag{3.4}$$

$$\begin{aligned}
&\|Z-Z'\|^2_{\mathbb{H}^2}+\|U-U'\|^2_{\mathbb{H}^2_N}\\
&\quad\le \hat{K}\Big(\mathbb{E}\big[e^{\rho T}|\xi-\xi'|^2\big]\\
&\qquad+\mathbb{E}\Big[\int_0^T e^{\rho t}\big|f\big(t,Y(t),Z(t),U(t)\big)-f'\big(t,Y(t),Z(t),U(t)\big)\big|^2dt\Big]\Big),
\end{aligned} \tag{3.5}$$

and

$$\begin{aligned}
&\|Y-Y'\|^2_{\mathbb{S}^2}\\
&\quad\le \hat{K}\Big(\mathbb{E}\big[e^{\rho T}|\xi-\xi'|^2\big]\\
&\qquad+\mathbb{E}\Big[\int_0^T e^{\rho t}\big|f\big(t,Y(t),Z(t),U(t)\big)-f'\big(t,Y'(t),Z'(t),U'(t)\big)\big|^2dt\Big]\Big),
\end{aligned} \tag{3.6}$$

$$\begin{aligned}
&\|Y-Y'\|^2_{\mathbb{S}^2}\\
&\quad\le \hat{K}\Big(\mathbb{E}\big[e^{\rho T}|\xi-\xi'|^2\big]\\
&\qquad+\mathbb{E}\Big[\int_0^T e^{\rho t}\big|f\big(t,Y(t),Z(t),U(t)\big)-f'\big(t,Y(t),Z(t),U(t)\big)\big|^2dt\Big]\Big),
\end{aligned} \tag{3.7}$$

where the constant $\hat{K}$ *depends on* $T, K, \rho > 0$.

Proof 1. Estimate (3.4). We apply the Itô's formula to $e^{\rho t}|Y(t)-Y'(t)|^2$ and we derive

$$e^{\rho t}\big|Y(t)-Y'(t)\big|^2+\rho\int_t^\tau e^{\rho s}\big|Y(s)-Y'(s)\big|^2ds+\int_t^\tau e^{\rho s}\big|Z(s)-Z'(s)\big|^2ds$$

$$
\begin{aligned}
&+\int_t^\tau \int_{\mathbb{R}} e^{\rho s}\big|U(s,z)-U'(s,z)\big|^2 Q(s,dz)\eta(s)ds \\
&= e^{\rho\tau}\big|Y(\tau)-Y'(\tau)\big|^2 \\
&\quad -2\int_t^\tau e^{\rho s}\big(Y(s)-Y'(s)\big) \\
&\quad \cdot\big(-f\big(s,Y(s),Z(s),U(s)\big)+f'\big(s,Y'(s),Z'(s),U'(s)\big)\big)ds \\
&\quad -2\int_t^\tau e^{\rho s}\big(Y(s-)-Y'(s-)\big)\big(Z(s)-Z'(s)\big)dW(s) \\
&\quad -2\int_t^\tau \int_{\mathbb{R}} e^{\rho s}\big(Y(s-)-Y'(s-)\big)\big(U(s,z)-U'(s,z)\big)\tilde{N}(ds,dz) \\
&\quad -\int_t^\tau \int_{\mathbb{R}} e^{\rho s}\big|U(s,z)-U'(s,z)\big|^2 \tilde{N}(ds,dz), \quad 0\le t\le\tau\le T. \qquad (3.8)
\end{aligned}
$$

We can notice that the stochastic integrals in (3.8) are local martingales, see Theorem 2.3.1. From (3.8) we get

$$
\begin{aligned}
&\bigg|2\int_t^\tau e^{\rho s}\big(Y(s)-Y'(s)\big)\big(Z(s)-Z'(s)\big)dW(s) \\
&\qquad +2\int_t^\tau \int_{\mathbb{R}} e^{\rho s}\big(Y(s)-Y'(s)\big)\big(U(s,z)-U'(s,z)\big)\tilde{N}(ds,dz) \\
&\qquad +\int_t^\tau \int_{\mathbb{R}} e^{\rho s}\big|U(s,z)-U'(s,z)\big|^2 \tilde{N}(ds,dz)\bigg| \\
&\quad = \bigg| e^{\rho\tau}\big|Y(\tau)-Y'(\tau)\big|^2 \\
&\qquad -2\int_t^\tau e^{\rho s}\big(Y(s)-Y'(s)\big)\big(-f\big(s,Y(s),Z(s),U(s)\big) \\
&\qquad +f'\big(s,Y'(s),Z'(s),U'(s)\big)\big)ds - e^{\rho t}\big|Y(t)-Y'(t)\big|^2 \\
&\qquad -\rho\int_t^\tau e^{\rho s}\big|Y(s)-Y'(s)\big|^2 ds - \int_t^\tau e^{\rho s}\big|Z(s)-Z'(s)\big|^2 ds \\
&\qquad -\int_t^\tau \int_{\mathbb{R}} e^{\rho s}\big|U(s,z)-U'(s,z)\big|^2 Q(s,dz)\eta(s)ds\bigg|,
\end{aligned}
$$

and from the growth conditions for the Lipschitz generators f, f' and the assumptions that $(Y,Z,U),(Y',Z',U')\in\mathbb{S}^2(\mathbb{R})\times\mathbb{H}^2(\mathbb{R})\times\mathbb{H}^2_N(\mathbb{R})$ we deduce that the stochastic integrals are uniformly bounded by an integrable random variable ζ, i.e.

$$
\bigg|2\int_t^\tau e^{\rho s}\big(Y(s)-Y'(s)\big)\big(Z(s)-Z'(s)\big)dW(s)
$$

$$+2\int_t^\tau\int_{\mathbb{R}} e^{\rho s}\big(Y(s)-Y'(s)\big)\big(U(s,z)-U'(s,z)\big)\tilde{N}(ds,dz)$$
$$+\int_t^\tau\int_{\mathbb{R}} e^{\rho s}\big|U(s,z)-U'(s,z)\big|^2\tilde{N}(ds,dz)\bigg|\leq\zeta,\quad 0\leq t\leq\tau\leq T.$$

Hence, the stochastic integrals in (3.8) are uniformly integrable martingales, see Theorem I.51 in Protter (2004). Taking the expected value of (3.8) and estimating the Lebesgue integral in (3.8) from above, we obtain

$$\begin{aligned}
&\mathbb{E}\big[e^{\rho t}\big|Y(t)-Y'(t)\big|^2\big]+\rho\mathbb{E}\bigg[\int_t^T e^{\rho s}\big|Y(s)-Y'(s)\big|^2 ds\bigg]\\
&\quad+\mathbb{E}\bigg[\int_t^T e^{\rho s}\big|Z(s)-Z'(s)\big|^2 ds\bigg]\\
&\quad+\mathbb{E}\bigg[\int_t^T\int_{\mathbb{R}} e^{\rho s}\big|U(s,z)-U'(s,z)\big|^2 Q(s,dz)\eta(s)ds\bigg]\\
&\leq\mathbb{E}\big[e^{\rho T}|\xi-\xi'|^2\big]+2\mathbb{E}\bigg[\int_t^T e^{\rho s}\big|Y(s)-Y'(s)\big|\\
&\quad\cdot\big|f\big(s,Y(s),Z(s),U(s)\big)-f'\big(s,Y'(s),Z'(s),U'(s)\big)\big|ds\bigg],\quad 0\leq t\leq T.
\end{aligned}\tag{3.9}$$

Notice that for any $\alpha>0$ we have the inequality

$$2|uv|\leq\frac{1}{\alpha}|u|^2+\alpha|v|^2.\tag{3.10}$$

Consequently, we have the estimate

$$\begin{aligned}
&2\big|Y(s)-Y'(s)\big|\big|f\big(s,Y(s),Z(s),U(s)\big)-f'\big(s,Y'(s),Z'(s),U'(s)\big)\big|\\
&\quad\leq\alpha\big|Y(s)-Y'(s)\big|^2+\frac{1}{\alpha}\big|f\big(s,Y(s),Z(s),U(s)\big)-f'\big(s,Y'(s),Z'(s),U'(s)\big)\big|^2.
\end{aligned}$$

If we choose $\alpha=\rho$ and $t=0$, then from (3.9) we deduce the estimate

$$\begin{aligned}
&\mathbb{E}\big[e^{\rho t}\big|Y(t)-Y'(t)\big|^2\big]+\|Z-Z'\|^2_{\mathbb{H}^2}+\|U-U'\|^2_{\mathbb{H}^2_N}\\
&\quad\leq\mathbb{E}\big[e^{\rho T}|\xi-\xi'|^2\big]\\
&\qquad+\frac{1}{\rho}\mathbb{E}\bigg[\int_0^T e^{\rho t}\big|f\big(t,Y(t),Z(t),U(t)\big)-f'\big(t,Y'(t),Z'(t),U'(t)\big)\big|^2 dt\bigg],
\end{aligned}\tag{3.11}$$

which proves (3.4).

2. Estimate (3.5). For any $\alpha > 0$ we have the inequality

$$\begin{aligned}
&2\big|Y(s)-Y'(s)\big|\big|f\big(s,Y(s),Z(s),U(s)\big)-f'\big(s,Y'(s),Z'(s),U'(s)\big)\big| \\
&\quad\le 2\big|Y(s)-Y'(s)\big|\big|f\big(s,Y(s),Z(s),U(s)\big)-f'\big(s,Y(s),Z(s),U(s)\big)\big| \\
&\qquad + 2\big|Y(s)-Y'(s)\big|\big|f'\big(s,Y(s),Z(s),U(s)\big)-f'\big(s,Y'(s),Z'(s),U'(s)\big)\big| \\
&\quad\le 2\alpha\big|Y(s)-Y'(s)\big|^2+\frac{1}{\alpha}\big|f\big(s,Y(s),Z(s),U(s)\big)-f'\big(s,Y(s),Z(s),U(s)\big)\big|^2 \\
&\qquad +\frac{K}{\alpha}\Big(\big|Y(s)-Y'(s)\big|^2+\big|Z(s)-Z'(s)\big|^2 \\
&\qquad +\int_{\mathbb{R}}\big|U(s,z)-U'(s,z)\big|^2 Q(s,dz)\eta(s)\Big).
\end{aligned}$$

If we now choose $t=0$, α such that $\alpha > K$ and $\rho = 2\alpha + \frac{K}{\alpha}$ in (3.9), then we are done. Such a choice of α is only possible if ρ is sufficiently large. Hence, we first succeed in establishing (3.5) for large ρ. Next, we can easily deduce that (3.5) holds for any ρ.

Starting from (3.9) and reasoning as above, we can also establish the following estimates

$$\begin{aligned}
&\mathbb{E}\big[e^{\rho t}\big|Y(t)-Y'(t)\big|^2\big] \\
&\quad\le \mathbb{E}\big[e^{\rho T}|\xi-\xi'|^2\big] \\
&\qquad +\frac{1}{\rho}\mathbb{E}\bigg[\int_0^T e^{\rho s}\big|f\big(s,Y(s),Z(s),U(s)\big)-f'\big(s,Y'(s),Z'(s),U'(s)\big)\big|^2 ds\bigg], \\
&\quad 0\le t\le T,
\end{aligned} \tag{3.12}$$

and

$$\begin{aligned}
&\mathbb{E}\big[e^{\rho t}\big|Y(t)-Y'(t)\big|^2\big] \\
&\quad\le \hat{K}\bigg(\mathbb{E}\big[e^{\rho T}|\xi-\xi'|^2\big] \\
&\qquad +\mathbb{E}\bigg[\int_0^T e^{\rho s}\big|f\big(s,Y(s),Z(s),U(s)\big)-f'\big(s,Y(s),Z(s),U(s)\big)\big|^2 ds\bigg]\bigg), \\
&\quad 0\le t\le T,
\end{aligned} \tag{3.13}$$

which we use in the next part of this proof.

3. Estimate (3.6). From (3.8) we get

$$e^{\rho t}\big|Y(t)-Y'(t)\big|^2+\rho\int_t^T e^{\rho s}\big|Y(s)-Y'(s)\big|^2 ds+\int_t^T e^{\rho s}\big|Z(s)-Z'(s)\big|^2 ds$$

$$
\begin{aligned}
&+\int_t^T \int_{\mathbb{R}} e^{\rho s} \left|U(s,z)-U'(s,z)\right|^2 N(ds,dz) \\
&= e^{\rho T}\left|\xi-\xi'\right|^2 \\
&\quad -2\int_t^T e^{\rho s}\big(Y(s)-Y'(s)\big) \\
&\quad \cdot \big(-f\big(s,Y(s),Z(s),U(s)\big)+f'\big(s,Y'(s),Z'(s),U'(s)\big)\big)ds \\
&\quad -2\int_t^T e^{\rho s}\big(Y(s-)-Y'(s-)\big)\big(Z(s)-Z'(s)\big)dW(s) \\
&\quad -2\int_t^T \int_{\mathbb{R}} e^{\rho s}\big(Y(s-)-Y'(s-)\big)\big(U(s,z)-U'(s,z)\big)\tilde{N}(ds,dz), \\
&0 \le t \le T.
\end{aligned} \tag{3.14}
$$

It is straightforward to derive the following estimate

$$
\begin{aligned}
&\sup_{t\in[0,T]} e^{\rho t}\left|Y(t)-Y'(t)\right|^2 \\
&\quad \le e^{\rho T}\left|\xi-\xi'\right|^2 \\
&\qquad +2\int_0^T e^{\rho s}\left|Y(s)-Y'(s)\right| \\
&\qquad \cdot \left|f\big(s,Y(s),Z(s),U(s)\big)-f'\big(s,Y'(s),Z'(s),U'(s)\big)\right|ds \\
&\qquad +2\sup_{t\in[0,T]}\left|\int_t^T e^{\rho s}\big(Y(s-)-Y'(s-)\big)\big(Z(s)-Z'(s)\big)dW(s)\right| \\
&\qquad +2\sup_{t\in[0,T]}\left|\int_t^T \int_{\mathbb{R}} e^{\rho s}\big(Y(s-)-Y'(s-)\big)\big(U(s,z)-U'(s,z)\big)\tilde{N}(ds,dz)\right| \\
&\quad \le e^{\rho T}\left|\xi-\xi'\right|^2 \\
&\qquad +2\int_0^T e^{\rho s}\left|Y(s)-Y'(s)\right| \\
&\qquad \cdot \left|f\big(s,Y(s),Z(s),U(s)\big)-f'\big(s,Y'(s),Z'(s),U'(s)\big)\right|ds \\
&\qquad +4\sup_{t\in[0,T]}\left|\int_0^t e^{\rho s}\big(Y(s-)-Y'(s-)\big)\big(Z(s)-Z'(s)\big)dW(s)\right| \\
&\qquad +4\sup_{t\in[0,T]}\left|\int_0^t \int_{\mathbb{R}} e^{\rho s}\big(Y(s-)-Y'(s-)\big)\big(U(s,z)-U'(s,z)\big)\tilde{N}(ds,dz)\right|.
\end{aligned} \tag{3.15}
$$

By the Burkholder-Davis-Gundy inequality and some classical estimates we can deduce that for any $\alpha > 0$ we have

$$
\begin{aligned}
&\mathbb{E}\Big[\sup_{t\in[0,T]} e^{\rho t}\big|Y(t)-Y'(t)\big|^2\Big] \\
&\quad\le \mathbb{E}\big[e^{\rho T}\big|\xi-\xi'\big|^2\big] \\
&\qquad + 2\mathbb{E}\Big[\int_0^T e^{\rho s}\big|Y(s)-Y'(s)\big| \\
&\qquad \cdot \big|f\big(s,Y(s),Z(s),U(s)\big)-f'\big(s,Y'(s),Z'(s),U'(s)\big)\big|ds\Big] \\
&\qquad + 2K_1\mathbb{E}\Big[\Big\{\int_0^T e^{2\rho s}\big|Y(s)-Y'(s)\big|^2\big|Z(s)-Z'(s)\big|^2 ds\Big\}^{1/2}\Big] \\
&\qquad + 2K_2\mathbb{E}\Big[\Big\{\int_0^T\int_{\mathbb{R}} e^{2\rho s}\big|Y(s-)-Y'(s-)\big|^2\big|U(s,z)-U'(s,z)\big|^2 N(ds,dz)\Big\}^{1/2}\Big] \\
&\quad\le \mathbb{E}\big[e^{\rho T}\big|\xi-\xi'\big|^2\big] \\
&\qquad + 2\mathbb{E}\Big[\int_0^T e^{\rho s}\big|Y(s)-Y'(s)\big| \\
&\qquad \cdot \big|f\big(s,Y(s),Z(s),U(s)\big)-f'\big(s,Y'(s),Z'(s),U'(s)\big)\big|ds\Big] \\
&\qquad + 2K_1\mathbb{E}\Big[\sup_{t\in[0,T]} e^{\frac{\rho}{2}t}\big|Y(t)-Y'(t)\big|\Big\{\int_0^T e^{\rho s}\big|Z(s)-Z'(s)\big|^2 ds\Big\}^{1/2}\Big] \\
&\qquad + 2K_2\mathbb{E}\Big[\sup_{t\in[0,T]} e^{\frac{\rho}{2}t}\big|Y(t)-Y'(t)\big| \\
&\qquad \cdot \Big\{\int_0^T\int_{\mathbb{R}} e^{\rho s}\big|U(s,z)-U'(s,z)\big|^2 N(ds,dz)\Big\}^{1/2}\Big] \\
&\quad\le \mathbb{E}\big[e^{\rho T}\big|\xi-\xi'\big|^2\big] \\
&\qquad + 2\mathbb{E}\Big[\int_0^T e^{\rho s}\big|Y(s)-Y'(s)\big| \\
&\qquad \cdot \big|f\big(s,Y(s),Z(s),U(s)\big)-f'\big(s,Y'(s),Z'(s),U'(s)\big)\big|ds\Big] \\
&\qquad + \frac{K_1+K_2}{\alpha}\mathbb{E}\Big[\sup_{t\in[0,T]} e^{\rho t}\big|Y(t)-Y'(t)\big|^2\Big] \\
&\qquad + K_1\alpha\mathbb{E}\Big[\int_0^T e^{\rho s}\big|Z(s)-Z'(s)\big|^2 ds\Big]
\end{aligned}
$$

$$
\begin{aligned}
&+ K_2\alpha\mathbb{E}\bigg[\int_0^T\int_{\mathbb{R}} e^{\rho s}\big|U(s,z)-U'(s,z)\big|^2 N(ds,dz)\bigg]\\
&= \mathbb{E}\big[e^{\rho T}\big|\xi-\xi'\big|^2\big]\\
&\quad + 2\mathbb{E}\bigg[\int_0^T e^{\rho s}\big|Y(s)-Y'(s)\big|\\
&\quad \cdot\big|f\big(s,Y(s),Z(s),U(s)\big)-f'\big(s,Y'(s),Z'(s),U'(s)\big)\big|ds\bigg]\\
&\quad + \frac{K_1+K_2}{\alpha}\mathbb{E}\Big[\sup_{t\in[0,T]} e^{\rho t}\big|Y(t)-Y'(t)\big|^2\Big]\\
&\quad + K_1\alpha\mathbb{E}\bigg[\int_0^T e^{\rho s}\big|Z(s)-Z'(s)\big|^2 ds\bigg]\\
&\quad + K_2\alpha\mathbb{E}\bigg[\int_0^T\int_{\mathbb{R}} e^{\rho s}\big|U(s,z)-U'(s,z)\big|^2 Q(s,dz)\eta(s)ds\bigg].
\end{aligned}
\tag{3.16}
$$

If we choose $\alpha > K_1 + K_2$, then we can derive

$$
\begin{aligned}
&\big\|Y-Y'\big\|_{\mathbb{S}^2}^2\\
&\quad\le \tilde{K}\bigg(\mathbb{E}\big[e^{\rho T}\big|\xi-\xi'\big|^2\big]+\big\|Z-Z'\big\|_{\mathbb{H}^2}^2+\big\|U-U'\big\|_{\mathbb{H}_N^2}^2\\
&\qquad + 2\mathbb{E}\bigg[\int_0^T e^{\rho s}\big|Y(s)-Y'(s)\big|\\
&\qquad \cdot\big|f\big(s,Y(s),Z(s),U(s)\big)-f'\big(s,Y'(s),Z'(s),U'(s)\big)\big|ds\bigg]\bigg),
\end{aligned}
$$

where the constant $\tilde{K}$ is independent of ρ. Estimates (3.10) and (3.11) yield

$$
\begin{aligned}
&\big\|Y-Y'\big\|_{\mathbb{S}^2}^2\\
&\quad\le \tilde{K}\bigg(2\mathbb{E}\big[e^{\rho T}\big|\xi-\xi'\big|^2\big]+\sqrt{\rho}\mathbb{E}\bigg[\int_0^T e^{\rho s}\big|Y(s)-Y'(s)\big|^2 ds\bigg]\\
&\qquad + \frac{1}{\sqrt{\rho}}\mathbb{E}\bigg[\int_0^T e^{\rho s}\big|f\big(s,Y(s),Z(s),U(s)\big)-f'\big(s,Y'(s),Z'(s),U'(s)\big)\big|^2 ds\bigg]\\
&\qquad + \frac{1}{\rho}\mathbb{E}\bigg[\int_0^T e^{\rho s}\big|f\big(s,Y(s),Z(s),U(s)\big)-f'\big(s,Y'(s),Z'(s),U'(s)\big)\big|^2 ds\bigg]\bigg).
\end{aligned}
$$

By (3.12) we finally get

$$
\big\|Y-Y'\big\|_{\mathbb{S}^2}^2 \le \tilde{K}\Big((2+\sqrt{\rho}T)\mathbb{E}\big[e^{\rho T}\big|\xi-\xi'\big|^2\big]
$$

$$+\left(\frac{1}{\rho}+\frac{T+1}{\sqrt{\rho}}\right)$$
$$\cdot\mathbb{E}\left[\int_0^T e^{\rho s}\left|f\big(s,Y(s),Z(s),U(s)\big)-f'\big(s,Y'(s),Z'(s),U'(s)\big)\right|^2ds\right]\Bigg), \tag{3.17}$$

which proves (3.6).

4. Estimate (3.7). From (3.17) we derive

$$\begin{aligned}
&\|Y-Y'\|_{\mathbb{S}^2}^2\\
&\quad\le\hat{K}\Bigg(\mathbb{E}\big[e^{\rho T}|\xi-\xi'|^2\big]\\
&\qquad+\mathbb{E}\left[\int_0^T e^{\rho s}\left|f\big(s,Y(s),Z(s),U(s)\big)-f'\big(s,Y(s),Z(s),U(s)\big)\right|^2ds\right]\\
&\qquad+\mathbb{E}\left[\int_0^T e^{\rho s}\left|f'\big(s,Y(s),Z(s),U(s)\big)-f'\big(s,Y'(s),Z'(s),U'(s)\big)\right|^2ds\right]\Bigg).
\end{aligned}$$

We get (3.7) by the Lipschitz property of f' and inequalities (3.5) and (3.13). □

In Sect. 4.2 we need a modification of an estimate from Lemma 3.1.1. From (3.9) we get

$$\begin{aligned}
&\mathbb{E}\big[e^{\rho t}\left|Y(t)-Y'(t)\right|^2\big]+\rho\mathbb{E}\left[\int_t^T e^{\rho s}\left|Y(s)-Y'(s)\right|^2ds\right]\\
&\qquad+\mathbb{E}\left[\int_t^T e^{\rho s}\left|Z(s)-Z'(s)\right|^2ds\right]\\
&\qquad+\mathbb{E}\left[\int_t^T\int_{\mathbb{R}} e^{\rho s}\left|U(s,z)-U'(s,z)\right|^2Q(s,dz)\eta(s)ds\right]\\
&\quad\le\mathbb{E}\big[e^{\rho T}|\xi-\xi'|^2\big]\\
&\qquad+2\mathbb{E}\Bigg[\int_t^T e^{\rho s}\left|Y(s)-Y'(s)\right|\\
&\qquad\cdot\left|f\big(s,Y(s),Z(s),U(s)\big)-f'\big(s,Y(s),Z(s),U(s)\big)\right|ds\Bigg]\\
&\qquad+2\mathbb{E}\Bigg[\int_t^T e^{\rho s}\left|Y(s)-Y'(s)\right|\\
&\qquad\cdot\left|f'\big(s,Y(s),Z(s),U(s)\big)-f'\big(s,Y'(s),Z'(s),U'(s)\big)\right|ds\Bigg].
\end{aligned}$$

By the Lipschitz property of f' and (3.10) we derive

$$\begin{aligned}
&\mathbb{E}\big[e^{\rho t}\big|Y(t)-Y'(t)\big|^2\big]+\rho\mathbb{E}\bigg[\int_t^T e^{\rho s}\big|Y(s)-Y'(s)\big|^2 ds\bigg]\\
&\quad+\mathbb{E}\bigg[\int_t^T e^{\rho s}\big|Z(s)-Z'(s)\big|^2 ds\bigg]\\
&\quad+\mathbb{E}\bigg[\int_t^T\int_{\mathbb{R}} e^{\rho s}\big|U(s,z)-U'(s,z)\big|^2 Q(s,dz)\eta(s)ds\bigg]\\
&\le \mathbb{E}\big[e^{\rho T}|\xi-\xi'|^2\big]\\
&\quad+2\mathbb{E}\bigg[\int_t^T e^{\rho s}\big|Y(s)-Y'(s)\big|\\
&\quad\cdot\big|f\big(s,Y(s),Z(s),U(s)\big)-f'\big(s,Y(s),Z(s),U(s)\big)\big|ds\bigg]\\
&\quad+\mathbb{E}\bigg[\int_t^T e^{\rho s}\Big(\alpha+\frac{K}{\alpha}\Big)\big|Y(s)-Y'(s)\big|^2 ds\bigg]\\
&\quad+\mathbb{E}\bigg[\int_t^T e^{\rho s}\frac{K}{\alpha}\big|Z(s)-Z'(s)\big|^2 ds\bigg]\\
&\quad+\mathbb{E}\bigg[\int_t^T\int_{\mathbb{R}} e^{\rho s}\frac{K}{\alpha}\big|U(s,z)-U'(s,z)\big|^2 Q(s,dz)\eta(s)ds\bigg].
\end{aligned}$$

We choose sufficiently large ρ and α such that $\alpha>K$ and $\alpha+\frac{K}{\alpha}=\rho$. We obtain the estimate

$$\begin{aligned}
&\mathbb{E}\big[e^{\rho t}\big|Y(t)-Y'(t)\big|^2\big]+\mathbb{E}\bigg[\int_t^T e^{\rho s}\big|Z(s)-Z'(s)\big|^2 ds\bigg]\\
&\quad+\mathbb{E}\bigg[\int_t^T\int_{\mathbb{R}} e^{\rho s}\big|U(s,z)-U'(s,z)\big|^2 Q(s,dz)\eta(s)ds\bigg]\\
&\le \hat{K}\bigg(\mathbb{E}\big[e^{\rho T}|\xi-\xi'|^2\big]+\mathbb{E}\bigg[\int_t^T e^{\rho s}\big|Y(s)-Y'(s)\big|\\
&\quad\cdot\big|f\big(s,Y(s),Z(s),U(s)\big)-f'\big(s,Y(s),Z(s),U(s)\big)\big|ds\bigg]\bigg),\quad 0\le t\le T.
\end{aligned}\tag{3.18}$$

The a priori estimates from Lemma 3.1.1 are very useful in the study of BSDEs, and they are often applied in this book.

We now prove the existence of a unique solution to the BSDE (3.1). The idea is to construct a sequence of solutions to simpler BSDEs for which the existence of a unique solution can be established by Proposition 3.1.2 and the predictable representation property. Next, we show convergence of the sequence by using the a priori estimates.

Theorem 3.1.1 *Assume that* (A1)–(A3) *hold. The BSDE* (3.1) *has a unique solution* $(Y, Z, U) \in \mathbb{S}^2(\mathbb{R}) \times \mathbb{H}^2(\mathbb{R}) \times \mathbb{H}^2_N(\mathbb{R})$.

Proof 1. The Picard iteration. Let $Y^0(t) = Z^0(t) = U^0(t,z) = 0$, $(t,z) \in [0,T] \times \mathbb{R}$ and consider the recursive equation

$$Y^{n+1}(t) = \xi + \int_t^T f\big(s, Y^n(s-), Z^n(s), U^n(s)\big)ds - \int_t^T Z^{n+1}(s)dW(s) - \int_t^T \int_{\mathbb{R}} U^{n+1}(s,z)\tilde{N}(ds,dz), \quad 0 \le t \le T. \tag{3.19}$$

Assume that $(Y^n, Z^n, U^n) \in \mathbb{S}^2(\mathbb{R}) \times \mathbb{H}^2(\mathbb{R}) \times \mathbb{H}^2_N(\mathbb{R})$. Then

$$\mathbb{E}\Big[\int_0^T \big|f\big(t, Y^n(t), Z^n(t), U^n(t)\big)\big|^2 dt\Big] \le 2\mathbb{E}\Big[\int_0^T |f(t,0,0,0)|^2 dt\Big] + 2K\big(T\|Y^n\|^2_{\mathbb{S}^2} + \|Z^n\|^2_{\mathbb{H}^2} + \|U^n\|^2_{\mathbb{H}^2_N}\big) < \infty.$$

Hence, by Proposition 3.1.2 there exists a unique solution $(Y^{n+1}, Z^{n+1}, U^{n+1}) \in \mathbb{S}^2(\mathbb{R}) \times \mathbb{H}^2(\mathbb{R}) \times \mathbb{H}^2_N(\mathbb{R})$ to the BSDE (3.19) with the generator f independent of $(Y^{n+1}, Z^{n+1}, U^{n+1})$.

2. *The convergence of the sequence* $(Y^n, Z^n, U^n)_{n\in\mathbb{N}}$. In step 1 we constructed a sequence $(Y^n, Z^n, U^n) \in \mathbb{S}^2(\mathbb{R}) \times \mathbb{H}^2(\mathbb{R}) \times \mathbb{H}^2_N(\mathbb{R})$. From (3.11) and (3.17) we deduce the inequality

$$\begin{aligned}
&\|Y^{n+1} - Y^n\|^2_{\mathbb{S}^2} + \|Z^{n+1} - Z^n\|^2_{\mathbb{H}^2} + \|U^{n+1} - U^n\|^2_{\mathbb{H}^2_N} \\
&\le \Big(\frac{\tilde{K}+1}{\rho} + \frac{\tilde{K}(T+1)}{\sqrt{\rho}}\Big) \\
&\quad \cdot \mathbb{E}\Big[\int_0^T e^{\rho t}\big|f\big(t, Y^n(t), Z^n(t), U^n(t)\big) - f\big(t, Y^{n-1}(t), Z^{n-1}(t), U^{n-1}(t)\big)\big|^2 dt\Big],
\end{aligned} \tag{3.20}$$

where $\tilde{K}$ is independent of ρ. By the Lipschitz property of f we obtain

$$\begin{aligned}
&\|Y^{n+1} - Y^n\|^2_{\mathbb{S}^2} + \|Z^{n+1} - Z^n\|^2_{\mathbb{H}^2} + \|U^{n+1} - U^n\|^2_{\mathbb{H}^2_N} \\
&\le \bar{K}\Big(\frac{1}{\rho} + \frac{1}{\sqrt{\rho}}\Big)\big(\|Y^n - Y^{n-1}\|^2_{\mathbb{S}^2} + \|Z^n - Z^{n-1}\|^2_{\mathbb{H}^2} + \|U^n - U^{n-1}\|^2_{\mathbb{H}^2_N}\big),
\end{aligned} \tag{3.21}$$

which proves the contraction property for sufficiently large ρ. Hence, there exists a unique limit $(Y, Z, U) \in \mathbb{S}^2(\mathbb{R}) \times \mathbb{H}^2(\mathbb{R}) \times \mathbb{H}^2_N(\mathbb{R})$ of the converging sequence $(Y^n, Z^n, U^n)_{n\in\mathbb{N}}$.

3. The existence and uniqueness of a solution. The unique limit (Y, Z, U) satisfies the BSDE (3.1). Indeed, it is easy to show

$$\lim_{n\to\infty} \mathbb{E}\Big[\sup_{t\in[0,T]}\Big|\int_t^T f\big(s, Y^n(s), Z^n(s), U^n(s)\big)ds - \int_t^T f\big(s, Y(s), Z(s), U(s)\big)ds\Big|^2\Big] = 0,$$

$$\lim_{n\to\infty} \mathbb{E}\Big[\sup_{t\in[0,T]}\Big|\int_t^T Z^n(s)dW(s) - \int_t^T Z(s)dW(s)\Big|^2\Big] = 0,$$

$$\lim_{n\to\infty} \mathbb{E}\Big[\sup_{t\in[0,T]}\Big|\int_t^T \int_{\mathbb{R}} U^n(s,z)\tilde{N}(ds,dz) - \int_t^T \int_{\mathbb{R}} U(s,z)\tilde{N}(ds,dz)\Big|^2\Big] = 0.$$

The uniqueness of a solution to (3.1) follows immediately from the a priori estimates (3.5) and (3.7). □

We point out that by setting $\xi' = f' = Y' = Z' = U' = 0$ in (3.4)–(3.7) we can derive some useful norm estimates for the solution (Y, Z, U).

In applications considered in this book we also deal with BSDEs of the form

$$\begin{aligned} Y(t) &= \xi + \int_t^T f\big(s, Y(s), Z(s), U(s,.)\big)ds + \int_t^T C(s)dW(s) \\ &\quad + \int_t^T \int_{\mathbb{R}} V(s,z)N(ds,dz) - \int_t^T Z(s)dW(s) \\ &\quad - \int_t^T \int_{\mathbb{R}} U(s,z)\tilde{N}(ds,dz), \quad 0 \le t \le T, \end{aligned} \tag{3.22}$$

where C and V are given. If we assume that $\mathbb{E}[\int_0^T |C(t)|^2 dt] < \infty$, $\mathbb{E}[\int_0^T \int_{\mathbb{R}} |V(t,z)|^2 Q(t,dz)\eta(t)dt] < \infty$, $\mathbb{E}[\int_0^T |\int_{\mathbb{R}} V(t,z)Q(t,dz)\eta(t)|^2 dt] < \infty$, then by the change of variables

$$\hat{Z}(t) = Z(t) - C(t), \qquad \hat{U}(t,z) = U(t,z) - V(t,z),$$

we can investigate the BSDE

$$\begin{aligned} Y(t) &= \xi + \int_t^T f\big(s, Y(s), \hat{Z}(s) + C(s), \hat{U}(s,.) + V(s,.)\big)ds \\ &\quad + \int_t^T \int_{\mathbb{R}} V(s,z)Q(s,dz)\eta(s)ds \\ &\quad - \int_t^T \hat{Z}(s)dW(s) - \int_t^T \int_{\mathbb{R}} \hat{U}(s,z)\tilde{N}(ds,dz), \quad 0 \le t \le T, \end{aligned}$$

which now fits (3.1).

In this book we investigate BSDEs with jumps with Lipschitz generators and Theorem 3.1.1 is sufficient for our investigation. However, BSDEs with quadratic and exponential generators also play an important role in the theory and applications, see Sects. 3.3, 3.4, 13.1 and Chap. 11.

3.2 Comparison Principles

We state two versions of a comparison principle which allows us to compare the solutions to BSDEs based on the terminal conditions and the generators.

The first comparison principle for BSDEs with jumps was established by Barles et al. (1997).

Theorem 3.2.1 *Consider the BSDE*

$$\begin{aligned} Y(t) = \xi &+ \int_t^T f\Big(s, Y(s), Z(s), \int_{\mathbb{R}} U(s,z)\delta(s,z)Q(s,dz)\eta(s)\Big)ds \\ &- \int_t^T Z(s)dW(s) - \int_t^T \int_{\mathbb{R}} U(s,z)\tilde{N}(ds,dz), \quad 0 \le t \le T. \end{aligned} \tag{3.23}$$

Assume that

(i) *the terminal value* $\xi \in \mathbb{L}^2(\mathbb{R})$,
(ii) *the generator* $f : \Omega \times [0,T] \times \mathbb{R} \times \mathbb{R} \times \mathbb{R} \to \mathbb{R}$ *is predictable and Lipschitz continuous in the sense that*

$$\big|f(\omega,t,y,z,u) - f\big(\omega,t,y',z',u'\big)\big| \le K\big(\big|y-y'\big| + \big|z-z'\big| + \big|u-u'\big|\big),$$

a.s., a.e. $(\omega,t) \in \Omega \times [0,T]$, *for all* $(y,z,u), (y',z',u') \in \mathbb{R}\times\mathbb{R}\times\mathbb{R}$,
(iii) *for each* $(t,y,z) \in [0,T]\times\mathbb{R}\times\mathbb{R}$ *the mapping* $u \mapsto f(t,y,z,u)$ *is nondecreasing,*
(iv) $\mathbb{E}[\int_0^T |f(t,0,0,0)|^2 dt] < \infty$,
(v) *the process* $\delta : \Omega \times [0,T] \times \mathbb{R} \to \mathbb{R}$ *is predictable, non-negative and the mapping* $t \mapsto \int_{\mathbb{R}} |\delta(t,z)|^2 Q(t,dz)\eta(t)$ *is bounded,*

and let (ξ', f') *satisfy the same set of assumptions. Let* $(Y,Z,U) \in \mathbb{S}^2(\mathbb{R}) \times \mathbb{H}^2(\mathbb{R}) \times \mathbb{H}^2_N(\mathbb{R})$ *and* $(Y',Z',U') \in \mathbb{S}^2(\mathbb{R}) \times \mathbb{H}^2(\mathbb{R}) \times \mathbb{H}^2_N(\mathbb{R})$ *be the unique solutions to the BSDE* (3.23) *with* (ξ, f) *and* (ξ', f'). *If*

- $\bar{\xi} = \xi - \xi' \ge 0$,
- $\bar{f}(t,y,z,u) = f(t,y,z,u) - f'(t,y,z,u) \ge 0$, $(t,y,z,u) \in [0,T]\times\mathbb{R}\times\mathbb{R}\times\mathbb{R}$,

then $Y(t) \ge Y'(t)$, $0 \le t \le T$. *In addition, if* $Y(t_0) = Y'(t_0)$ *a.s. for some* $t_0 \in [0,T]$, *then* $Y(t) = Y'(t)$, $t_0 \le t \le T$.

Proof The existence of unique solutions (Y, Z, U), (Y', Z', U') follows from Theorem 3.1.1. We next define

$$\bar{Y}(t) = Y(t) - Y'(t), \qquad \bar{Z}(t) = Z(t) - Z'(t), \qquad \bar{U}(t) = U(t) - U'(t),$$

and

$$\begin{aligned}
&\Delta_y f(t) \\
&\quad = \biggl(f\biggl(t, Y(t-), Z(t), \int_{\mathbb{R}} U(t,z)\delta(t,z)Q(t,dz)\eta(t)\biggr) \\
&\qquad - f\biggl(t, Y'(t-), Z(t), \int_{\mathbb{R}} U(t,z)\delta(t,z)Q(t,dz)\eta(t)\biggr)\biggr)/\bigl(Y(t-) - Y'(t-)\bigr) \\
&\qquad \cdot \mathbf{1}\bigl\{\bar{Y}(t-) \neq 0\bigr\}, \\
&\Delta_z f(t) \\
&\quad = \biggl(f\biggl(t, Y'(t-), Z(t), \int_{\mathbb{R}} U(t,z)\delta(t,z)Q(t,dz)\eta(t)\biggr) \\
&\qquad - f\biggl(t, Y'(t-), Z'(t), \int_{\mathbb{R}} U(t,z)\delta(t,z)Q(t,dz)\eta(t)\biggr)\biggr)/\bigl(Z(t) - Z'(t)\bigr) \\
&\qquad \cdot \mathbf{1}\bigl\{\bar{Z}(t) \neq 0\bigr\}, \\
&\Delta_u f(t) \\
&\quad = \biggl(f\biggl(t, Y'(t-), Z'(t), \int_{\mathbb{R}} U(t,z)\delta(t,z)Q(t,dz)\eta(t)\biggr) \\
&\qquad - f\biggl(t, Y'(t-), Z'(t), \int_{\mathbb{R}} U'(t,z)\delta(t,z)Q(t,dz)\eta(t)\biggr)\biggr) \\
&\qquad /\biggl(\int_{\mathbb{R}} \bigl(U(t) - U'(t)\bigr)\delta(t,z)Q(t,dz)\eta(t)\biggr)\mathbf{1}\bigl\{\bar{U}(t) \neq 0\bigr\}.
\end{aligned}$$

Notice that the processes Δ_y, Δ_z, Δ_u are bounded by the Lipschitz property of f. We get the BSDE

$$\begin{aligned}
&\bar{Y}(t) \\
&\quad = \bar{\xi} \\
&\qquad + \int_t^T \biggl(\Delta_y f(s)\bar{Y}(s-) + \Delta_z f(s)\bar{Z}(s) + \Delta_u f(s)\int_{\mathbb{R}} \bar{U}(s,z)\delta(s,z)Q(s,dz)\eta(s) \\
&\qquad + \bar{f}\biggl(s, Y'(s-), Z'(s), \int_{\mathbb{R}} U'(s,z)\delta(s,z)Q(s,dz)\eta(s)\biggr)\biggr)ds \\
&\qquad - \int_t^T \bar{Z}(s)dW(s) - \int_t^T \int_{\mathbb{R}} \bar{U}(s,z)\tilde{N}(ds,dz), \quad 0 \leq t \leq T,
\end{aligned}$$

and we derive

$$\begin{aligned}
\bar{Y}(t) &= \bar{\xi} e^{\int_t^T \Delta_y f(s)ds} \\
&+ \int_t^T \bar{f}\Big(s, Y'(s), Z'(s), \int_{\mathbb{R}} U'(s,z)\delta(z,s)Q(s,dz)\eta(s)\Big) e^{\int_t^s \Delta_y f(u)du} ds \\
&- \int_t^T \bar{Z}(s) e^{\int_t^s \Delta_y f(u)du} dW(s) + \int_t^T \bar{Z}(s) e^{\int_t^s \Delta_y f(u)du} \Delta_z f(s) ds \\
&- \int_t^T \int_{\mathbb{R}} \bar{U}(s,z) e^{\int_t^s \Delta_y f(u)du} \tilde{N}(ds,dz) \\
&+ \int_t^T \int_{\mathbb{R}} \bar{U}(s,z) e^{\int_t^s \Delta_y f(u)du} \Delta_u f(s)\delta(s,z)Q(s,z)\eta(s)ds, \quad 0 \le t \le T.
\end{aligned}$$

Let us introduce an equivalent probability measure $\mathbb{Q}$ by the Radon-Nikodym derivative

$$\begin{aligned}
&\frac{d\mathbb{Q}}{d\mathbb{P}}\Big|_{\mathscr{F}_t} = M(t), \quad 0 \le t \le T, \\
&dM(t) = M(t-)\Big(\Delta_z f(t)dW(t) + \int_{\mathbb{R}} \Delta f_u(t)\delta(t,z)\tilde{N}(dt,dz)\Big).
\end{aligned}$$

Since the process δ is non-negative and the generator f is non-decreasing in u, the kernel $\Delta f_u(t)\delta(t,z)$ is non-negative. Since the mappings $t \mapsto |\Delta_z f(t)|^2$ and $t \mapsto \int_{\mathbb{R}} |\Delta f_u(t)|^2 |\delta(t,z)|^2 Q(t,dz)\eta(t)$ are bounded, by Proposition 2.5.1 the process M is a square integrable martingale. Hence, M defines an equivalent probability measure. By the change of measure, see Theorem 2.5.1, we derive the equation

$$\begin{aligned}
\bar{Y}(t) &= \bar{\xi} e^{\int_t^T \Delta_y f(s)ds} \\
&+ \int_t^T \bar{f}\Big(s, Y'(s), Z'(s), \int_{\mathbb{R}} U'(s,z)\delta(s,z)Q(s,dz)\eta(s)\Big) e^{\int_t^s \Delta_y f(u)du} ds \\
&- \int_t^T \bar{Z}(s) e^{\int_t^s \Delta_y f(u)du} dW^{\mathbb{Q}}(s) \\
&- \int_t^T \int_{\mathbb{R}} \bar{U}(s,z) e^{\int_t^s \Delta_y f(u)du} \tilde{N}^{\mathbb{Q}}(ds,dz), \quad 0 \le t \le T. \qquad (3.24)
\end{aligned}$$

We conclude that the stochastic integrals in (3.24) are $\mathbb{Q}$-local martingales. By the Burkholder-Davis-Gundy inequality, the Cauchy-Schwarz inequality, boundedness of $\Delta_y f$ and Theorem 2.3.2 we obtain

$$\mathbb{E}^{\mathbb{Q}}\Big[\sup_{t\in[0,T]}\Big|\int_0^t \bar{Z}(s) e^{\int_t^s \Delta_y f(u)du} dW^{\mathbb{Q}}(s)\Big|\Big]$$

$$\leq K\mathbb{E}^{\mathbb{Q}}\Big[\sqrt{\int_0^T \big|\bar{Z}(s)e^{\int_0^s \Delta_y f(u)du}\big|^2 ds}\Big] \leq K\mathbb{E}\Big[\Big|\frac{d\mathbb{Q}}{d\mathbb{P}}\Big|^2\Big]\mathbb{E}\Big[\int_0^T \big|\bar{Z}(s)\big|^2 ds\Big] < \infty, \tag{3.25}$$

and

$$\begin{aligned}
&\mathbb{E}^{\mathbb{Q}}\Big[\sup_{t\in[0,T]}\Big|\int_0^t \int_{\mathbb{R}} \bar{U}(s,z)e^{\int_t^s \Delta_y f(u)du}\tilde{N}^{\mathbb{Q}}(ds,dz)\Big|\Big] \\
&\quad \leq K\mathbb{E}^{\mathbb{Q}}\Big[\sqrt{\int_0^T \int_{\mathbb{R}} \big|\bar{U}(s,z)e^{\int_t^s \Delta_y f(u)du}\big|^2 N(ds,dz)}\Big] \\
&\quad \leq K\mathbb{E}\Big[\Big|\frac{d\mathbb{Q}}{d\mathbb{P}}\Big|^2\Big]\mathbb{E}\Big[\int_0^T \int_{\mathbb{R}} \big|\bar{U}(s,z)\big|^2 N(ds,dz)\Big] \\
&\quad = K\mathbb{E}\Big[\Big|\frac{d\mathbb{Q}}{d\mathbb{P}}\Big|^2\Big]\mathbb{E}\Big[\int_0^T \int_{\mathbb{R}} \big|\bar{U}(s,z)\big|^2 Q(s,dz)\eta(s)ds\Big] < \infty.
\end{aligned} \tag{3.26}$$

From the last two estimates we can conclude that the stochastic integrals in (3.24) are true $\mathbb{Q}$-martingales, see Theorem I.51 in Protter (2004). Taking the conditional expectation of (3.24), we get

$$\begin{aligned}
\bar{Y}(t) = \mathbb{E}^{\mathbb{Q}}\Big[&\bar{\xi}e^{\int_t^T \Delta_y f(s)ds} \\
&+ \int_t^T \bar{f}\Big(s, Y'(s), Z'(s), \int_{\mathbb{R}} U'(s,z)\delta(s,z)Q(s,dz)\eta(s)\Big) \\
&\cdot e^{\int_t^s \Delta_y f(u)du}ds|\mathscr{F}_t\Big].
\end{aligned} \tag{3.27}$$

The first assertion of our theorem is proved by (3.27). We now prove the second assertion. From (3.27) and the assumptions $\bar{Y}(t_0) = 0$, $\bar{\xi} \geq 0$, $\bar{f}(s, Y'(s), Z'(s), \int_{\mathbb{R}} U'(s,z)\delta(s,z)Q(s,dz)\eta(s)) \geq 0$ we deduce that $\bar{\xi} = 0$ a.s. and $\bar{f}(s, Y'(s), Z'(s), \int_{\mathbb{R}} U'(s,z)\delta(s,z)Q(s,dz)\eta(s)) = 0$, a.s., a.e. $t_0 \leq s \leq T$. Hence, from (3.24) we obtain

$$\begin{aligned}
\bar{Y}(t) = &-\int_t^T \bar{Z}(s)e^{\int_t^s \Delta_y f(u)du}dW^{\mathbb{Q}}(s) \\
&-\int_t^T \int_{\mathbb{R}} \bar{U}(s,z)e^{\int_t^s \Delta_y f(u)du}\tilde{N}^{\mathbb{Q}}(ds,dz), \quad t_0 \leq t \leq T,
\end{aligned}$$

and the second assertion is proved by taking the conditional expectation. □

The comparison principle from Theorem 3.2.1 holds for a special case of a BSDE under strong assumptions on the generator. In order to weaken the assumptions on the generator, Royer (2006) proved a different version of a comparison principle.

Theorem 3.2.2 *Consider the BSDE* (3.1). *Assume that*

(i) *the terminal value* $\xi \in \mathbb{L}^2(\mathbb{R})$,
(ii) *the generator* $f : \Omega \times [0,T] \times \mathbb{R} \times \mathbb{R} \times L^2_Q(\mathbb{R}) \to \mathbb{R}$ *is predictable and satisfies*

$$\begin{aligned}
&\left|f(\omega,t,y,z,u) - f\big(\omega,t,y',z',u\big)\right| \le K\big(|y-y'| + |z-z'|\big),\\
&f(\omega,t,y,z,u) - f\big(\omega,t,y,z,u'\big)\\
&\quad\le \int_{\mathbb{R}} \delta^{y,z,u,u'}(t,x)\big(u(x) - u'(x)\big)Q(t,dx)\eta(t),
\end{aligned}$$

a.s., a.e. $(\omega,t) \in \Omega \times [0,T]$, *for all* $(y,z,u), (y'z',u), (y,z,u') \in \mathbb{R} \times \mathbb{R} \times L^2_Q(\mathbb{R})$, *where* $\delta^{y,z,u,u'} : \Omega \times [0,T] \times \mathbb{R} \to (-1,\infty)$ *is a predictable process such that the mapping* $t \mapsto \int_{\mathbb{R}} |\delta^{y,z,u,u'}(t,x)|^2 Q(t,dx)\eta(t)$ *is uniformly bounded in* (y,z,u,u'),
(iii) $\mathbb{E}[\int_0^T |f(t,0,0,0)|^2 dt] < \infty$,

and let (ξ', f') *satisfy* (A1)–(A3). *Let* $(Y,Z,U) \in \mathbb{S}^2(\mathbb{R}) \times \mathbb{H}^2(\mathbb{R}) \times \mathbb{H}^2_N(\mathbb{R})$ *and* $(Y',Z',U') \in \mathbb{S}^2(\mathbb{R}) \times \mathbb{H}^2(\mathbb{R}) \times \mathbb{H}^2_N(\mathbb{R})$ *be the unique solutions to the BSDE* (3.1) *with* (ξ,f) *and* (ξ',f'). *If*

- $\bar{\xi} = \xi - \xi' \ge 0$,
- $\bar{f}(t,y,z,u) = f(t,y,z,u) - f'(t,y,z,u) \ge 0$, $(t,y,z,u) \in [0,T] \times \mathbb{R} \times \mathbb{R} \times \mathbb{R}$,

then $Y(t) \ge Y'(t)$, $0 \le t \le T$. *In addition, if* $Y(t_0) = Y'(t_0)$ *a.s. for some* $t_0 \in [0,T]$, *then* $Y(t) = Y'(t)$, $t_0 \le t \le T$.

Proof From assumption (ii) we deduce that the generator f also satisfies the following inequality

$$f(\omega,t,y,z,u) - f\big(\omega,t,y,z,u'\big) \ge \int_{\mathbb{R}} \tilde{\delta}^{y,z,u,u'}(t,x)\big(u(x) - u'(x)\big)Q(t,dx)\eta(t), \tag{3.28}$$

a.s., a.e. $(\omega,t) \in \Omega \times [0,T]$, for all $(y,z,u), (y,z,u') \in \mathbb{R} \times \mathbb{R} \times L^2_Q(\mathbb{R})$, where $\tilde{\delta}^{y,z,u,u'} : \Omega \times [0,T] \times \mathbb{R} \to (-1,\infty)$ is a predictable process such that $t \mapsto \int_{\mathbb{R}} |\tilde{\delta}^{y,z,u,u'}(t,x)|^2 Q(t,dx)\eta(t)$ is uniformly bounded in (y,z,u,u'). Hence, the generator f is Lipschitz continuous in u in the sense of (A2). We conclude that there exists a unique solution (Y,Z,U) to (3.1) with (ξ,f). The existence of a unique solution (Y',Z',U') follows from Theorem 3.1.1.

Arguing as in the first part of the proof of Theorem 3.2.1, we get

$$\begin{aligned}
\bar{Y}(t) = \bar{\xi}e^{\int_t^T \Delta_y f(s)ds} &+ \int_t^T \bar{f}\big(s, Y'(s), Z'(s), U'(s)\big)e^{\int_t^s \Delta_y f(u)du}ds\\
&- \int_t^T \bar{Z}(s)e^{\int_t^s \Delta_y f(u)du}dW(s) + \int_t^T \bar{Z}(s)e^{\int_t^s \Delta_y f(u)du}\Delta_z f(s)ds
\end{aligned}$$

$$
\begin{aligned}
&- \int_t^T \int_{\mathbb{R}} \bar{U}(s,z) e^{\int_t^s \Delta_y f(u)du} \tilde{N}(ds,dz) \\
&+ \int_t^T \big(f\big(s, Y'(s), Z'(s), U(s)\big) \\
&- f\big(s, Y'(s), Z'(s), U'(s)\big)\big) e^{\int_t^s \Delta_y f(u)du} ds, \quad 0 \le t \le T. \qquad (3.29)
\end{aligned}
$$

Applying (3.28), we derive the inequality

$$
\begin{aligned}
\bar{Y}(t) \ge{}& \bar{\xi} e^{\int_t^T \Delta_y f(s)ds} + \int_t^T \bar{f}\big(s, Y'(s), Z'(s), U'(s)\big) e^{\int_t^s \Delta_y f(u)du} ds \\
&- \int_t^T \bar{Z}(s) e^{\int_t^s \Delta_y f(u)du} dW(s) + \int_t^T \bar{Z}(s) e^{\int_t^s \Delta_y f(u)du} \Delta_z f(s) ds \\
&- \int_t^T \int_{\mathbb{R}} \bar{U}(s,z) e^{\int_t^s \Delta_y f(u)du} \tilde{N}(ds,dz) \\
&+ \int_t^T \int_{\mathbb{R}} \bar{U}(s,z) e^{\int_t^s \Delta_y f(u)du} \tilde{\delta}(s,z) Q(s,dz) \eta(s) ds, \quad 0 \le t \le T,
\end{aligned}
$$

and by the change of measure we obtain

$$
\begin{aligned}
\bar{Y}(t) \ge{}& \bar{\xi} e^{\int_t^T \Delta_y f(s)ds} + \int_t^T \bar{f}\big(s, Y'(s), Z'(s), U'(s)\big) e^{\int_t^s \Delta_y f(u)du} ds \\
&- \int_t^T \bar{Z}(s) e^{\int_t^s \Delta_y f(u)du} dW^{\mathbb{Q}}(s) \\
&- \int_t^T \int_{\mathbb{R}} \bar{U}(s,z) e^{\int_t^s \Delta_y f(u)du} \tilde{N}^{\mathbb{Q}}(ds,dz), \quad 0 \le t \le T, \qquad (3.30)
\end{aligned}
$$

where we introduce

$$
\begin{aligned}
&\frac{d\mathbb{Q}}{d\mathbb{P}}\Big|_{\mathscr{F}_t} = M(t), \quad 0 \le t \le T, \\
&dM(t) = M(t-)\Big(\Delta_z f(t) dW(t) + \int_{\mathbb{R}} \tilde{\delta}(t,z) \tilde{N}(dt,dz) \Big).
\end{aligned}
$$

Taking the conditional expected value of (3.30) under $\mathbb{Q}$ and taking into account the assumptions on the generators and the terminal conditions, we obtain $\bar{Y}(t) \ge 0$, $0 \le t \le T$, and the first assertion of our theorem is proved, see the proof of Theorem 3.2.1 for details. We prove the second assertion. From (3.30) we get

$$
\begin{aligned}
\bar{Y}(t_0) \ge \mathbb{E}^{\mathbb{Q}}\Big[& \bar{\xi} e^{\int_{t_0}^T \Delta_y f(s)ds} \\
&+ \int_{t_0}^T \bar{f}\big(s, Y'(s), Z'(s), U'(s)\big) e^{\int_{t_0}^s \Delta_y f(u)du} ds \,|\, \mathscr{F}_{t_0} \Big], \qquad (3.31)
\end{aligned}
$$

which together with $\bar{Y}(t_0)=0$, $\bar{\xi}\geq 0$, $\bar{f}(s, Y'(s), Z'(s), \int_{\mathbb{R}} U'(s,z)\delta(s,z)\times Q(s,dz)\eta(s))\geq 0$ yield that $\bar{\xi}=0$ a.s. and $\bar{f}(s, Y'(s), Z'(s), \int_{\mathbb{R}} U'(s,z)\delta(s,z)\times Q(s,dz)\eta(s))=0$, a.s., a.e. $t_0\leq s\leq T$. From (3.29) and the assumption on the generator f we now deduce the inequality

$$\begin{aligned}\bar{Y}(t) \leq &-\int_t^T \bar{Z}(s)e^{\int_t^s \Delta_y f(u)du}dW(s)+\int_t^T \bar{Z}(s)e^{\int_t^s \Delta_y f(u)du}\Delta_z f(s)ds\\&-\int_t^T\int_{\mathbb{R}} \bar{U}(s,z)e^{\int_t^s \Delta_y f(u)du}\tilde{N}(ds,dz)\\&+\int_t^T \bar{U}(s,z)e^{\int_t^s \Delta_y f(u)du}\delta(s,z)Q(s,dz)\eta(s)ds,\quad t_0\leq t\leq T.\end{aligned}$$

By the change of measure we obtain $\bar{Y}(t)\leq 0$, $t_0\leq t\leq T$. Hence, $\bar{Y}(t)=0$, $t_0\leq t\leq T$. □

A comparison principle is an important tool in the study of BSDEs. In this book we apply both versions of a comparison principle from Theorem 3.2.1 and Theorem 3.2.2. In the financial context the comparison principle implies desirable properties of pricing equations. For example, it implies that a more severe claim ($\xi'\geq\xi$) must have a higher price ($Y'\geq Y$).

We point out that the comparison principle must be applied with care. Notice that the assumptions of Theorems 3.2.1 and 3.2.2 are more restrictive than the assumptions of Theorem 3.1.1. We remark that under the assumptions of Theorem 3.1.1 a comparison does not hold unless we deal with the natural Brownian filtration and a BSDE driven by a Brownian motion.

Example 3.1 Consider the BSDE

$$Y(t)=\xi-2\int_t^T U(s)ds-\int_t^T Z(s)dW(s)-\int_t^T U(s)\tilde{N}(ds),\quad 0\leq t\leq T,$$

driven by the random measure N generated by the Poisson process with intensity 1. Choose $\xi=\int_0^T N(ds)$, then $Y(t)=\int_0^t N(ds)-(T-t)$, $Z(t)=0$, $U(t)=1$ is the unique solution to the BSDE. Choose $\xi'=0$, then $Y'(t)=0$, $Z'(t)=0$, $U'(t)=0$ is the unique solution to the BSDE. Even though $\xi\geq\xi'$, the inequality $Y(t)\geq Y'(t)$ may fail.

A financial example in which a price process solves a BSDE with jumps but does not satisfy the comparison principle and the property of monotonicity with respect to the claim is presented in Sect. 10.4.

3.3 Examples of Linear and Nonlinear BSDEs Without Jumps

In this chapter we consider the natural Brownian filtration $\mathscr{F}^W$ and we investigate three important types of BSDEs driven by a Brownian motion. Our goal is to illustrate some useful techniques for BSDEs and for this reason we omit jumps in the equations. In particular, we show that some BSDEs can be reduce to a BSDE with zero generator and, consequently, finding the solutions to those BSDEs can be reduced to finding the predictable representation of a random variable. BSDEs with jumps are investigated in the next chapter.

Let us introduce a linear backward stochastic differential equation. A linear BSDE is an equation of the form

$$Y(t) = \xi + \int_t^T \alpha(s)Y(s)ds + \int_t^T \beta(s)Z(s)ds - \int_t^T Z(s)dW(s), \quad 0 \leq t \leq T. \tag{3.32}$$

Proposition 3.3.1 *Let ξ be an $\mathscr{F}^W$-measurable random variable such that $\xi \in \mathbb{L}^2(\mathbb{R})$, and let α and β be $\mathscr{F}^W$-predictable, bounded processes. There exists a unique solution $(Y, Z) \in \mathbb{S}^2(\mathbb{R}) \times \mathbb{H}^2(\mathbb{R})$ to the BSDE* (3.32). *The process Y has the representation*

$$Y(t) = \mathbb{E}^{\mathbb{Q}}\big[e^{\int_t^T \alpha(s)ds)}\xi|\mathscr{F}_t^W\big], \quad 0 \leq t \leq T,$$

and the control process Z is derived from

$$Z(t) = e^{-\int_0^t \alpha(s)ds}\mathscr{Z}(t), \quad 0 \leq t \leq T,$$

and the representation

$$e^{\int_0^T \alpha(s)ds}\xi = \mathbb{E}^{\mathbb{Q}}\big[e^{\int_0^T \alpha(s)ds}\xi\big] + \int_0^T \mathscr{Z}(s)dW^{\mathbb{Q}}(s),$$

where the equivalent probability measure $\mathbb{Q}$ is defined by

$$\frac{d\mathbb{Q}}{d\mathbb{P}}\Big|\mathscr{F}_T^W = e^{\int_0^T \beta(s)dW(s) - \frac{1}{2}\int_0^T \beta^2(s)ds}.$$

Proof The existence of a unique solution (Y, Z) follows from Theorem 3.1.1. We prove the representation of the solution. Recalling Proposition 2.5.1 and (2.15), we notice that $\mathbb{Q}$ is an equivalent probability measure. We introduce the processes

$$\hat{Y}(t) = Y(t)e^{\int_0^t \alpha(s)ds}, \quad \hat{Z}(t) = Z(t)e^{\int_0^t \alpha(s)ds}, \quad 0 \leq t \leq T.$$

By the Itô's formula the unique solution (Y, Z) to (3.32) must satisfies the BSDE

$$\hat{Y}(t) = e^{\int_0^T \alpha(s)ds}\xi - \int_t^T \hat{Z}(s)dW^{\mathbb{Q}}(s), \quad 0 \leq t \leq T,$$

or

$$\hat{Y}(t) = \hat{Y}(0) + \int_0^t \hat{Z}(s)dW^{\mathbb{Q}}(s), \quad 0 \le t \le T.$$

Since the process $\hat{Z}$ is a.s. square integrable, we first deduce that $\hat{Y}$ is a $\mathbb{Q}$-local martingale, and following (3.25) we next deduce that $\hat{Y}$ is a $\mathbb{Q}$-martingale. Hence, we obtain the representation $\hat{Y}(t) = \mathbb{E}^{\mathbb{Q}}[\hat{Y}(T)|\mathscr{F}_t^W]$. The process $\hat{Z}$ is now derived from the predictable representation of the $\mathbb{Q}$-martingale $\hat{Y}$. □

We will observe that linear BSDEs arise when we investigate pricing and hedging problems in complete markets and when we deal with quadratic pricing and hedging in incomplete markets.

Next, we consider the backward stochastic differential equation

$$Y(t) = \xi + \int_t^T \beta(s)\big|Z(s)\big|ds - \int_t^T Z(s)dW(s), \quad 0 \le t \le T. \tag{3.33}$$

Proposition 3.3.2 *Let ξ be an $\mathscr{F}^W$-measurable random variable such that $\xi \in \mathbb{L}^2(\mathbb{R})$, and let β be an $\mathscr{F}^W$-predictable, positive, bounded process. There exists a unique solution $(Y, Z) \in \mathbb{S}^2(\mathbb{R}) \times \mathbb{H}^2(\mathbb{R})$ to the BSDE* (3.33). *The process Y has the representation*

$$Y(t) = \sup_{\mathbb{Q}\in\mathscr{Q}} \mathbb{E}^{\mathbb{Q}}\big[\xi|\mathscr{F}_t^W\big], \quad 0 \le t \le T,$$

where

$$\mathscr{Q} = \bigg\{\mathbb{Q} \sim \mathbb{P} : \frac{d\mathbb{Q}}{d\mathbb{P}}\Big|\mathscr{F}_T^W = e^{\int_0^T \phi(s)dW(s) - \frac{1}{2}\int_0^T |\phi(s)|^2 ds},$$
$$\phi \text{ is } \mathscr{F}^W\text{-predictable}, \ \big|\phi(t)\big| \le \beta(t)\bigg\}.$$

Proof The existence of a unique solution (Y, Z) follows from Theorem 3.1.1. We prove the representation of the solution. Let us deal with the BSDE

$$Y^{\phi}(t) = \xi + \int_t^T \phi(s)Z^{\phi}(s)ds - \int_t^T Z^{\phi}(s)dW(s), \quad 0 \le t \le T, \tag{3.34}$$

with a predictable process ϕ such that $|\phi(t)| \le \beta(t)$. By Theorem 3.1.1 there exists a unique solution $(Y^{\phi}, Z^{\phi}) \in \mathbb{S}^2(\mathbb{R}) \times \mathbb{H}^2(\mathbb{R})$ to (3.34). Since $|\phi(t)| \le \beta(t)$, we have $\phi(t)z \le \beta(t)|z|$, $(t, z) \in [0, T] \times \mathbb{R}$, and from the comparison principle we deduce that $Y^{\phi}(t) \le Y(t)$, $0 \le t \le T$. We get

$$\sup_{|\phi(t)|\le\beta(t)} Y^{\phi}(t) \le Y(t), \quad 0 \le t \le T. \tag{3.35}$$

If we choose $\phi^*(t) = \frac{|Z(t)|}{Z(t)}\beta(t)\mathbf{1}\{Z(t) \neq 0\}$, then by uniqueness of solution to (3.34) we must have $Z^{\phi^*}(t) = Z(t)$, $Y^{\phi^*}(t) = Y(t)$, $0 \le t \le T$. We also get

$$Y(t) = Y^{\phi^*}(t) \le \sup_{|\phi(t)| \le \beta(t)} Y^{\phi}(t), \quad 0 \le t \le T. \tag{3.36}$$

We can now deduce from (3.35)–(3.36) that $Y(t) = \sup_{|\phi(t)|\le\beta(t)} Y^{\phi}(t)$, $0 \le t \le T$. Finally, Proposition 3.3.1 yields the representation

$$Y^{\phi}(t) = \mathbb{E}^{\mathbb{Q}^{\phi}}\big[\xi|\mathscr{F}_t^W\big], \quad 0 \le t \le T,$$

under the equivalent probability measure

$$\frac{d\mathbb{Q}^{\phi}}{d\mathbb{P}}\Big|\mathscr{F}_T^W = e^{\int_0^T \phi(s)dW(s) - \frac{1}{2}\int_0^T |\phi(s)|^2 ds}. \qquad \square$$

We will observe that nonlinear BSDEs of the form (3.33) arise when we deal with an ambiguity risk measure and no-good-deal pricing. If the control process Z does not change its sign, then the nonlinear BSDE (3.33) reduces to a linear BSDE.

Finally, let us investigate a backward stochastic differential equation with a quadratic generator. We consider the following equation

$$Y(t) = \xi + \frac{1}{2}\int_t^T \beta\big|Z(s)\big|^2 ds - \int_t^T Z(s)dW(s), \quad 0 \le t \le T. \tag{3.37}$$

Notice that the quadratic BSDE (3.37) does not fit directly into the framework of Theorem 3.1.1 which only deals with Lipschitz generators.

Proposition 3.3.3 *Let ξ be an $\mathscr{F}^W$-measurable, bounded random variable, and let $\beta \neq 0$. There exists a unique solution $(Y, Z) \in \mathbb{S}^2(\mathbb{R}) \times \mathbb{H}^2(\mathbb{R})$ to the BSDE* (3.37). *The process Y has the representation*

$$Y(t) = \frac{1}{\beta}\ln \mathbb{E}\big[e^{\beta\xi}|\mathscr{F}_t^W\big], \quad 0 \le t \le T,$$

and the control process Z is derived from

$$Z(t) = \frac{\mathscr{Z}(t)}{\beta\mathbb{E}[e^{\beta\xi}|\mathscr{F}_t^W]}, \quad 0 \le t \le T,$$

and the representation

$$e^{\beta\xi} = \mathbb{E}\big[e^{\beta\xi}\big] + \int_0^T \mathscr{Z}(s)dW(s).$$

Proof We change the variables

$$\hat{Y}(t) = e^{\beta Y(t)}, \quad \hat{Z}(t) = \beta\hat{Y}(t)Z(t), \quad 0 \le t \le T.$$

By the Itô's formula any solution to (3.37) must also satisfy the BSDE

$$\hat{Y}(t) = e^{\beta\xi} - \int_t^T \hat{Z}(t)dW(t), \quad 0 \le t \le T. \tag{3.38}$$

By Theorem 3.1.1 there exists a unique solution $(\hat{Y}, \hat{Z}) \in \mathbb{S}^2(\mathbb{R}) \times \mathbb{H}^2(\mathbb{R})$ to the BSDE (3.38). Hence, the BSDE (3.37) has at most one solution. We also have $\hat{Y}(t) = \mathbb{E}[e^{\beta\xi}|\mathscr{F}_t^W]$, $0 \le t \le T$. Since $|\xi| \le K$, we deduce $0 < k' \le \hat{Y}(t) \le K'$, $0 \le t \le T$. We obtain that $(Y, Z) \in \mathbb{S}^2(\mathbb{R}) \times \mathbb{H}^2(\mathbb{R})$, and the proof is complete. □

Notice that square integrability of ξ is not sufficient for having a unique solution to the quadratic BSDE (3.37). In this book we do not deal with quadratic BSDEs but we will point out that BSDEs with quadratic generators may arise when we deal with the entropic risk measure (the exponential premium principle) and optimization problems for an exponential utility.

3.4 Examples of Linear and Nonlinear BSDEs with Jumps

The techniques discussed in Sect. 3.3 can also be applied to BSDEs with jumps. However, some additional assumptions and modifications have to be introduced. We now consider the natural filtration $\mathscr{F}^J$ generated by a pure jump càdlàg process J. Let N denote the jump measure of J.

First, we consider the linear BSDE

$$\begin{aligned} Y(t) = \xi &+ \int_t^T \alpha(s)Y(s)ds + \int_t^T \int_{\mathbb{R}} \gamma(s,z)U(s,z)Q(s,dz)\eta(s)ds \\ &- \int_t^T \int_{\mathbb{R}} U(s,z)\tilde{N}(ds,dz), \quad 0 \le t \le T. \end{aligned} \tag{3.39}$$

Proposition 3.4.1 *Let ξ be an $\mathscr{F}^J$-measurable random variable such that $\xi \in \mathbb{L}^2(\mathbb{R})$, and let α and γ be $\mathscr{F}^J$-predictable processes such that*

$$|\alpha(t)| \le K, \quad \int_{\mathbb{R}} |\gamma(t,z)|^2 Q(t,dz)\eta(t) \le K, \quad 0 \le t \le T.$$

There exists a unique solution $(Y, U) \in \mathbb{S}^2(\mathbb{R}) \times \mathbb{H}_N^2(\mathbb{R})$ to the BSDE (3.39). Moreover, let $\gamma(t,z) > -1$, $0 \le t \le T$, $z \in \mathbb{R}$. The process Y has the representation

$$Y(t) = \mathbb{E}^{\mathbb{Q}}\big[e^{\int_t^T \alpha(s)ds}\xi|\mathscr{F}_t^J\big], \quad 0 \le t \le T,$$

and the control process U is derived from

$$U(t,z) = e^{-\int_0^t \alpha(s)ds}\mathscr{U}(t,z), \quad 0 \le t \le T, \ z \in \mathbb{R},$$

and the representation

$$e^{\int_0^T \alpha(s)ds}\xi = \mathbb{E}^{\mathbb{Q}}\big[e^{\int_0^T \alpha(s)ds}\xi\big] + \int_0^T \int_{\mathbb{R}} \mathscr{U}(s,z)\tilde{N}^{\mathbb{Q}}(ds,dz),$$

where the equivalent probability measure $\mathbb{Q}$ *is defined by*

$$\frac{d\mathbb{Q}}{d\mathbb{P}}\Big|\mathscr{F}_T^J = M(T),$$
$$\frac{dM(t)}{M(t-)} = \int_{\mathbb{R}} \gamma(t,z)\tilde{N}(dt,dz), \quad M(0) = 1.$$

The proof of Proposition 3.4.1 is analogous to the proof of Proposition 3.3.1. We remark that in order to change the measure and establish the representation of the solution as the expectation under an equivalent probability measure we have to assume that $\gamma(t,z) > -1$.

Next, we consider the nonlinear BSDE

$$\begin{aligned} Y(t) = \xi &+ \int_t^T \int_{\mathbb{R}} \gamma(s,z)\big|U(s,z)\big|Q(s,dz)\eta(s)ds \\ &- \int_t^T \int_{\mathbb{R}} U(s,z)\tilde{N}(ds,dz), \quad 0 \le t \le T. \end{aligned} \tag{3.40}$$

Proposition 3.4.2 *Let* ξ *be an* $\mathscr{F}^J$*-measurable random variable such that* $\xi \in \mathbb{L}^2(\mathbb{R})$, *and let* γ *be an* $\mathscr{F}^J$*-predictable process such that*

$$\int_{\mathbb{R}} \big|\gamma(t,z)\big|^2 Q(t,dz)\eta(t) \le K, \quad 0 \le t \le T.$$

There exists a unique solution $(Y,U) \in \mathbb{S}^2(\mathbb{R}) \times \mathbb{H}_N^2(\mathbb{R})$ *to the BSDE* (3.40). *Moreover, let* $0 < \gamma(t,z) < 1$, $0 \le t \le T$, $z \in \mathbb{R}$. *The process* Y *has the representation*

$$Y(t) = \sup_{\mathbb{Q}\in\mathscr{Q}} \mathbb{E}^{\mathbb{Q}}\big[\xi|\mathscr{F}_t^J\big], \quad 0 \le t \le T,$$

where

$$\mathscr{Q} = \Big\{\mathbb{Q} \sim \mathbb{P} : \frac{d\mathbb{Q}}{d\mathbb{P}}\Big|\mathscr{F}_T^J = M(T), \ \frac{dM(t)}{M(t-)} = \int_{\mathbb{R}} \kappa(t,z)\tilde{N}(dt,dz), \ M(0) = 1,$$
$$\kappa \text{ is } \mathscr{F}^J\text{-predictable}, \ \big|\kappa(t,z)\big| \le \gamma(t,z)\Big\}.$$

The proof of Proposition 3.4.2 is analogous to the proof of Proposition 3.3.2. We remark that in order to apply the comparison principle from Theorem 3.2.2 and establish the representation of the solution as the robust expectation with respect to a set of equivalent probability measures we have to assume that $0 < \gamma(t,z) < 1$.

If the control process U does not change its sign, then the nonlinear BSDE (3.40) reduces to a linear BSDE.

Finally, we investigate a BSDE with an exponential generator. We consider the following equation

$$\begin{aligned} Y(t) = \xi &+ \int_t^T \int_{\mathbb{R}} \left(\frac{1}{\gamma} \left(e^{\gamma U(s,z)} - 1 \right) - U(s,z) \right) Q(s,dz)\eta(s)ds \\ &- \int_t^T \int_{\mathbb{R}} U(s,z)\tilde{N}(ds,dz), \quad 0 \le t \le T. \end{aligned} \tag{3.41}$$

Proposition 3.4.3 *Let ξ be an $\mathscr{F}^J$-measurable, bounded random variable, and let $\gamma \neq 0$. There exists a unique solution $(Y,U) \in \mathbb{S}^2(\mathbb{R}) \times \mathbb{H}^2_N(\mathbb{R})$ to the BSDE* (3.41). *The process Y has the representation*

$$Y(t) = \frac{1}{\gamma} \ln \mathbb{E}\left[e^{\gamma\xi} | \mathscr{F}_t^J\right], \quad 0 \le t \le T,$$

and the control process U is derived from

$$U(t,z) = \frac{1}{\gamma} \ln\left(1 + \mathscr{U}(t,z)e^{-\gamma Y(t-)}\right), \quad 0 \le t \le T,\ z \in \mathbb{R}$$

and the representation

$$e^{\gamma\xi} = \mathbb{E}\left[e^{\gamma\xi}\right] + \int_0^T \int_{\mathbb{R}} \mathscr{U}(s,z)\tilde{N}(ds,dz).$$

Proof The proof is analogous to the proof of Proposition 3.3.3. We only comment on the property that $U \in \mathbb{H}^2_N(\mathbb{R})$. By the Itô's formula the process $\hat{Y}(t) = e^{\gamma Y(t)}$ satisfies the BSDE

$$\hat{Y}(t) = e^{\gamma\xi} - \int_t^T \int_{\mathbb{R}} \mathscr{U}(s,z)\tilde{N}(ds,dz), \quad 0 \le t \le T, \tag{3.42}$$

where

$$\mathscr{U}(t,z) = \hat{Y}(t)\left(e^{\gamma U(t,z)} - 1\right), \quad 0 \le t \le T,\ z \in \mathbb{R}.$$

We immediately get the solution $\hat{Y}(t) = \mathbb{E}[e^{\gamma\xi} | \mathscr{F}_t^J]$, $0 \le t \le T$. Moreover, we have $0 < k' \le \hat{Y}(t) \le K'$, $0 \le t \le T$. From the dynamics (3.42) we can deduce that $\int_{\mathbb{R}} \mathscr{U}(t,z)N(\{t\},dz) = \hat{Y}(t) - \hat{Y}(t-)$. Hence, $\mathscr{U}(t,z) = \hat{Y}^{t,z}(t) - \hat{Y}(t-)$ where $\hat{Y}^{t,z}(t)$ denotes the expectation $\mathbb{E}[e^{\gamma\xi} | \mathscr{F}_t^J]$ given there is a jump of size z at time t. Consequently, $1 + \mathscr{U}(t,z)e^{-\gamma Y(t-)} = 1 + \frac{\mathscr{U}(t,z)}{\hat{Y}(t-)} = \frac{\hat{Y}^{t,z}(t)}{\hat{Y}(t-)}$ is bounded away from zero and bounded from above, and $|U(t,z)| \le K|\mathscr{U}(t,z)|$. Since $\mathscr{U} \in \mathbb{H}^2_N(\mathbb{R})$, we can conclude that $U \in \mathbb{H}^2_N(\mathbb{R})$. □

It is interesting to note that in the jump setting the entropic risk measure, i.e. the process $Y(t)=\frac{1}{\gamma}\ln\mathbb{E}[e^{\gamma\xi}|\mathscr{F}_t]$, leads to an exponential BSDE, whereas in the diffusion setting the entropic risk measure leads to a quadratic BSDE.

3.5 Malliavin Differentiability of Solution

Let us consider the natural filtration generated by a Lévy process. In this chapter we investigate the BSDE

$$\begin{aligned} Y(t) &= \xi + \int_t^T f\left(s, Y(s), Z(s), \int_{\mathbb{R}} U(s,z)\delta(z)\nu(dz)\right)ds \\ &\quad - \int_t^T Z(s)dW(s) - \int_t^T\int_{\mathbb{R}} U(s,z)\tilde{N}(ds,dz), \quad 0\le t\le T, \quad (3.43) \end{aligned}$$

where $\tilde{N}$ is a compensated Poisson random measure with a Lévy measure ν, and $\delta:[0,T]\times\mathbb{R}\to\mathbb{R}$ is a measurable function satisfying

$$|\delta(z)|\le Kz, \quad z\in\mathbb{R}.$$

As in Sect. 2.2, we define

$$\begin{aligned} &m(A)=\int_A z^2\nu(dz), \quad \upsilon(t,A)=\int_0^t\int_A m(dz)ds+\int_0^t \mathbf{1}\{0\in A\}\sigma^2 ds, \\ &\Upsilon^c(t)=\sigma W(t), \quad \Upsilon^d(t,A)=\int_0^t\int_A z\tilde{N}(ds,dz), \quad 0\le t\le T, A\in\mathscr{B}(\mathbb{R}). \end{aligned}$$

Since we follow the exposition from Solé et al. (2007), let us redefine

$$Z(t):=\frac{Z(t)}{\sigma}, \qquad U(t,z):=\frac{U(t,z)}{z}, \qquad \delta(z):=\frac{\delta(z)}{z}, \quad (3.44)$$

and we deal with the following BSDE

$$\begin{aligned} Y(t) &= \xi + \int_t^T f\left(s, Y(s), Z(s), \int_{\mathbb{R}} U(s,z)\delta(z)m(dz)\right)ds \\ &\quad - \int_t^T Z(s)d\Upsilon^c(s) - \int_t^T\int_{\mathbb{R}} U(s,z)\Upsilon^d(ds,dz), \quad 0\le t\le T. \quad (3.45) \end{aligned}$$

By $\mathbb{H}^2_{\Upsilon^d}(\mathbb{R})$ we denote the space of predictable processes $U:\Omega\times[0,T]\times\mathbb{R}\to\mathbb{R}$ satisfying

$$\mathbb{E}\left[\int_0^T\int_{\mathbb{R}}|U(t,z)|^2 m(dz)dt\right]<\infty.$$

We study Malliavin differentiability of the solution to the BSDE (3.45).

Theorem 3.5.1 *Assume that*

(i) *the filtration $\mathscr{F}$ is the natural filtration generated by a Lévy process,*
(ii) *the terminal value $\xi \in \mathbb{D}^{1,2}(\mathbb{R})$ satisfies*

$$\mathbb{E}\Big[\int_{[0,T]\times\mathbb{R}} |D_{s,z}\xi|^2 \upsilon(ds,dz)\Big] < \infty,$$

$$\lim_{\varepsilon\downarrow 0}\mathbb{E}\Big[\int_0^T\int_{|z|\leq\varepsilon} |D_{s,z}\xi|^2 m(dz)ds\Big] = 0,$$

(iii) *the function $\delta : \mathbb{R} \to \mathbb{R}$ is measurable and bounded,*
(iv) *the generator $f : \Omega \times [0,T] \times \mathbb{R} \times \mathbb{R} \times \mathbb{R} \to \mathbb{R}$ is predictable and Lipschitz continuous in the sense that*

$$\big|f(\omega,t,y,z,u) - f\big(\omega,t,y',z',u'\big)\big| \leq K\big(\big|y-y'\big| + \big|z-z'\big| + \big|u-u'\big|\big),$$

a.s., a.e. $(\omega,t) \in \Omega \times [0,T]$, for all $(y,z,u), (y',z',u') \in \mathbb{R}\times\mathbb{R}\times\mathbb{R}$,
(v) $\mathbb{E}[\int_0^T |f(t,0,0,0)|^2 dt] < \infty$,
(vi) *a.s., a.e. $(\omega,t) \in \Omega \times [0,T]$ the mapping $(y,z,u) \mapsto f(\omega,t,y,z,u)$ is continuously differentiable with uniformly bounded and continuous partial derivatives f_y, f_z, f_u,*
(vii) *for each $(t,y,z,u) \in [0,T]\times\mathbb{R}\times\mathbb{R}\times\mathbb{R}$ we have $f(\omega,t,y,z,u) \in \mathbb{D}^{1,2}(\mathbb{R})$, and*

$$\mathbb{E}\Big[\int_{[0,T]\times\mathbb{R}}\int_0^T \big|D_{s,z}f(\omega,t,0,0,0)\big|^2 dt\upsilon(ds,dz)\Big] < \infty,$$

(viii) *υ-a.e. $(s,z) \in [0,T]\times\mathbb{R}$ we have*

$$\begin{aligned}&\big|D_{s,z}f(\omega,t,\tilde{y},\tilde{z},\tilde{u}) - |D_{s,z}f\big(\omega,t,y',z',u'\big)\big|\\ &\quad\leq K\big(\big|\tilde{y}-y'\big| + \big|\tilde{z}-z'\big| + \big|\tilde{u}-u'\big|\big),\end{aligned}$$

a.s., a.e. $(\omega,t) \in \Omega \times [0,T]$, for all $(\tilde{y},\tilde{z},\tilde{u}), (y',z',u') \in \mathbb{R}\times\mathbb{R}\times\mathbb{R}$.

(a) *There exists a unique solution $(Y,Z,U) \in \mathbb{S}^2(\mathbb{R}) \times \mathbb{H}^2(\mathbb{R}) \times \mathbb{H}^2_{\Upsilon^d}(\mathbb{R})$ to the BSDE* (3.45).
(b) *There exists a unique solution $(Y^{s,0}, Z^{s,0}, U^{s,0}) \in \mathbb{S}^2(\mathbb{R}) \times \mathbb{H}^2(\mathbb{R}) \times \mathbb{H}^2_{\Upsilon^d}(\mathbb{R})$ to the BSDE*

$$\begin{aligned}Y^{s,0}(t) &= D_{s,0}\xi + \int_t^T f^{s,0}(r)dr - \int_t^T Z^{s,0}(r)d\Upsilon^c(r)\\ &\quad - \int_t^T\int_{\mathbb{R}} U^{s,0}(r,y)\Upsilon^d(dr,dy), \quad 0 \leq s \leq t \leq T, \quad (3.46)\end{aligned}$$

where

$$f^{s,0}(r) = D_{s,0}f\Big(\omega,r,Y(r),Z(r),\int_{\mathbb{R}} U(r,y)\delta(y)m(dy)\Big)$$

$$+ f_y\left(\omega, r, Y(r), Z(r), \int_{\mathbb{R}} U(r,y)\delta(y)m(dy)\right)Y^{s,0}(r)$$

$$+ f_z\left(\omega, r, Y(r), Z(r), \int_{\mathbb{R}} U(r,y)\delta(y)m(dy)\right)Z^{s,0}(r)$$

$$+ f_u\left(\omega, r, Y(r), Z(r), \int_{\mathbb{R}} U(r,y)\delta(y)m(dy)\right)\int_{\mathbb{R}} U^{s,0}(r,y)\delta(y)m(dy). \tag{3.47}$$

(c) *There exists a unique solution* $(Y^{s,z}, Z^{s,z}, U^{s,z}) \in \mathbb{S}^2(\mathbb{R}) \times \mathbb{H}^2(\mathbb{R}) \times \mathbb{H}^2_{\Upsilon^d}(\mathbb{R})$ *to the BSDE*

$$Y^{s,z}(t) = D_{s,z}\xi + \int_t^T f^{s,z}(r)dr - \int_t^T Z^{s,z}(r)d\Upsilon^c(r)$$
$$- \int_t^T\int_{\mathbb{R}} U^{s,z}(r,y)\Upsilon^d(dr,dy), \quad 0 \le s \le t \le T,\ z \neq 0, \tag{3.48}$$

where

$$f^{s,z}(r)$$
$$= \Bigg\{ f\Big(\omega^{s,z}, r, zY^{s,z}(r) + Y(r),$$
$$zZ^{s,z}(r) + Z(r), z\int_{\mathbb{R}} U^{s,z}(r,y)\delta(y)m(dy) + \int_{\mathbb{R}} U(r,y)\delta(y)m(dy)\Big)$$
$$- f\left(\omega, r, Y(r), Z(r), \int_{\mathbb{R}} U(r,y)\delta(y)m(dy)\right)\Bigg\}/z. \tag{3.49}$$

(d) *Set*

$$Y^{s,z}(t) = Z^{s,z}(t) = U^{s,z}(t,y) = 0, \quad (y,z) \in \mathbb{R}\times\mathbb{R},\ t < s \le T.$$

We have $(Y, Z, U) \in \mathbb{L}^{1,2}(\mathbb{R}) \times \mathbb{L}^{1,2}(\mathbb{R}) \times \mathbb{L}^{1,2}(\mathbb{R})$ *and* $(Y^{s,z}(t), Z^{s,z}(t), U^{s,z}(t,y))_{0\le s,t\le T,(y,z)\in\mathbb{R}\times\mathbb{R}}$ *is a version of the Malliavin derivative* $(D_{s,z}Y(t), D_{s,z}Z(t), D_{s,z}U(t,y))_{0\le s,t\le T,(y,z)\in\mathbb{R}\times\mathbb{R}}$.

Proof 1. The existence and uniqueness of solutions. By (3.44) and Theorem 3.1.1 there exists a unique solution (Y, Z, U) to (3.45), and for υ-a.e. $(s,z) \in [0,T]\times\mathbb{R}$ there exists a unique solution $(Y^{s,z}, Z^{s,z}, U^{s,z})$ to (3.46) and (3.48).

2. *The Picard iteration and the property that* $(Y^n, Z^n, U^n) \in \mathbb{L}^{1,2}(\mathbb{R}) \times \mathbb{L}^{1,2}(\mathbb{R}) \times \mathbb{L}^{1,2}(\mathbb{R})$. Referring to Theorem 3.1.1, we consider a sequence $(Y^n, Z^n, U^n)_{n\in\mathbb{N}}$, constructed by the Picard procedure, which converges to (Y, Z, U). We show that $(Y^n, Z^n, U^n) \in \mathbb{L}^{1,2}(\mathbb{R}) \times$

$\mathbb{L}^{1,2}(\mathbb{R}) \times \mathbb{L}^{1,2}(\mathbb{R})$ implies $(Y^{n+1}, Z^{n+1}, U^{n+1}) \in \mathbb{L}^{1,2}(\mathbb{R}) \times \mathbb{L}^{1,2}(\mathbb{R}) \times \mathbb{L}^{1,2}(\mathbb{R})$ and from $\mathbb{E}[\int_{[0,T]} \sup_{t\in[0,T]} |D_{s,z}Y^n(t)|^2 \upsilon(ds,dz)] < \infty$ we can deduce $\mathbb{E}[\int_{[0,T]\times\mathbb{R}} \sup_{t\in[0,T]} |D_{s,z}Y^{n+1}(t)|^2 \upsilon(ds,dz)] < \infty$. For that purpose we study the iterations

$$\begin{aligned} Y^{n+1}(t) &= \xi + \int_t^T f^n(r)dr \\ &\quad - \int_t^T Z^{n+1}(r)d\Upsilon^c(r) - \int_t^T \int_{\mathbb{R}} U^{n+1}(r,y)\Upsilon^d(dr,dy), \quad 0 \le t \le T, \end{aligned} \tag{3.50}$$

where

$$f^n(r) = f\left(r, Y^n(r), Z^n(r), \int_{\mathbb{R}} U^n(r,y)\delta(y)m(dy)\right).$$

We establish differentiability of all terms in (3.50). From Proposition 2.6.5 we conclude that $\int_{\mathbb{R}} U^n(r,y)\delta(y)m(dy)$ is Malliavin differentiable. By the chain rule from Proposition 2.6.4 the generator f^n is Malliavin differentiable and we derive

$$\begin{aligned} D_{s,0}f^n(r) &= D_{s,0}f\left(\omega, r, Y^n(r), Z^n(r), \int_{\mathbb{R}} U^n(r,y)\delta(y)m(dy)\right) \\ &\quad + f_y\left(\omega, r, Y^n(r), Z^n(r), \int_{\mathbb{R}} U^n(r,y)\delta(y)m(dy)\right) D_{s,0}Y^n(r) \\ &\quad + f_z\left(\omega, r, Y^n(r), Z^n(r), \int_{\mathbb{R}} U^n(r,y)\delta(y)m(dy)\right) D_{s,0}Z^n(r) \\ &\quad + f_u\left(\omega, r, Y^n(r), Z^n(r), \int_{\mathbb{R}} U^n(r,y)\delta(y)m(dy)\right) \\ &\quad \cdot \int_{\mathbb{R}} D_{s,0}U^n(r,y)\delta(y)m(dy), \end{aligned} \tag{3.51}$$

and for $z \neq 0$

$$\begin{aligned} D_{s,z}f^n(r) = \Bigg\{ f\Bigg(&\omega^{s,z}, r, zD_{s,z}Y^n(r) + Y^n(r), zD_{s,z}Z^n(r) + Z^n(r), \\ &z\int_{\mathbb{R}} D_{s,z}U^n(r,y)\delta(y)m(dy) + \int_{\mathbb{R}} U^n(r,y)\delta(y)m(dy)\Bigg) \\ &- f\left(\omega, r, Y^n(r), Z^n(r), \int_{\mathbb{R}} U^n(r,y)\delta(y)m(dy)\right)\Bigg\}/z. \end{aligned} \tag{3.52}$$

Applying Proposition 2.6.5 again, we deduce that $\int_t^T f^n(r)dr$ is Malliavin differentiable. We obtain the derivative

$$D_{s,z}\left(\xi + \int_t^T f^n(r)dr\right)$$
$$= D_{s,z}\xi + \int_t^T D_{s,z}f^n(r)dr, \quad \upsilon\text{-a.e. } (s,z) \in [0,T] \times \mathbb{R},\ s \le t \le T.$$

By Proposition 2.6.3 we have

$$Y^{n+1}(t) = \mathbb{E}\left[\xi + \int_t^T f^n(r)dr|\mathscr{F}_t\right] \in \mathbb{D}^{1,2}(\mathbb{R}), \quad 0 \le t \le T,$$

and we conclude from (3.50) that

$$\int_t^T Z^{n+1}(r)d\Upsilon^c(r) \in \mathbb{D}^{1,2}(\mathbb{R}), \quad 0 \le t \le T, \tag{3.53}$$

and

$$\int_t^T \int_{\mathbb{R}} U^{n+1}(r,y)\Upsilon^d(dr,dy) \in \mathbb{D}^{1,2}(\mathbb{R}), \quad 0 \le t \le T. \tag{3.54}$$

Proposition 2.6.6 shows that $(Z^{n+1}, U^{n+1}) \in \mathbb{L}^{1,2}(\mathbb{R}) \times \mathbb{L}^{1,2}(\mathbb{R})$. We still have to investigate Y^{n+1}. We can differentiate the recursive equation (3.50) and by Proposition 2.6.6 we obtain for υ-a.e. $(s,z) \in [0,T] \times \mathbb{R}$

$$D_{s,z}Y^{n+1}(t) = D_{s,z}\xi + \int_t^T D_{s,z}f^n(r)dr - \int_t^T D_{s,z}Z^{n+1}(r)d\Upsilon^c(r)$$
$$- \int_t^T \int_{\mathbb{R}} D_{s,z}U^{n+1}(r,y)\Upsilon^d(dr,dy), \quad s \le t \le T. \tag{3.55}$$

The BSDE (3.55) with the generator (3.51) or (3.52) satisfies the assumptions of Theorem 3.1.1 and for υ-a.e. $(s,z) \in [0,T] \times \mathbb{R}$ there exists a unique solution $(D_{s,z}Y^{n+1}, D_{s,z}Z^{n+1}, D_{s,z}U^{n+1}) \in \mathbb{S}^2(\mathbb{R}) \times \mathbb{H}^2(\mathbb{R}) \times \mathbb{H}^2_{\Upsilon^d}(\mathbb{R})$ to (3.55). By the a priori estimates (3.5) and (3.7), the Lipschitz properties of f and $D_{s,z}f$ and boundedness of derivatives f_y, f_z, f_u we can derive the inequality

$$\|D_{s,z}Y^{n+1}\|^2_{\mathbb{S}^2} + \|D_{s,z}Z^{n+1}\|^2_{\mathbb{H}^2} + \|D_{s,z}U^{n+1}\|^2_{\mathbb{H}^2_{\Upsilon^d}}$$
$$\le K\mathbb{E}\left[|D_{s,z}\xi|^2 + \int_s^T |D_{s,z}f^n(r)|^2 dr\right]$$
$$\le K\left(\mathbb{E}[|D_{s,z}\xi|^2] + \mathbb{E}\left[\int_0^T |D_{s,z}f(\omega,r,0,0,0)|^2 dr\right]\right.$$

$$+ \|Y^n\|_{\mathbb{S}^2}^2 + \|Z^n\|_{\mathbb{H}^2}^2 + \|U^n\|_{\mathbb{H}^2_{\gamma d}}^2$$

$$+ \|D_{s,z}Y^n\|_{\mathbb{S}^2}^2 + \|D_{s,z}Z^n\|_{\mathbb{H}^2}^2 + \|D_{s,z}U^n\|_{\mathbb{H}^2_{\gamma d}}^2\Big). \tag{3.56}$$

This in turn yields $\mathbb{E}[\int_{[0,T]\times\mathbb{R}} \sup_{t\in[0,T]} |D_{s,z}Y^{n+1}(t)|^2 \upsilon(ds,dz)] < \infty$, and $Y^{n+1} \in \mathbb{L}^{1,2}(\mathbb{R})$.

3. The square integrability of the solution $Y^{s,z}(t)$, $Z^{s,z}(t)$, $U^{s,z}(t,y)$ with respect to the product measure υ. Consider the unique solution $(Y^{s,z}, Z^{s,z}, U^{s,z}) \in \mathbb{S}^2(\mathbb{R}) \times \mathbb{H}^2(\mathbb{R}) \times \mathbb{H}^2_{\gamma d}(\mathbb{R})$ to (3.46) or (3.48). The a priori estimates (3.5) and (3.7) and the Lipschitz property of $D_{s,z}f$ yield the inequality

$$\begin{aligned}
&\|Y^{s,z}\|_{\mathbb{S}^2}^2 + \|Z^{s,z}\|_{\mathbb{H}^2}^2 + \|U^{s,z}\|_{\mathbb{H}^2_{\gamma d}}^2 \\
&\leq K\mathbb{E}\Big[|D_{s,z}\xi|^2 + \int_s^T \Big|D_{s,z}f\Big(\omega, r, Y(r), Z(r), \int_{\mathbb{R}} U(r,y)\delta(y)m(dy)\Big)\Big|^2 dr\Big] \\
&\leq K\Big(\mathbb{E}[|D_{s,z}\xi|^2] + \mathbb{E}\Big[\int_0^T |D_{s,z}f(\omega, r, 0, 0, 0,)|^2 dr\Big] \\
&\quad + \|Y\|_{\mathbb{S}^2}^2 + \|Z\|_{\mathbb{H}^2}^2 + \|U\|_{\mathbb{H}^2_{\gamma d}}^2\Big),
\end{aligned}$$

and we also derive

$$\mathbb{E}\Big[\int_{([0,T]\times\mathbb{R})^2} |Y^{s,z}(t)|^2 \upsilon(dt,dy)\upsilon(ds,dz)\Big] < \infty,$$

$$\mathbb{E}\Big[\int_{([0,T]\times\mathbb{R})^2} |Z^{s,z}(t)|^2 \upsilon(dt,dy)\upsilon(ds,dz)\Big] < \infty,$$

$$\mathbb{E}\Big[\int_{([0,T]\times\mathbb{R})^2} |U^{s,z}(t,y)|^2 \upsilon(dt,dy)\upsilon(ds,dz)\Big] < \infty.$$

4. The convergence of $(Y^n, Z^n, U^n)_{n\in\mathbb{N}}$ in $\mathbb{L}^{1,2}(\mathbb{R}) \times \mathbb{L}^{1,2}(\mathbb{R}) \times \mathbb{L}^{1,2}(\mathbb{R})$. From the proof of Theorem 3.1.1 we already know that $(Y^n, Z^n, U^n)_{n\in\mathbb{N}}$ converges in $\mathbb{S}^2(\mathbb{R}) \times \mathbb{H}^2(\mathbb{R}) \times \mathbb{H}^2_{\gamma d}(\mathbb{R})$. We have to prove that the corresponding Malliavin derivatives converge. First, we prove the limit

$$\begin{aligned}
\lim_{n\to\infty} \int_0^T \big(&\|Y^{s,0} - D_{s,0}Y^{n+1}\|_{\mathbb{S}^2}^2 \\
&+ \|Z^{s,0} - D_{s,0}Z^{n+1}\|_{\mathbb{H}^2}^2 + \|U^{s,0} - D_{s,0}U^{n+1}\|_{\mathbb{H}^2_{\gamma d}}^2\big)ds = 0.
\end{aligned} \tag{3.57}$$

By the a priori estimates (3.4), (3.6) and the Cauchy-Schwarz inequality we derive

$$\|Y^{s,0} - D_{s,0}Y^{n+1}\|_{\mathbb{S}^2}^2 + \|Z^{s,0} - D_{s,0}Z^{n+1}\|_{\mathbb{H}^2}^2 + \|U^{s,0} - D_{s,0}U^{n+1}\|_{\mathbb{H}^2_{\gamma d}}^2$$

$$
\begin{aligned}
&\leq K\mathbb{E}\bigg[\int_s^T e^{\rho r}\big|f^{s,0}(r)-D_{s,0}f^n(r)\big|^2 dr\bigg]\\
&\leq K\mathbb{E}\bigg[\int_0^T e^{\rho r}\bigg|D_{s,0}f\Big(r,Y(r),Z(r),\int_{\mathbb{R}}U(r,y)\delta(y)m(dy)\Big)\\
&\quad -D_{s,0}f\Big(r,Y^n(r),Z^n(r),\int_{\mathbb{R}}U^n(r,y)\delta(y)m(dy)\Big)\bigg|^2 dr\bigg]\\
&\quad +K\mathbb{E}\bigg[\int_0^T e^{\rho r}\bigg|\bigg(f_y\Big(r,Y(r),Z(r),\int_{\mathbb{R}}U(r,y)\delta(y)m(dy)\Big)\\
&\quad -f_y\Big(r,Y^n(r),Z^n(r),\int_{\mathbb{R}}U^n(r,y)\delta(y)m(dy)\Big)\bigg)Y^{s,0}(r)\\
&\quad +\bigg(f_z\Big(r,Y(r),Z(r),\int_{\mathbb{R}}U(r,y)\delta(y)m(dy)\Big)\\
&\quad -f_z\Big(r,Y^n(r),Z^n(r),\int_{\mathbb{R}}U^n(r,y)\delta(y)m(dy)\Big)\bigg)Z^{s,0}(r)\\
&\quad +\bigg(f_u\Big(r,Y(r),Z(r),\int_{\mathbb{R}}U(r,y)\delta(y)m(dy)\Big)\\
&\quad -f_u\Big(r,Y^n(r),Z^n(r),\int_{\mathbb{R}}U^n(r,y)\delta(y)m(dy)\Big)\bigg)\\
&\quad \cdot\int_{\mathbb{R}}U^{s,0}(r,y)\delta(y)m(dy)\bigg|^2 dr\bigg]\\
&\quad +K\mathbb{E}\bigg[\int_0^T e^{\rho r}\bigg|f_y\Big(r,Y^n(r),Z^n(r),\int_{\mathbb{R}}U^n(r,y)\delta(y)m(dy)\Big)\\
&\quad \cdot\big(Y^{s,0}(r)-D_{s,0}Y^n(r)\big)\\
&\quad +f_z\Big(r,Y^n(r),Z^n(r),\int_{\mathbb{R}}U^n(r,y)\delta(y)m(dy)\Big)\big(Z^{s,0}(r)-D_{s,0}Z^n(r)\big)\\
&\quad +f_u\Big(r,Y^n(r),Z^n(r),\int_{\mathbb{R}}U^n(r,y)\delta(y)m(dy)\Big)\\
&\quad \cdot\int_{\mathbb{R}}\big(U^{s,0}(r,y)-D_{s,0}U^n(r,y)\big)\delta(y)m(dy)\bigg|^2 dr\bigg]. \qquad (3.58)
\end{aligned}
$$

The first expected value in (3.58) converges to zero by the Lipschitz property of $D_{s,0}f$ and the convergence of (Y^n, Z^n, U^n) to (Y, Z, U). The second expected value converges to zero by the dominated convergence theorem, continuity of derivatives f_y, f_z, f_u and the convergence of (Y^n, Z^n, U^n) to (Y, Z, U). Since the first two expected values in (3.58) converge to zero, for any small $\varepsilon_0 > 0$ we can find sufficiently large n_0 such that for all $n \geq n_0$ we have

$$\int_0^T \big(\|Y^{s,0} - D_{s,0}Y^{n+1}\|_{\mathbb{S}^2}^2 + \|Z^{s,0} - D_{s,0}Z^{n+1}\|_{\mathbb{H}^2}^2 + \|U^{s,0} - D_{s,0}U^{n+1}\|_{\mathbb{H}^2_{\gamma d}}^2\big)ds$$
$$< \varepsilon_0 + \tilde{K}\int_0^T \big(\|Y^{s,0} - D_{s,0}Y^{n}\|_{\mathbb{H}^2}^2 + \|Z^{s,0} - D_{s,0}Z^{n}\|_{\mathbb{H}^2}^2$$
$$+ \|U^{s,0} - D_{s,0}U^{n}\|_{\mathbb{H}^2_{\gamma d}}^2\big)ds,$$

where we use boundedness of derivatives f_y, f_z, f_u. From estimates (3.11) and (3.17) we can conclude that the constant K in (3.58) can be made sufficiently small by taking ρ sufficiently large. Consequently, we can choose ρ such that $\tilde{K} < 1$. By recursion we derive for $n \geq n_0$

$$\int_0^T \big(\|Y^{s,0} - D_{s,0}Y^{n+1}\|_{\mathbb{S}^2}^2 + \|Z^{s,0} - D_{s,0}Z^{n+1}\|_{\mathbb{H}^2}^2 + \|U^{s,0} - D_{s,0}U^{n+1}\|_{\mathbb{H}^2_{\gamma d}}^2\big)ds$$
$$< \tilde{K}\int_0^T \big(\|Y^{s,0} - D_{s,0}Y^{n}\|_{\mathbb{S}^2}^2 + \|Z^{s,0} - D_{s,0}Z^{n}\|_{\mathbb{H}^2}^2 + \|U^{s,0} - D_{s,0}U^{n}\|_{\mathbb{H}^2_{\gamma d}}^2\big)ds$$
$$+ \varepsilon_0$$
$$< \tilde{K}^{n-n_0}\int_0^T \big(\|Y^{s,0} - D_{s,0}Y^{n_0}\|_{\mathbb{S}^2}^2 + \|Z^{s,0} - D_{s,0}Z^{n_0}\|_{\mathbb{H}^2}^2$$
$$+ \|U^{s,0} - D_{s,0}U^{n_0}\|_{\mathbb{H}^2_{\gamma d}}^2\big)ds + \frac{\varepsilon_0}{1-\tilde{K}}.$$

We can conclude that the limit (3.57) holds. Next, we prove the following limit

$$\lim_{n\to\infty}\int_{[0,T]\times(\mathbb{R}\setminus\{0\})} \big(\|Y^{s,z} - D_{s,z}Y^{n+1}\|_{\mathbb{S}^2}^2$$
$$+ \|Z^{s,z} - D_{s,z}Z^{n+1}\|_{\mathbb{H}^2}^2 + \|U^{s,z} - D_{s,z}U^{n+1}\|_{\mathbb{H}^2_{\gamma d}}^2\big)m(dz)ds = 0. \quad (3.59)$$

By the a priori estimates (3.4) and (3.6), the Lipschitz property of f and the Cauchy-Schwarz inequality we derive

$$\|Y^{s,z} - D_{s,z}Y^{n+1}\|_{\mathbb{S}^2}^2 + \|Z^{s,z} - D_{s,z}Z^{n+1}\|_{\mathbb{H}^2}^2 + \|U^{s,z} - D_{s,z}U^{n+1}\|_{\mathbb{H}^2_{\gamma d}}^2$$
$$\leq K\mathbb{E}\Big[\int_s^T e^{\rho r}\big|f^{s,z}(r) - D_{s,z}f^n(r)\big|^2 dr\Big]$$
$$\leq \tilde{K}\Big(\|Y^{s,z} - D_{s,z}Y^{n}\|_{\mathbb{S}^2}^2 + \|Z^{s,z} - D_{s,z}Z^{n}\|_{\mathbb{H}^2}^2 + \|U^{s,z} - D_{s,z}U^{n}\|_{\mathbb{H}^2_{\gamma d}}^2$$
$$+ \frac{\|Y^n - Y\|_{\mathbb{S}^2}^2 + \|Z^n - Z\|_{\mathbb{H}^2}^2 + \|U^n - U\|_{\mathbb{H}^2_{\gamma d}}^2}{z^2}\Big). \quad (3.60)$$

We notice that

$$\int_{[0,T]\times(\mathbb{R}\setminus\{0\})}\Big(\big\|Y^{s,z}-D_{s,z}Y^{n+1}\big\|_{\mathbb{S}^2}^2 + \big\|Z^{s,z}-D_{s,z}Z^{n+1}\big\|_{\mathbb{H}^2}^2+\big\|U^{s,z}-D_{s,z}U^{n+1}\big\|_{\mathbb{H}^2_{\gamma^d}}^2\Big)m(dz)ds$$
$$=\lim_{\varepsilon\downarrow 0}\int_0^T\int_{|z|>\varepsilon}\Big(\big\|Y^{s,z}-D_{s,z}Y^{n+1}\big\|_{\mathbb{H}^2}^2 + \big\|Z^{s,z}-D_{s,z}Z^{n+1}\big\|_{\mathbb{H}^2}^2+\big\|U^{s,z}-D_{s,z}U^{n+1}\big\|_{\mathbb{H}^2_{\gamma^d}}^2\Big)m(dz)ds, \quad (3.61)$$

and we can prove that this convergence is uniform in n, see Theorem 4.1 in Delong and Imkeller (2010b). We fix $\varepsilon>0$. From (3.60) we obtain

$$\int_0^T\int_{|z|>\varepsilon}\Big(\big\|Y^{s,z}-D_{s,z}Y^{n+1}\big\|_{\mathbb{S}^2}^2 + \big\|Z^{s,z}-D_{s,z}Z^{n+1}\big\|_{\mathbb{H}^2}^2+\big\|U^{s,z}-D_{s,z}U^{n+1}\big\|_{\mathbb{H}^2_{\gamma^d}}^2\Big)m(dz)ds$$
$$\le\tilde{K}\bigg\{\int_0^T\int_{|z|>\varepsilon}\Big(\big\|Y^{s,z}-D_{s,z}Y^{n}\big\|_{\mathbb{S}^2}^2 + \big\|Z^{s,z}-D_{s,z}Z^{n}\big\|_{\mathbb{H}^2}^2+\big\|U^{s,z}-D_{s,z}U^{n}\big\|_{\mathbb{H}^2_{\gamma^d}}^2\Big)m(dz)ds$$
$$+\Big(\big\|Y^n-Y\big\|_{\mathbb{S}^2}^2+\big\|Z^n-Z\big\|_{\mathbb{H}^2}^2+\big\|U^n-U\big\|_{\mathbb{H}^2_{\gamma^d}}^2\Big)T\int_{|z|>\varepsilon}\nu(dz)\bigg\}.$$

From estimates (3.11) and (3.17) we can conclude that the constant K in (3.60) can be made sufficiently small by taking ρ sufficiently large. Hence, we can choose $\tilde{K}<1$. Since $(Y^n,Z^n,U^n)_{n\in\mathbb{N}}$ converges, for any small $\varepsilon_0>0$ we can find sufficiently large n_0 such that for all $n\ge n_0$ we have

$$\int_0^T\int_{|z|>\varepsilon}\Big(\big\|Y^{s,z}-D_{s,z}Y^{n+1}\big\|_{\mathbb{S}^2}^2 + \big\|Z^{s,z}-D_{s,z}Z^{n+1}\big\|_{\mathbb{H}^2}^2+\big\|U^{s,z}-D_{s,z}U^{n+1}\big\|_{\mathbb{H}^2_{\gamma^d}}^2\Big)m(dz)ds$$
$$<\tilde{K}\int_0^T\int_{|z|>\varepsilon}\Big(\big\|Y^{s,z}-D_{s,z}Y^{n}\big\|_{\mathbb{S}^2}^2 + \big\|Z^{s,z}-D_{s,z}Z^{n}\big\|_{\mathbb{H}^2}^2+\big\|U^{s,z}-D_{s,z}U^{n}\big\|_{\mathbb{H}^2_{\gamma^d}}^2\Big)m(dz)ds+\varepsilon_0.$$

By recursion we derive for $n\ge n_0$

$$\int_0^T\int_{|z|>\varepsilon}\Big(\big\|Y^{s,z}-D_{s,z}Y^{n+1}\big\|_{\mathbb{S}^2}^2$$

$$+ \|Z^{s,z} - D_{s,z}Z^{n+1}\|^2_{\mathbb{H}^2} + \|U^{s,z} - D_{s,z}U^{n+1}\|^2_{\mathbb{H}^2_{\gamma d}}\big)m(dz)ds$$

$$< \tilde{K}^{n-n_0} \int_0^T \int_{|z|>\varepsilon} \big(\|Y^{s,z} - D_{s,z}Y^{n_0}\|^2_{\mathbb{S}^2}$$

$$+ \|Z^{s,z} - D_{s,z}Z^{n_0}\|^2_{\mathbb{H}^2} + \|U^{s,z} - D_{s,z}U^{n_0}\|^2_{\mathbb{H}^2_{\gamma d}}\big)m(dz)ds + \frac{\varepsilon_0}{1-\tilde{K}},$$

and we end up with the limit

$$\lim_{n\to\infty} \int_0^T \int_{|z|>\varepsilon} \big(\|Y^{s,z} - D_{s,z}Y^{n+1}\|^2_{\mathbb{S}^2}$$

$$+ \|Z^{s,z} - D_{s,z}Z^{n+1}\|^2_{\mathbb{H}^2} + \|U^{s,z} - D_{s,z}U^{n+1}\|^2_{\mathbb{H}^2_{\gamma d}}\big)m(dz)ds = 0, \quad \varepsilon > 0.$$

The limit (3.59) follows by interchanging the limits in n and ε in (3.61). Combining (3.57) and (3.59) we get

$$\lim_{n\to\infty} \int_{[0,T]\times\mathbb{R}} \big(\|Y^{s,z} - D_{s,z}Y^{n+1}\|^2_{\mathbb{S}^2}$$

$$+ \|Z^{s,z} - D_{s,z}Z^{n+1}\|^2_{\mathbb{H}^2} + \|U^{s,z} - D_{s,z}U^{n+1}\|^2_{\mathbb{H}^2_{\gamma d}}\big)\upsilon(ds, dz) = 0.$$

5. The Malliavin derivative process. Since the space $\mathbb{L}^{1,2}(\mathbb{R})$ is a Hilbert space and the Malliavin derivative is a closed operator, the assertions that $(Y, Z, U) \in \mathbb{L}^{1,2}(\mathbb{R}) \times \mathbb{L}^{1,2}(\mathbb{R}) \times \mathbb{L}^{1,2}(\mathbb{R})$ and $(Y^{s,z}(t), Z^{s,z}(t), U^{s,z}(t, y))_{0\le s,t\le T,(y,z)\in(\mathbb{R})\mathbb{R}}$ is a version of the derivative $(D_{s,z}Y(t), D_{s,z}Z(t), D_{s,z}U(t, y))_{0\le s,t\le T,(y,z)\in\mathbb{R}\times\mathbb{R}}$ hold. □

The key consequence of Theorem 3.5.1 is that we can interpret the solution (Z, U) as the Malliavin derivative of Y. Before we formally state this result, we have to define the predictable projection of a process.

Definition 3.5.1 Let V be a measurable process such that for any predictable time τ the random variable $V(\tau)\mathbf{1}\{\tau < \infty\}$ is integrable with respect to the filtration $\mathscr{F}_{\tau-}$. There exists a unique predictable process $V^{\mathscr{P}}$ satisfying

$$\mathbb{E}\big[V(\tau)\mathbf{1}\{\tau < \infty\}|\mathscr{F}_{\tau-}\big] = V^{\mathscr{P}}(\tau)\mathbf{1}\{\tau < \infty\},$$

which is called the predictable projection of V.

We remark that for an adapted, càdlàg, quasi-left continuous process we have $V^{\mathscr{P}}(t) = V(t-)$, see Theorem 4.23 in He et al. (1992).

Proposition 3.5.1 *Under the assumptions of Theorem* 3.5.1 *we have*

$((D_{t,0}Y)^{\mathscr{P}}(t))_{0\le t\le T}$ *is a version of* $(Z(t))_{0\le t\le T}$,

$((D_{t,z}Y)^{\mathscr{P}}(t))_{0\le t\le T, z\in(\mathbb{R}\setminus\{0\})}$ *is a version of* $(U(t,z))_{0\le t\le T, z\in(\mathbb{R}\setminus\{0\})}$,

where (Y,Z,U) *solves the BSDE* (3.45).

Proof The solution to (3.45) satisfies

$$Y(s) = Y(0) - \int_0^s f\left(r, Y(r), Z(r), \int_{\mathbb{R}} U(r,y)\delta(y)m(dy)\right)dr \\ + \int_0^s Z(r)d\Upsilon^c(r) + \int_0^s \int_{\mathbb{R}} U(r,y)\Upsilon^d(dr,dy), \quad 0 \le s \le T. \tag{3.62}$$

Differentiating (3.62) in line with Proposition 2.6.6, we obtain for υ-a.e. $(u,z) \in [0,T]\times\mathbb{R}$

$$D_{u,0}Y(s) = Z(u) - \int_u^s D_{u,0}f(r)dr + \int_u^s D_{u,0}Z(r)d\Upsilon^c(r) \\ + \int_u^s \int_{\mathbb{R}} D_{u,0}U(r,y)\Upsilon^d(dr,dy), \quad 0 \le u \le s \le T,$$

$$D_{u,z}Y(s) = U(u,z) - \int_u^s D_{u,z}f(r)dr + \int_u^s D_{u,z}Z(r)d\Upsilon^c(r) \\ + \int_u^s \int_{\mathbb{R}} D_{u,z}U(r,y)\Upsilon^d(dr,dy), \quad 0 \le u \le s \le T,\ z \neq 0,$$

where the derivative operators $D_{u,z}$ are defined by (3.47) and (3.49). Since the mappings $s \mapsto \int_u^s D_{u,z}f(r)dr$, $s \mapsto \int_u^s D_{u,z}Z(r)d\Upsilon^c(r)$ are continuous and the mapping $s \mapsto \int_u^s \int_{\mathbb{R}} D_{u,z}U(r,y)\Upsilon^d(dr,dy)$ is càdlàg, letting $s \downarrow u$ we get the following relations

$$D_{u,0}Y(u) = Z(u), \quad \text{a.s., a.e. } (\omega,u) \in \Omega \times [0,T],$$
$$D_{u,z}Y(u) = U(u,z), \quad \text{a.s., } \upsilon\text{-a.e. } (\omega,u,z) \in \Omega \times [0,T] \times \big(\mathbb{R}\setminus\{0\}\big).$$

Some measurability considerations complete the proof, see Corollary 4.1 in Delong and Imkeller (2010b). □

Proposition 3.5.2 *Under the assumptions of Theorem* 3.5.1 *and with* $|\delta(z)| \le Kz$, $z \in \mathbb{R}$, *we have*

$((D_tY)^{\mathscr{P}}(t))_{0\le t\le T}$ *is a version of* $(Z(t))_{0\le t\le T}$,
$((zD_{t,z}Y)^{\mathscr{P}}(t))_{0\le t\le T, z\in(\mathbb{R}\setminus\{0\})}$ *is a version of* $(U(t,z))_{0\le t\le T, z\in(\mathbb{R}\setminus\{0\})}$,

where (Y,Z,U) *solves the BSDE* (3.43).

We recall that D_t is the classical Malliavin derivative on the Wiener space, see Sect. 2.2.

Proposition 3.5.2 shows an important application of the Malliavin calculus to BSDEs. The formulas from Proposition 3.5.2 can be used to derive the control processes (Z, U) provided that we already have the solution Y. Examples are considered in Sect. 8.2. In Chap. 4 we use the Malliavin representations of (Z, U) to establish explicit formulas for the control processes of a general Markovian BSDE. In the financial context, Proposition 3.5.2 characterizes a hedging strategy, which usually depends on (Z, U), as the Malliavin derivative of a price process or a replicating portfolio Y. The Malliavin representations from Proposition 3.5.2 can also be used in numerical algorithms.

We now use Proposition 3.5.2 to gain an insight into the predictable representation property. We recover the Clark-Ocone formula.

Theorem 3.5.2 *Assume that* $\xi \in \mathbb{D}^{1,2}(\mathbb{R})$. *We have the representation*

$$\xi = \mathbb{E}[\xi] + \int_0^T \mathscr{Z}^{\mathscr{P}}(s)dW(s) + \int_0^T \int_{\mathbb{R}} \mathscr{U}^{\mathscr{P}}(s,z)\tilde{N}(ds,dz),$$

where $\mathscr{Z}(s) = \mathbb{E}[D_s\xi|\mathscr{F}_s]$ *and* $\mathscr{U}(s,z) = z\mathbb{E}[D_{s,z}\xi|\mathscr{F}_s]$.

Proof Finding the predictable representation of ξ is equivalent to solving the BSDE

$$Y(t) = \xi - \int_t^T Z(s)dW(s) - \int_t^T \int_{\mathbb{R}} U(s,z)\tilde{N}(ds,dz), \quad 0 \le t \le T.$$

Clearly, $Y(t) = \mathbb{E}[\xi|\mathscr{F}_t]$. Moreover, $Y \in \mathbb{D}^{1,2}(\mathbb{R})$ by Proposition 2.6.3. We can prove our result by following the reasoning from the proof of Proposition 3.5.1 and applying Propositions 2.6.3 and 2.6.6. If $\sup_{(t,z)\in[0,T]\times\mathbb{R}} \mathbb{E}[|D_{t,z}\xi|^2] < \infty$, then the result follows directly from Proposition 3.5.2. □

The Clark-Ocone formula gives a direct method for deriving the predictable representation of a random variable. We notice that the Malliavin derivatives determine the integrators in the predictable representation. We now present two examples.

Example 3.2 Let $\xi = (K - e^{\sigma W(T)-\frac{1}{2}\sigma^2 T})^+$. In Example 2.12 we show that $D_{t,0}\xi = -e^{\sigma W(T)-\frac{1}{2}\sigma^2 T}\mathbf{1}\{e^{\sigma W(T)-\frac{1}{2}\sigma^2 T} < K\}$. By Proposition 2.6.1 we get $D_t\xi = -\sigma e^{\sigma W(T)-\frac{1}{2}\sigma^2 T}\mathbf{1}\{e^{\sigma W(T)-\frac{1}{2}\sigma^2 T} < K\}$. We can conclude that there exists a measurable function $\varphi : [0,T] \times \mathbb{R} \to \mathbb{R}$ such that

$$\varphi\big(t, W(t)\big) = -\sigma\mathbb{E}\big[e^{\sigma W(T)-\frac{1}{2}\sigma^2 T}\mathbf{1}\big\{e^{\sigma W(T)-\frac{1}{2}\sigma^2 T} < K\big\}|\mathscr{F}_t\big], \quad 0 \le t \le T,$$

and the predictable representation of ξ is given by the formula

$$\xi = \mathbb{E}[\xi] + \int_0^T \varphi\big(s, W(s)\big)dW(s).$$

Example 3.3 Let $\xi = (J(T) - K)^+$ where J is a compound Poisson process with jump size distribution q. We assume that q is supported on $(0, \infty)$ and satisfies $\int_0^\infty y^2 q(dy) < \infty$. In Example 2.14 we show that $zD_{t,z}\xi = (J(T) + z - K)^+ - (J(T) - K)^+$. We can conclude that there exists a measurable function $\varphi : [0, T] \times [0, \infty) \to \mathbb{R}$ such that

$$\varphi\big(t, J(t)\big) = \mathbb{E}\big[\big(J(T) - K\big)^+ | \mathscr{F}_t\big], \quad 0 \le t \le T,$$

and the predictable representation of ξ is given by the formula

$$\xi = \mathbb{E}[\xi] + \int_0^T \int_0^\infty \big(\varphi\big(s, J(s-) + z\big) - \varphi\big(s, J(s-)\big)\big)\tilde{N}(ds, dz),$$

where we use the predictable projection of the càdlàg, quasi-left continuous process J.

In Sects. 3.3 and 3.4 we point out that the nonlinear BSDE (3.33) or (3.40) can be reduced to a linear BSDE if the control process does not change its sign. We may deduce the sign of the control process by the applying Malliavin calculus.

Example 3.4 Let us recall Example 1.3 and the expectation

$$Y(t) = \sup_{\mathbb{Q} \in \mathscr{Q}} \mathbb{E}^{\mathbb{Q}}\big[\big(J(T) - K\big)^+ | \mathscr{F}_t\big], \quad 0 \le t \le T,$$

where J is a compound Poisson process under $\mathbb{P}$ with jump size distribution q and intensity λ, and

$$\mathscr{Q} = \bigg\{\mathbb{Q} \sim \mathbb{P} : \frac{d\mathbb{Q}}{d\mathbb{P}} | \mathscr{F}_T^J = M(T), \ \frac{dM(t)}{M(t-)} = \int_{\mathbb{R}} \kappa(t, z)\tilde{N}(dt, dz), \ M(0) = 1,$$
$$\kappa \text{ is } \mathscr{F}^J\text{-predictable}, \ \big|\kappa(t, z)\big| \le \delta < 1\bigg\}.$$

Under an equivalent probability measure $\mathbb{Q} \in \mathscr{Q}$ determined by κ the process J has the jump size distribution and the intensity

$$q^{\mathbb{Q}}(t, dz) = \frac{1 + \kappa(t, z)}{\int_{\mathbb{R}}(1 + \kappa(t, z))q(dz)} q(dz), \quad 0 \le t \le T, \ z \in \mathbb{R},$$
$$\lambda^{\mathbb{Q}}(t) = \int_{\mathbb{R}} \big(1 + \kappa(t, z)\big)q(dz)\lambda, \quad 0 \le t \le T.$$

Let us assume that the distribution q is supported on $(0, \infty)$ and satisfies $\int_0^\infty z^2 q(dz) < \infty$. By Proposition 3.4.2 the process Y solves the nonlinear BSDE

$$Y(t) = \big(J(T) - K\big)^+ + \int_t^T \int_0^\infty \delta\big|U(s, z)\big|\lambda q(dz)ds$$

$$-\int_t^T\int_0^\infty U(s,z)\tilde{N}(ds,dz),\quad 0\le t\le T. \tag{3.63}$$

We now consider the linear BSDE

$$\mathscr{Y}(t)=\big(J(T)-K\big)^+ +\int_t^T\int_0^\infty \delta\mathscr{U}(s,z)\lambda q(dz)ds$$
$$-\int_t^T\int_0^\infty \mathscr{U}(s,z)\tilde{N}(ds,dz),\quad 0\le t\le T. \tag{3.64}$$

By Proposition 3.4.1 the solution $\mathscr{U}$ is derived from the predictable representation

$$\big(J(T)-K\big)^+=\mathbb{E}^{\mathbb{Q}^*}\big[\big(J(T)-K\big)^+\big]+\int_0^T\int_0^\infty \mathscr{U}(s,z)\tilde{N}^{\mathbb{Q}^*}(ds,dz),$$

where $\mathbb{Q}^*$ is the equivalent probability measure under which J has the jump size distribution q and the intensity $(1+\delta)\lambda$. Recalling the results from Example 3.3, we can conclude that the control process $\mathscr{U}(s,z)=\varphi(s,J(s-)+z)-\varphi(s,J(s-))$ from the representation of the martingale $\varphi(t,J(t))=\mathbb{E}^{\mathbb{Q}^*}[(J(T)-K)^+|\mathscr{F}_t]$ is non-negative. By uniqueness of solution to (3.63) we deduce that (Y,U) and $(\mathscr{Y},\mathscr{U})$ must coincide. Consequently, the nonlinear BSDE (3.63) reduces to the linear BSDE (3.64).

Bibliographical Notes BSDEs were introduced by Bismut (1973) and Pardoux and Peng (1990). The classical reference on the theory and financial applications of BSDEs with Lipschitz generators driven by Brownian motions is El Karoui et al. (1997b). The theory and financial applications of BSDEs driven by Brownian motions are also discussed in Chap. 8 in Carmona (2008), and the theory and financial applications of BSDEs with jumps are presented in Chap. 8 in Rong (2005). The proofs of Lemma 3.1.1, Theorems 3.1.1 and 3.2.1 are inspired by El Karoui et al. (1997b) and extended to cover the case of a random measure, see also Barles et al. (1997) and Royer (2006) who consider a Poisson random measure and Becherer (2006) who considers a general random measure. In Theorem 3.1.1 we deal with Lipschitz generators. It is worth pointing out papers which deal with non-Lipschitz generators. The existence of solution for BSDEs driven by Brownian motions with continuous generators having a linear and a superlinear-quadratic growth was proved by Lepeltier and Martin (1997) and Lepeltier and Martin (1998), and the existence of solution for BSDEs driven by Brownian motions with continuous, monotonic generators having a general growth in Y was established by Fan and Jiang (2010). Quadratic BSDEs driven by Brownian motions are investigated in the monograph by Dos Reis (2011). The classical result on the existence of a solution to a quadratic BSDE driven by a Brownian motion is due to Kobylanski (2000). Other existence results for quadratic BSDEs driven by Brownian motions (continuous martingales) can be found in Barrieu and El Karoui (2013), Tevzadze (2008), Delbaen et al. (2011). BSDEs with Poison jumps with continuous generators having a linear growth were studied in Yin and Mao (2008). The existence of solution for BSDEs driven by Brownian motions and Poisson random measures with continuous generators satisfying a weak monotone condition was proved by Rong (1997). Various existence results for quadratic BSDEs driven by Brownian motions and random measures can be found in Ankirchner et al. (2010), Morlais (2009), Lim and Quenez (2011), Jiao et al. (2013), Kharroubi and Lim (2012). In the proof of Theorem 3.2.2 we closely follow the arguments from Royer (2006). Example 3.1 is taken from Barles et al. (1997). For other versions of a comparison principle we refer to Royer (2006), Cohen (2011), Yin and Mao (2008), Quenez and Sulem (2012). For a comparison principle for the control processes (a so-called comonotonic property) we refer to Chen et al. (2005). A converse comparison principle is studied in Coquet et al. (2001) and De Scheemaekere (2011). In the proof of Proposition 3.3.1 we follow the arguments from El Karoui et al. (1997b), the proof of Proposition 3.3.2 is inspired by El Karoui and Quenez (1995), and the exponential change of variable used in Proposition 3.3.3 is taken from Kobylanski (2000). The proofs of Theorem 3.5.1 and Proposition 3.5.1 are taken (with appropriate adaptations) from Delong and Imkeller (2010b). The Clark-Ocone formula for Lévy processes can be found in Di Nunno et al. (2009). For $\mathbb{L}^p$, $p \in (1,2)$, solutions to BSDEs we refer to Pardoux and Peng (1990) and Briand et al. (2003). For the notion of a weak solution to a BSDE we refer to Buckdahn et al. (2004).

Chapter 4
Forward-Backward Stochastic Differential Equations

Abstract We investigate a backward SDE with a generator and a terminal condition which depend on the state of a Markov process solving a forward SDE driven by a Brownian motion and a compensated Poisson random measure. Such an equation is called a forward-backward SDE. In the Markovian setting we show that the unique solution to a backward SDE can be written as a function of a forward state process. We derive formulas for the control processes by applying the Itô's formula and the Malliavin calculus. We establish the connection between the solution to a BSDE and the viscosity solution to a partial integro-differential equation. A generalization of the Feynman-Kac formula is given. We also deal with a coupled forward-backward SDE in which a solution to the backward component also affects the forward component.

In Chap. 3 we study BSDEs with random terminal conditions and generators which depend on a Brownian motion and a random measure in an arbitrary way. In this chapter we assume that the randomness of the terminal condition and the generator comes from the state of a Markov process solving a forward SDE. We deal with forward-backward stochastic differential equations driven by a Brownian motion and a compensated Poisson random measure. We derive crucial results, such as the connection between BSDEs and PIDEs and the nonlinear Feynman-Kac formula. Let us remark that in majority of applications we deal with FBSDEs where the forward SDE models a risk factor.

4.1 The Markovian Structure of FBSDEs

We consider the forward stochastic differential equation

$$\begin{aligned}\mathscr{X}(s) = x &+ \int_t^s \mu\big(\mathscr{X}(r-)\big)dr + \int_t^s \sigma\big(\mathscr{X}(r-)\big)dW(r)\\ &+ \int_t^s \int_{\mathbb{R}} \gamma\big(\mathscr{X}(r-), z\big)\tilde{N}(dr, dz), \quad 0 \leq t \leq s \leq T. \end{aligned} \tag{4.1}$$

Ł. Delong, *Backward Stochastic Differential Equations with Jumps and Their Actuarial and Financial Applications*, EAA Series, DOI 10.1007/978-1-4471-5331-3_4,

The process $\mathscr{X}$ is called a state process. By $(\mathscr{X}^{t,x}(s), t \le s \le T)$ we denote a solution to (4.1) parametrized by the initial data (t,x). We set $\mathscr{X}^{t,x}(s) = x$, $s \le t$. Let us assume that

(B1) the random measure N is a Poisson random measure generated by a Lévy process with a Lévy measure ν,
(B2) the functions $\mu : \mathbb{R} \to \mathbb{R}$ and $\sigma : \mathbb{R} \to \mathbb{R}$ are Lipschitz continuous,
(B3) the function $\gamma : \mathbb{R} \times \mathbb{R} \to \mathbb{R}$ is measurable and satisfies

$$\begin{aligned} \left|\gamma(x,z)\right| &\le K\left(1 \wedge |z|\right), \quad (x,z) \in \mathbb{R} \times \mathbb{R}, \\ \left|\gamma(x,z) - \gamma\left(x',z\right)\right| &\le K\left|x - x'\right|\left(1 \wedge |z|\right), \quad (x,z), \left(x',z\right) \in \mathbb{R} \times \mathbb{R}. \end{aligned}$$

We deal with a time-homogenous dynamics of the state process but time-dependence can be easily introduced. We also restrict our study to a Poisson random measure generated by a Lévy process, which is the most common case investigated in the literature. An extension to cover a general random measure is possible by using formula (8.17) or considering an additional state process.

We state some properties of a solution to the forward SDE (4.1).

Theorem 4.1.1 *Assume that* (B1)–(B3) *hold.*

(a) *For each* $(t,x) \in [0,T] \times \mathbb{R}$ *there exists a unique adapted, càdlàg solution* $\mathscr{X}^{t,x} := (\mathscr{X}^{t,x}(s), t \le s \le T)$ *to* (4.1).
(b) *The solution* $\mathscr{X}$ *is a homogenous Markov process.*
(c) *For all* $(t,x), (t,x')(t',x') \in [0,T] \times \mathbb{R}$ *and* $p \ge 2$ *we have*

$$\begin{aligned} \mathbb{E}\Big[\sup_{t \le s \le T} \left|\mathscr{X}^{t,x}(s) - x\right|^p\Big] &\le K\left(1 + |x|^p\right)(T - t), \\ \mathbb{E}\Big[\sup_{t \le s \le T} \left|\mathscr{X}^{t,x}(s) - \mathscr{X}^{t,x'}(s)\right|^p\Big] &\le K\left|x - x'\right|^p, \\ \mathbb{E}\Big[\sup_{t \le s \le T} \left|\mathscr{X}^{t,x}(s) - \mathscr{X}^{t',x'}(s)\right|^p\Big] &\le K\left(\left|x - x'\right|^p + \left(1 + \left|x \vee x'\right|^p\right)\left|t - t'\right|\right). \end{aligned} \tag{4.2}$$

Proof Case (a) follows from Theorem 6.2.9. in Applebaum (2004), case (b) follows from Theorem 6.4.6 in Applebaum (2004). The first and the second estimates from case (c) are taken from Proposition 1.1 in Barles et al. (1997), the third estimate can be proved by classical techniques for SDEs, see Lemma II.2.2 in Gihman and Skorohod (1979). □

Theorem 4.1.2 *Let* (B1)–(B3) *hold.*

(a) *Assume that the functions* μ *and* σ *are continuously differentiable and* γ *is continuously differentiable in the first variable* x. *Then, the process* $\mathscr{X}^{t,x}$ *is Malliavin differentiable and we have*

$$\sup_{(s,z) \in [0,T] \times \mathbb{R}} \mathbb{E}\Big[\sup_{r \in [s,T]} \left|D_{s,z}\mathscr{X}^{t,x}(r)\right|^2\Big] < \infty. \tag{4.3}$$

(b) *Assume that*

(i) *the functions* μ *and* σ *are twice continuously differentiable and the first derivatives satisfy* (B2),

(ii) *the function* γ *is twice continuously differentiable in the first variable* x *and the first derivative satisfies* (B3),

(iii) *the function* $(x, z) \mapsto \gamma_x(x, z)$ *is continuous,*

(iv) $\sigma(x) \neq 0$ *and* $1 + \gamma_x(x, z) \neq 0$ *for* $(x, z) \in \mathbb{R} \times \mathbb{R}$.

Then, the probability law of $\mathscr{X}^{t,x}(s)$, *for* $s > t$, *is absolutely continuous with respect to the Lebesgue measure.*

Proof Assertion (a) follows from Theorem 3 in Petrou (2008). We prove assertion (b). Choose $s \in (t, T]$. By Propositions 2.6.1, 2.6.5 and 2.6.6 we derive

$$\begin{aligned} D_u \mathscr{X}^{t,x}(s) \\ &= \sigma\big(\mathscr{X}^{t,x}(u)\big) + \int_u^s \mu'\big(\mathscr{X}^{t,x}(r-)\big) D_u \mathscr{X}^{t,x}(r-)dr \\ &\quad + \int_u^s \sigma'\big(\mathscr{X}^{t,x}(r-)\big) D_u \mathscr{X}^{t,x}(r-)dW(r) \\ &\quad + \int_u^s \int_{\mathbb{R}} \gamma'\big(\mathscr{X}^{t,x}(r-), z\big) D_u \mathscr{X}^{t,x}(r-)\tilde{N}(dr, dz), \quad 0 \le t \le u \le s \le T. \end{aligned} \tag{4.4}$$

It is clear that we have

$$\begin{aligned} D_u \mathscr{X}^{t,x}(s) \\ &= \sigma\big(\mathscr{X}^{t,x}(u)\big) e^{\int_u^s \mu'(\mathscr{X}^{t,x}(r))dr + \int_u^s \sigma'(\mathscr{X}^{t,x}(r-))dW(r) + \int_u^s \int_{\mathbb{R}} \gamma'(\mathscr{X}^{t,x}(r-),z)\tilde{N}(dr,dz)} \\ &\quad \cdot e^{\int_u^s \int_{|z|\le 1} (\ln(1+\gamma'(\mathscr{X}^{t,x}(r),z)) - \gamma'(\mathscr{X}^{t,x}(r),z))\nu(dz)dr - \frac{1}{2}\int_u^s |\sigma'(\mathscr{X}^{t,x}(r))|^2 dr} \\ &\quad \cdot \Pi_{\{u<r\le s:|\Delta J(r)|>1\}} \big(1 + \gamma'\big(\mathscr{X}^{t,x}(r-), \Delta J(r)\big) e^{-\gamma'(\mathscr{X}^{t,x}(r-),\Delta J(r))} \\ &\quad \cdot e^{\int_u^s \int_{|z|\le 1} (\ln(1+\gamma'(\mathscr{X}^{t,x}(r-),z)) - \gamma'(\mathscr{X}^{t,x}(r-),z))\tilde{N}(dr,dz)}, \quad 0 \le t \le u \le s \le T. \end{aligned}$$

From the càdlàg property of the integrals and the process $\mathscr{X}^{t,x}$ and assumption (iv) we can deduce that $u \mapsto D_u \mathscr{X}^{t,x}(s)$ is a.s. bounded away from zero on $[t, s]$. Hence, $\int_t^s |D_u \mathscr{X}^{t,x}(s)|^2 du$ is a.s. invertible. We can show that $\mathscr{X}^{t,x}(s)$ and $D_u \mathscr{X}^{t,x}(s)$ have moments of all orders. We can also show that $D_u \mathscr{X}^{t,x}(s)$ is Malliavin differentiable and $D_v D_u \mathscr{X}^{t,x}(s)$ has moments of all orders. Absolute continuity of the probability law of $\mathscr{X}^{t,x}(s)$ now follows from Theorem 18.3 in Øksendal and Sulem (2004). □

We investigate a backward SDE with a terminal condition and a generator which depend on the state process solving the forward SDE (4.1). We deal with the backward stochastic differential equation

$$\begin{aligned} Y^{t,x}(s) &= g\big(\mathscr{X}^{t,x}(T)\big) \\ &\quad + \int_s^T f\bigg(r, \mathscr{X}^{t,x}(r-), Y^{t,x}(r-), Z^{t,x}(r), \int_{\mathbb{R}} U^{t,x}(r,z)\delta(z)\nu(dz)\bigg)dr \\ &\quad - \int_s^T Z^{t,x}(r)dW(r) - \int_s^T \int_{\mathbb{R}} U^{t,x}(r,z)\tilde{N}(dr,dz), \quad t \le s \le T. \end{aligned} \tag{4.5}$$

The form of the generator considered in (4.5) is most common in applications, see also Sects. 3.2 and 3.5. The system of equations (4.1) and (4.5) is called a forward-backward stochastic differential equation. A solution to the system (4.1), (4.5) is a quadruple $(\mathscr{X}, Y, Z, U) \in \mathbb{S}^2(\mathbb{R}) \times \mathbb{S}^2(\mathbb{R}) \times \mathbb{H}^2(\mathbb{R}) \times \mathbb{H}^2_N(\mathbb{R})$.

We assume

(B4) the function $f : [0,T] \times \mathbb{R} \times \mathbb{R} \times \mathbb{R} \to \mathbb{R}$ is continuous and satisfies the Lipschitz condition

$$\begin{aligned} &\big|f(t,x,y,z,u) - f\big(t,x',y',z',u'\big)\big| \\ &\quad \le K\big(\big|x-x'\big| + \big|y-y'\big| + \big|z-z'\big| + \big|u-u'\big|\big), \end{aligned}$$

for all $(t,x,z,u), (t,x',z',u') \in [0,T] \times \mathbb{R} \times \mathbb{R} \times \mathbb{R}$,

(B5) the functions $\delta : \mathbb{R} \to \mathbb{R}$ and $g : \mathbb{R} \mapsto \mathbb{R}$ are measurable and they satisfy

$$\begin{aligned} \big|\delta(z)\big| &\le K\big(1 \wedge |z|\big), \quad z \in \mathbb{R}, \\ \big|g(x) - g\big(x'\big)\big| &\le K\big|x - x'\big|, \quad x, x' \in \mathbb{R}, \end{aligned}$$

(B6) for each $(t,x,y,z) \in [0,T] \times \mathbb{R} \times \mathbb{R} \times \mathbb{R}$ the mapping $u \mapsto f(t,x,y,z,u)$ is non-decreasing, and $\delta(z) \ge 0$, $z \in \mathbb{R}$.

Assumptions (B1)–(B5) are standard in the theory of FBSDEs. Assumption (B6) is needed in Sect. 4.2 where we apply a comparison principle.

Existence of a unique solution to the FBSDE can be immediately deduced from Theorems 3.1.1 and 4.1.1.

Theorem 4.1.3 *Under assumptions* (B1)–(B5) *there exists a unique solution* $(\mathscr{X}, Y, Z, U) \in \mathbb{S}^2(\mathbb{R}) \times \mathbb{S}^2(\mathbb{R}) \times \mathbb{H}^2(\mathbb{R}) \times \mathbb{H}^2_N(\mathbb{R})$ *to the FBSDE* (4.1), (4.5).

Intuitively, in our Markovian setting we expect to represent the solution to the BSDE (4.5) as a function of time and the state process (4.1). The first result gives such a representation for the process Y, see Corollary 2.3 and Remark 2.4 in Barles et al. (1997).

Proposition 4.1.1 *Under the assumptions of Theorem* 4.1.3 *there exists a measurable functions* $u : [0,T] \times \mathbb{R} \to \mathbb{R}$ *such that*

$$Y^{t,x}(s) = u\big(s, \mathscr{X}^{t,x}(s)\big), \quad 0 \le t \le s \le T.$$

In particular, $Y^{t,x}(t) = u(t,x)$.

We prove a technical, but useful, result which shows that the function $u(t,x)$ is Lipschitz continuous in x and locally $\frac{1}{2}$-Hölder continuous in t.

Lemma 4.1.1 *Under the assumptions of Theorem* 4.1.3 *the function* $u(t,x) = Y^{t,x}(t)$ *is continuous on* $[0,T]\times\mathbb{R}$. *Moreover, the function* u *satisfies*

$$\begin{aligned}&\left|u(t,x)-u(t',x')\right|^2\\&\quad\leq K\left(\left|x-x'\right|^2+\left(1+\left|x\vee x'\right|^2\right)\left|t-t'\right|\right),\quad (t,x),(t',x')\in[0,T]\times\mathbb{R}.\end{aligned}$$

Proof Consider the parametrized BSDE

$$\begin{aligned}&Y^{t,x}(s)\\&\quad=g\left(\mathscr{X}^{t,x}(T)\right)\\&\qquad+\int_s^T \mathbf{1}\{t\leq r\leq T\}f\left(r,\mathscr{X}^{t,x}(r),Y^{t,x}(r),Z^{t,x}(r),\int_{\mathbb{R}}U^{t,x}(r,z)\delta(z)\nu(dz)\right)dr\\&\qquad-\int_s^T Z^{t,x}(r)dW(r)-\int_s^T\int_{\mathbb{R}}U^{t,x}(r,z)\tilde{N}(dr,dz),\quad 0\leq s\leq T.\end{aligned}$$

We can define $Y^{t,x}(r)=Y^{t,x}(t)=u(t,x)$ and $Z^{t,x}(r)=U^{t,x}(r,z)=0$ for $r\leq t$. By the a priori estimate (3.7), assumptions (B4)–(B5) and the moment estimates (4.2) we derive

$$\begin{aligned}\left|u(t,x)\right|^2&=\left|Y^{t,x}(t)\right|^2\leq\left\|Y^{t,x}\right\|_{\mathbb{S}^2}\\&\leq K\mathbb{E}\left[\left|g\left(\mathscr{X}^{t,x}(T)\right)\right|^2+\int_0^T\mathbf{1}\{t\leq r\leq T\}\left|f\left(r,\mathscr{X}^{t,x}(r),0,0,0\right)\right|^2dr\right]\\&\leq K\mathbb{E}\left[1+\sup_{0\leq r\leq T}\left|\mathscr{X}^{t,x}(r)\right|^2\right]\leq K\left(1+|x|^2\right).\end{aligned}\tag{4.6}$$

Let $t'\geq t$. The a priori estimate (3.7) and assumptions (B4)–(B5) yield

$$\begin{aligned}&\left|Y^{t,x}(t)-Y^{t',x'}(t')\right|^2\\&\quad=\left|Y^{t,x}(0)-Y^{t',x'}(0)\right|^2\leq\mathbb{E}\left[\sup_{s\in[0,T]}\left|Y^{t,x}(s)-Y^{t',x'}(s)\right|^2\right]\\&\quad\leq K\mathbb{E}\left[\left|g\left(\mathscr{X}^{t,x}(T)\right)-g\left(\mathscr{X}^{t',x'}(T)\right)\right|^2\right.\\&\qquad+\int_0^T\left|\mathbf{1}_{t\leq r\leq T}f\left(r,\mathscr{X}^{t,x}(r),Y^{t',x'}(r),Z^{t',x'}(r),\int_{\mathbb{R}}U^{t',x'}(r,z)\delta(z)\nu(dz)\right)\right.\\&\qquad\left.\left.-\mathbf{1}_{t'\leq r\leq T}f\left(r,\mathscr{X}^{t',x'}(r),Y^{t',x'}(r),Z^{t',x'}(r),\int_{\mathbb{R}}U^{t',x'}(r,z)\delta(z)\nu(dz)\right)\right|^2dr\right]\end{aligned}$$

$$\leq K\mathbb{E}\bigg[\big|\mathscr{X}^{t,x}(T)-\mathscr{X}^{t',x'}(T)\big|^2+\int_{t'}^{T}\big|\mathscr{X}^{t,x}(r)-\mathscr{X}^{t',x'}(r)\big|^2dr$$

$$+\int_t^{t'}\bigg|f\bigg(r,\mathscr{X}^{t,x}(r),Y^{t',x'}(r),Z^{t',x'}(r),\int_{\mathbb{R}}U^{t',x'}(r,z)\delta(z)\nu(dz)\bigg)\bigg|^2dr\bigg]$$

$$\leq K\mathbb{E}\bigg[\sup_{0\leq r\leq T}\big|\mathscr{X}^{t,x}(r)-\mathscr{X}^{t',x'}(r)\big|^2$$

$$+\int_t^{t'}\bigg(1+\big|\mathscr{X}^{t,x}(r)\big|^2+\big|Y^{t',x'}(r)\big|^2$$

$$+\big|Z^{t',x'}(r)\big|^2+\int_{\mathbb{R}}\big|U^{t',x'}(r,z)\big|^2\nu(dz)\bigg)dr\bigg].$$

Since $Y^{t',x'}(r)=u(t',x')$ and $Z^{t',x'}(r)=U^{t',x'}(r,z)=0$ for $r\leq t'$, by (4.6) and (4.2) we get

$$\big|Y^{t,x}(t)-Y^{t',x'}(t')\big|^2\leq K\mathbb{E}\bigg[\sup_{0\leq r\leq T}\big|\mathscr{X}^{t,x}(r)-\mathscr{X}^{t',x'}(r)\big|^2$$

$$+\int_t^{t'}\Big(1+|x|^2+\sup_{0\leq r\leq T}\big|\mathscr{X}^{t,x}(r)-x\big|^2+\big|u(t',x')\big|^2\Big)dr\bigg]$$

$$\leq K\big(|x-x'|^2+\big(1+|x\vee x'|^2\big)|t'-t|\big).$$

□

We now characterize the control processes (Z,U).

Theorem 4.1.4 *Assume that* (B1)–(B5) *hold. Let the generator f satisfy the assumptions of Theorem* 3.5.1 *and let the functions μ and σ be continuously differentiable and γ be continuously differentiable in the first variable x. Consider the function u defined in Proposition* 4.1.1. *If $u\in\mathscr{C}^{0,1}([0,T]\times\mathbb{R})$ or the probability law of $\mathscr{X}^{t,x}(s)$, for $s>t$, is absolutely continuous with respect to the Lebesgue measure, then*

$$Z^{t,x}(s)=u_x\big(s,\mathscr{X}^{t,x}(s-)\big)\sigma\big(\mathscr{X}^{t,x}(s-)\big),\quad t\leq s\leq T,$$

$$U^{t,x}(s,z)=u\big(s,\mathscr{X}^{t,x}(s-)+\gamma\big(\mathscr{X}^{t,x}(s-),z\big)\big)$$

$$-u\big(s,\mathscr{X}^{t,x}(s-)\big),\quad t\leq s\leq T,\ z\in\mathbb{R}.$$

Proof By Proposition 4.1.1 we have $Y^{t,x}(s)=u(s,\mathscr{X}^{t,x}(s))$. From Theorem 4.1.2 we know that $\mathscr{X}$ is Malliavin differentiable. Property (4.3) yields that the assumptions required for the terminal condition from Theorem 3.5.1 are satisfied. Hence, from Theorem 3.5.1 and Proposition 3.5.2 we deduce the representations

$$Z^{t,x}(s)=\big(D_su\big(s,\mathscr{X}^{t,x}(s)\big)\big)^{\mathscr{P}},\quad U^{t,x}(s,z)=z\big(D_{s,z}u\big(s,\mathscr{X}^{t,x}(s)\big)\big)^{\mathscr{P}}.\quad (4.7)$$

Since $x \mapsto u(t,x)$ is Lipschitz continuous, we can apply the chain rule from Proposition 2.6.4 to (4.7). The derivative $D_s \mathscr{X}^{t,x}(s) = \sigma(\mathscr{X}^{t,x}(s))$ can be immediately deduced from (4.4). In a similar way to (4.4), we can prove that $zD_{s,z}\mathscr{X}^{t,x}(s) = \gamma(\mathscr{X}^{t,x}(s), z)$. The representations for Z and U follow by taking the predictable projections. □

We remark that u_x denote first derivative of the function $u(t,x)$ with respect to the second state variable and $\mathscr{X}^{t,x}$ denote the solution to the forward SDE (4.1) with the initial condition $\mathscr{X}(t) = x$. For assumptions under which $u \in \mathscr{C}^{0,1}([0,T] \times \mathbb{R})$ we refer to Bouchard and Elie (2008). Assumptions which guarantee that $u \in \mathscr{C}^{1,2}([0,T] \times \mathbb{R})$ can be found in Theorem 4.3.1. Assumptions under which the law of $\mathscr{X}$ is absolutely continuous are given in Theorem 4.1.2.

We point out that Theorem 4.1.4 gives direct formulas for the control processes (Z,U) of a BSDE, assuming that we already have the function u which characterizes the solution Y. The formulas from Theorem 4.1.4 are of great importance and they are used in applications (including in numerical algorithms). Examples are considered in Sect. 8.2. In the financial context, Theorem 4.1.4 yields delta-hedging strategies.

The question remains: how to find the function u and its derivative u_x which characterize the solution to a FBSDE. In the next chapter we show that u and u_x can be derived by solving a partial integro-differential equation. Before we move to the next chapter which deals with PIDEs, let us consider a FBSDE with zero generator. For this special FBSDE we can show that the function u and its derivative u_x can be represented as conditional expectations of the state process. Such representations allow us to calculate the solution to a FBSDE with zero generator or find the predictable representation of a pay-off contingent on a Markov process by using analytical formulas for expectations or applying Monte Carlo simulations.

Proposition 4.1.2 *We investigate the FBSDE*

$$\begin{aligned}
\mathscr{X}^{t,x}(s) &= x + \int_t^s \mu\big(\mathscr{X}^{t,x}(r-)\big)dr + \int_t^s \sigma\big(\mathscr{X}^{t,x}(r-)\big)dW(r) \\
&\quad + \int_t^s \int_{\mathbb{R}} \gamma\big(\mathscr{X}(r-), z\big)\tilde{N}(dr,dz), \quad 0 \le t \le s \le T, \\
Y^{t,x}(s) &= g\big(\mathscr{X}^{t,x}(T)\big) \\
&\quad - \int_s^T Z^{t,x}(r)dW(r) - \int_s^T \int_{\mathbb{R}} U^{t,x}(r,z)\tilde{N}(dr,dz), \quad 0 \le t \le s \le T.
\end{aligned} \tag{4.8}$$

Let the assumptions of Theorem 4.1.2 *hold and let the function* $g : \mathbb{R} \to \mathbb{R}$ *be Lipschitz continuous. Consider the function* u *defined in Proposition* 4.1.1. *We have the representations*

$$u(t,x) = \mathbb{E}\big[g\big(\mathscr{X}^{t,x}(T)\big)\big], \quad (t,x) \in [0,T] \times \mathbb{R}, \tag{4.9}$$

and

$$\begin{aligned} u_x(t,x) &= \frac{\partial}{\partial x}\mathbb{E}\big[g\big(\mathscr{X}^{t,x}(T)\big)\big] \\ &= \mathbb{E}\big[g'\big(\mathscr{X}^{t,x}(T)\big)\mathscr{Y}^{t,x}(T)\big], \quad (t,x) \in [0,T) \times \mathbb{R}, \end{aligned} \tag{4.10}$$

where the process $\mathscr{Y}^{t,x}$ *satisfies the forward SDE*

$$\begin{aligned} \mathscr{Y}^{t,x}(s) = 1 &+ \int_t^s \mathscr{Y}^{t,x}(r-)\mu'\big(\mathscr{X}^{t,x}(r-)\big)dr \\ &+ \int_t^s \mathscr{Y}^{t,x}(r-)\sigma'\big(\mathscr{X}^{t,x}(r-)\big)dW(r) \\ &+ \int_t^s \int_{\mathbb{R}} \mathscr{Y}^{t,x}(r-)\gamma'\big(\mathscr{X}^{t,x}(r-),z\big)\tilde{N}(dr,dz), \quad 0 \le t \le s \le T, \end{aligned} \tag{4.11}$$

and $\mathscr{Y}^{t,x}(\omega,s) = \frac{\partial}{\partial x}\mathscr{X}^{t,x}(\omega,s)$, $(\omega,s) \in \Omega \times [t,T]$.

Proof By Theorem 4.1.3 there exists a unique solution $(\mathscr{X}, Y, Z, U)$ to (4.8). Representation (4.9) is obvious. We prove (4.10). From Theorem 3.4 in Kunita (2004) we recall that $x \mapsto \mathscr{X}^{t,x}(\omega,s)$ is continuously differentiable and the derivative $\frac{\partial}{\partial x}\mathscr{X}^{t,x}(\omega,s)$ satisfies (4.11). By Theorem 4.1.2 the random variable $\mathscr{X}^{t,x}(T)$ has an absolutely continuous probability law with respect to the Lebesgue measure. Consequently, $x \mapsto g(\mathscr{X}^{t,x}(T))$ is a.s. differentiable. We now justify the interchange of the limit and the expectation. We fix $(t,x) \in [0,T) \times \mathbb{R}$. We consider a family $(A_\alpha)_{\alpha\in\mathbb{R}}$ defined by

$$A_\alpha = \left|\frac{g(\mathscr{X}^{t,x}(T)) - g(\mathscr{X}^{t,\alpha}(T))}{x-\alpha}\right|, \quad \alpha \in \mathbb{R}. \tag{4.12}$$

The Lipschitz property of g and the moment estimate (4.2) yield

$$\mathbb{E}\left[\left|\frac{g(\mathscr{X}^{t,x}(T)) - g(\mathscr{X}^{t,\alpha}(T))}{x-\alpha}\right|^2\right] \le K\frac{\mathbb{E}[|\mathscr{X}^{t,x}(T) - \mathscr{X}^{t,\alpha}(T)|^2]}{|x-\alpha|^2} \le K, \quad \alpha \in \mathbb{R}$$

which proves that the family $(A_\alpha)_{\alpha\in\mathbb{R}}$ is uniformly integrable. Hence, we get

$$\begin{aligned} &\lim_{\alpha\to x}\mathbb{E}\left[\frac{g(\mathscr{X}^{t,x}(T)) - g(\mathscr{X}^{t,\alpha}(T))}{x-\alpha}\right] \\ &= \mathbb{E}\left[\lim_{\alpha\to x}\frac{g(\mathscr{X}^{t,x}(T)) - g(\mathscr{X}^{t,\alpha}(T))}{x-\alpha}\right] = \mathbb{E}\big[g'\big(\mathscr{X}^{t,x}(T)\big)\mathscr{Y}^{t,x}(T)\big]. \end{aligned}$$

□

We give an example illustrating the application of Theorem 4.1.4 and Proposition 4.1.2.

Example 4.1 We are interested in finding the predictable representation of $\xi = (K - \mathscr{X}(T))^+$ where $d\mathscr{X}(t) = \mathscr{X}(t)\sigma dW(t)$, $X(0) = x_0 > 0$. Let $\mathscr{Y}^{t,x}(s) = \frac{\partial}{\partial x}\mathscr{X}^{t,x}(s)$. The process $\mathscr{Y}^{t,x}$ satisfies the equation

$$\mathscr{Y}^{t,x}(s) = 1 + \int_t^s \mathscr{Y}^{t,x}(r)\sigma dW(r), \quad 0 \leq t \leq s \leq T,$$

and we immediately deduce that $\mathscr{Y}^{t,x}(s) = \frac{\mathscr{X}^{t,x}(s)}{x}$, $t \leq s \leq T$. Let $u(t,x) = \mathbb{E}[(K - \mathscr{X}^{t,x}(T))^+]$. By Proposition 4.1.2 we derive

$$u_x(t,x) = -\mathbb{E}\big[\mathbf{1}\{\mathscr{X}^{t,x}(T) < K\}\mathscr{X}^{t,x}(T)\big]\frac{1}{x}, \quad (t,x) \in [0,T) \times (0,\infty). \quad (4.13)$$

Recalling the formula for the control process from Theorem 4.1.4, we conclude that the predictable representation of ξ takes the form

$$\xi = \mathbb{E}[\xi] + \int_0^T \varphi\big(t, \mathscr{X}^{0,x_0}(t)\big)dW(t),$$

where

$$\begin{aligned}\varphi(t,x) &= u_x(t,x)\sigma x \\ &= -\sigma\mathbb{E}\big[\mathbf{1}\{\mathscr{X}^{t,x}(T) < K\}\mathscr{X}^{t,x}(T)\big], \quad (t,x) \in [0,T] \times (0,\infty).\end{aligned}$$

Clearly, the representation coincides with the representation from Example 3.2 where we apply the Malliavin calculus. We remark that the approach based on the Malliavin calculus can be used for a general non-Markovian BSDE, whereas Theorem 4.1.4 and Proposition 4.1.2 only hold for Markovian BSDEs. It is well-known that

$$\varphi(t,x) = -x\sigma\Phi\left(-\frac{\ln(\frac{x}{K}) + \sigma^2/2(T-t)}{\sigma\sqrt{T-t}}\right), \quad (t,x) \in [0,T] \times (0,\infty),$$

where Φ denote the distribution function of the standard normal random variable.

4.2 The Feynman-Kac Formula and the Connection with Partial Integro-Differential Equations

In Proposition 4.1.1 and Theorem 4.1.4 we characterize the solution to a FBSDE with a function u. For a FBSDE with zero generator and for a FBSDE with generator independent of (Y, Z, U) we can use Proposition 4.1.2 to find the function u. For a general FBSDE the function u can be derived by solving a partial integro-differential equation. In this chapter we consider the PIDE

$$
\begin{aligned}
&-u_t(t,x) - \mathscr{L}u(t,x)\\
&\quad -f\big(t,x,u(t,x),u_x(t,x)\sigma(x),\mathscr{J}u(t,x)\big) = 0, \quad (t,x)\in[0,T)\times\mathbb{R}, \qquad (4.14)\\
&u(T,x) = g(x), \quad x\in\mathbb{R},
\end{aligned}
$$

where we introduce two operators

$$
\begin{aligned}
\mathscr{L}u(t,x) &= \mu(x)u_x(t,x) + \frac{1}{2}\sigma^2(x)u_{xx}(t,x)\\
&\quad + \int_{\mathbb{R}} \big(u\big(t,x+\gamma(x,z)\big) - u(t,x) - \gamma(x,z)u_x(t,x)\big)\nu(dz),\\
\mathscr{J}u(t,x) &= \int_{\mathbb{R}} \big(u\big(t,x+\gamma(x,z)\big) - u(t,x)\big)\delta(z)\nu(dz),
\end{aligned}
$$

which are defined for $u \in \mathscr{C}^{1,2}([0,T]\times\mathbb{R})$, see Sect. 12.2 in Cont and Tankov (2004).

First, we give a generalization the Feynman-Kac formula.

Theorem 4.2.1 *Assume that* (B1)–(B5) *hold. Let* $u \in \mathscr{C}^{1,2}([0,T]\times\mathbb{R})$ *satisfy the PIDE* (4.14) *and the linear growth conditions*

$$
\big|u(t,x)\big| \le K\big(1+|x|\big), \quad \big|u_x(t,x)\big| \le K\big(1+|x|\big), \quad (t,x)\in[0,T]\times\mathbb{R}.
$$

Then

$$
\begin{aligned}
Y^{t,x}(s) &= u\big(s,\mathscr{X}^{t,x}(s)\big), \quad t\le s\le T,\\
Z^{t,x}(s) &= u_x\big(s,\mathscr{X}^{t,x}(s-)\big)\sigma\big(\mathscr{X}^{t,x}(s-)\big), \quad t\le s\le T,\\
U^{t,x}(s,z) &= u\big(s,\mathscr{X}^{t,x}(s-)+\gamma\big(\mathscr{X}^{t,x}(s-),z\big)\big)\\
&\quad - u\big(s,\mathscr{X}^{t,x}(s-)\big), \quad t\le s\le T,\ z\in\mathbb{R},
\end{aligned}
\qquad (4.15)
$$

is the unique solution to the BSDE (4.5). *Moreover, we have the representation*

$$
\begin{aligned}
Y^{t,x}(t) &= u(t,x)\\
&= \mathbb{E}\Big[g\big(\mathscr{X}^{t,x}(T)\big)\\
&\qquad + \int_t^T f\Big(r,\mathscr{X}^{t,x}(r),Y^{t,x}(r),Z^{t,x}(r),\int_{\mathbb{R}} U^{t,x}(r)\delta(z)\nu(dz)\Big)dr\Big],\\
&\qquad 0\le t\le T.
\end{aligned}
\qquad (4.16)
$$

Proof By the Itô's formula we obtain

$$\begin{aligned}
&u\big(T, \mathscr{X}^{t,x}(T)\big) \\
&\quad = u\big(s, \mathscr{X}^{t,x}(s)\big) + \int_s^T u_t\big(r, \mathscr{X}^{t,x}(r-)\big)dr \\
&\qquad + \int_s^T u_x\big(r, \mathscr{X}^{t,x}(r-)\big)d\mathscr{X}^{t,x}(r) \\
&\qquad + \int_s^T \frac{1}{2}u_{xx}\big(r, \mathscr{X}^{t,x}(r-)\big)\sigma^2\big(\mathscr{X}^{t,x}(r-)\big)dr \\
&\qquad + \int_s^T \int_{\mathbb{R}} \Big(u\big(r, \mathscr{X}^{t,x}(r-) + \gamma\big(\mathscr{X}^{t,x}(r-), z\big)\big) \\
&\qquad - u\big(r, \mathscr{X}^{t,x}(r-)\big) - u_x\big(r, \mathscr{X}^{t,x}(r-)\big)\gamma\big(\mathscr{X}^{t,x}(r-), z\big)\Big)N(dr, dz), \\
&\qquad t \le s \le T.
\end{aligned}$$

Since the process $\mathscr{X}$ satisfies (4.1) and the function u satisfies (4.14), we derive

$$\begin{aligned}
&g\big(\mathscr{X}^{t,x}(T)\big) = u\big(s, \mathscr{X}^{t,x}(s)\big) - \int_s^T f\big(r, \mathscr{X}^{t,x}(r-), u\big(r, \mathscr{X}^{t,x}(r-)\big), \\
&u_x\big(r, \mathscr{X}^{t,x}(r-)\big)\sigma\big(\mathscr{X}^{t,x}(r-)\big), \mathscr{J}u\big(r, \mathscr{X}^{t,x}(r-)\big)\big)dr \\
&\quad + \int_s^T u_x\big(r, \mathscr{X}^{t,x}(r-)\big)\sigma\big(\mathscr{X}^{t,x}(r-)\big)dW(r) \\
&\quad + \int_s^T \int_{\mathbb{R}} \Big(u\big(r, \mathscr{X}^{t,x}(r-) + \gamma\big(\mathscr{X}^{t,x}(r-), z\big)\big) \\
&\quad - u\big(r, \mathscr{X}^{t,x}(r-)\big)\Big)\tilde{N}(dr, dz), \quad t \le s \le T.
\end{aligned} \tag{4.17}$$

Hence, the candidate solution (4.15) satisfies the BSDE (4.5). By the growth conditions for σ, γ, u, u_x and the moment estimates for $\mathscr{X}$ we can show that $(Y^{t,x}, Z^{t,x}, U^{t,x}) \in \mathbb{S}^2(\mathbb{R}) \times \mathbb{H}^2(\mathbb{R}) \times \mathbb{H}^2_N(\mathbb{R})$. By uniqueness of solution the candidate solution (4.15) is the unique solution to the BSDE (4.5). Representation (4.16) can be established by taking the conditional expectation of (4.17). □

Theorem 4.2.1 establishes a probabilistic representation of a solution to the PIDE (4.14). The result of that theorem shows that we can find the unique solution to the BSDE (4.5), which has the representation (4.16), by finding the unique solution to the PIDE (4.14). In the case when the generator f is independent of (Y, Z, U) we recover the classical Feynman-Kac formula, which is well-known in financial mathematics and option pricing. In the general case, we have the non-linear Feynman-Kac formula.

Theorem 4.2.1 relies on the smoothness assumption for u. We cannot always expect that a smooth function u solves (4.14), see Sect. 12.2 in Cont and Tankov

(2004) and the discussion at the end of Sect. 4.3. In order to relax smoothness assumption, we can work with the concept of a viscosity solution, see Definition 3.1 and Lemma 3.3 in Barles et al. (1997).

Definition 4.2.1 We say that $u \in \mathscr{C}([0,T] \times \mathbb{R})$ is

(i) a viscosity subsolution to (4.14) if

$$u(T,x) \le g(x), \quad x \in \mathbb{R},$$

and for any $\varphi \in \mathscr{C}^{1,2}([0,T] \times \mathbb{R})$, whenever $(t,x) \in [0,T] \times \mathbb{R}$ is a global maximum point of $u - \varphi$, we have

$$-\varphi_t(t,x) - \mathscr{L}\varphi(t,x) - f\big(t,x,u(t,x),\varphi_x(t,x)\sigma(t,x),\mathscr{J}\varphi(t,x)\big) \le 0.$$

(ii) a viscosity supersolution to (4.14) if

$$u(T,x) \ge g(x), \quad x \in \mathbb{R},$$

and for any $\varphi \in \mathscr{C}^{1,2}([0,T] \times \mathbb{R})$, whenever $(t,x) \in [0,T] \times \mathbb{R}$ is a global minimum point of $u - \varphi$, we have

$$-\varphi_t(t,x) - \mathscr{L}\varphi(t,x) - f\big(t,x,u(t,x),\varphi_x(t,x)\sigma(t,x),\mathscr{J}\varphi(t,x)\big) \ge 0.$$

(iii) a viscosity solution to (4.14) if it is both viscosity sub and supersolution to (4.14).

We now prove that the unique solution to the BSDE (4.5) is linked to the unique viscosity solution to the PIDE (4.14).

Theorem 4.2.2 *Assume that* (B1)–(B6) *hold. Let $Y^{t,x}$ be the unique solution to the FBSDE* (4.1), (4.5). *The function $u(t,x) = Y^{t,x}(t)$ is the unique viscosity solution to the PIDE* (4.14) *in the class of solutions that satisfy the growth condition*

$$\lim_{|x|\to\infty} \big|u(t,x)\big| e^{-c\log^2(|x|)} = 0, \quad 0 \le t \le T,\ c > 0.$$

Proof By Theorem 4.1.3 and Proposition 4.1.1 there exist a unique solution $(Y^{t,x}, Z^{t,x}, U^{t,x})$ to (4.5) and a measurable function u such that $Y^{t,x}(s) = u(s, \mathscr{X}^{t,x}(s)), t \le s \le T$. We prove that $u(t,x) = Y^{t,x}(t)$ is a subsolution to (4.14). By Lemma 4.1.1 u is continuous on $[0,T] \times \mathbb{R}$. Choose $(t,x) \in [0,T] \times \mathbb{R}$ and $\varphi \in \mathscr{C}^{1,2}([0,T] \times \mathbb{R})$ such that $\varphi(t,x) = u(t,x)$ and $\varphi \ge u$ on $[0,T] \times \mathbb{R}$. For any $\varphi \in \mathscr{C}^{1,2}([0,T] \times \mathbb{R})$ we can find a sequence of continuously differentiable functions with bounded derivatives $\varphi_n \in \mathscr{C}_b^\infty([0,T] \times \mathbb{R})$ such that $(\varphi_n)_{n\ge 1}$ (its first and second derivatives) converges to φ (its first and second derivatives) uniformly on compacts, see the proof of Theorem 4.2 in El Karoui et al. (1997b). Hence, we can prove the subsolution property for $\varphi_n \in \mathscr{C}_b^\infty([0,T] \times \mathbb{R})$ and then take the limit. In the sequel w.l.o.g. we consider $\varphi \in \mathscr{C}_b^\infty([0,T] \times \mathbb{R})$.

1. Initial estimates. Choose $h > 0$ such that $t + h \le T$. The solution to (4.5) satisfies the equation

$$Y(s) = u\big(t+h, \mathscr{X}^{t,x}(t+h)\big)$$
$$+ \int_s^{t+h} f\Big(r, \mathscr{X}^{t,x}(r-), Y(r-), Z(r-), \int_{\mathbb{R}} U(r,z)\delta(z)\nu(dz)\Big)dr$$
$$- \int_s^{t+h} Z(r)dW(r) - \int_s^{t+h}\int_{\mathbb{R}} U(r,z)\tilde{N}(dr,dz), \quad t \le s \le t+h. \tag{4.18}$$

Consider the BSDE

$$\bar{Y}(s) = \varphi\big(t+h, \mathscr{X}^{t,x}(t+h)\big)$$
$$+ \int_s^{t+h} f\Big(r, \mathscr{X}^{t,x}(r-), \bar{Y}(r-), \bar{Z}(r), \int_{\mathbb{R}} \bar{U}(r,z)\delta(z)\nu(dz)\Big)dr$$
$$- \int_s^{t+h} \bar{Z}(r)dW(r) - \int_s^{t+h}\int_{\mathbb{R}} \bar{U}(r,z)\tilde{N}(dr,dz), \quad t \le s \le t+h. \tag{4.19}$$

Since $\varphi \in \mathscr{C}_b^\infty([0,T]\times\mathbb{R})$ and $\mathscr{X}^{t,x}$ satisfies (4.2), from Theorem 4.1.3 we conclude that there exists a unique solution $(\bar{Y}, \bar{Z}, \bar{U}) \in \mathbb{S}^2(\mathbb{R}) \times \mathbb{H}^2(\mathbb{R}) \times \mathbb{H}^2_N(\mathbb{R})$ to (4.19). Since $\varphi \ge u$, by Theorem 3.2.1 and the comparison between (4.18) and (4.19) we obtain

$$\bar{Y}(s) \ge Y(s), \quad t \le s \le t+h. \tag{4.20}$$

Define

$$\Theta(t,x) = \varphi_t(t,x) + \mathscr{L}\varphi(t,x), \quad (t,x) \in [0,T]\times\mathbb{R},$$
$$\Gamma(t,x,z) = \varphi\big(t, x+\gamma(x,z)\big) - \varphi(t,x), \quad (t,x) \in [0,T]\times\mathbb{R}.$$

From assumptions (B1)–(B3) and $\varphi \in \mathscr{C}_b^\infty([0,T]\times\mathbb{R})$ we can deduce the following growth conditions

$$\big|\Theta(t,x)\big| \le K\big(1+|x|^2\big), \quad \big|\Gamma(t,x,z)\big| \le K(1\wedge z), \quad (t,x,z) \in [0,T]\times\mathbb{R}\times\mathbb{R}. \tag{4.21}$$

We now define

$$\hat{Y}(s) = \bar{Y}(s) - \varphi\big(s, \mathscr{X}^{t,x}(s)\big), \quad t \le s \le t+h,$$
$$\hat{Z}(s) = \bar{Z}(s) - \varphi_x\big(s, \mathscr{X}^{t,x}(s)\big)\sigma\big(\mathscr{X}^{t,x}(s-)\big), \quad t \le s \le t+h,$$
$$\hat{U}(s,z) = \bar{U}(s,z) - \Gamma\big(s, \mathscr{X}^{t,x}(s-), z\big), \quad t \le s \le t+h,$$

and by the Itô's formula we have

$$\varphi\big(s, \mathscr{X}^{t,x}(s)\big)$$
$$= \varphi\big(t+h, \mathscr{X}^{t,x}(t+h)\big) - \int_s^{t+h} \Theta\big(r, \mathscr{X}^{t,x}(r-)\big)dr$$

$$
\begin{aligned}
&- \int_s^{t+h} \varphi_x\big(r, \mathscr{X}^{t,x}(r-)\big)\sigma\big(\mathscr{X}^{t,x}(r-)\big)dW(r) \\
&- \int_s^{t+h} \int_{\mathbb{R}} \Gamma\big(r, \mathscr{X}^{t,x}(r-), z\big)\tilde{N}(dr, dz), \quad t \le s \le t+h. \qquad (4.22)
\end{aligned}
$$

From (4.19) and (4.22) it is easy to observe that $(\hat{Y}, \hat{Z}, \hat{U}) \in \mathbb{S}^2(\mathbb{R}) \times \mathbb{H}^2(\mathbb{R}) \times \mathbb{H}^2_N(\mathbb{R})$ is the unique solution to the BSDE

$$
\begin{aligned}
\hat{Y}(s) = &\int_s^{t+h} \Big\{ \Theta\big(r, \mathscr{X}^{t,x}(r-)\big) + f\Big(r, \mathscr{X}^{t,x}(r-), \varphi\big(r, \mathscr{X}^{t,x}(r-)\big) \\
&+ \hat{Y}(r-), \varphi_x\big(r, \mathscr{X}^{t,x}(r-)\big)\sigma\big(\mathscr{X}^{t,x}(r-)\big) + \hat{Z}(r), \\
&\int_{\mathbb{R}} \big(\Gamma\big(r, \mathscr{X}^{t,x}(r-), z\big) + \hat{U}(r,z)\big)\delta(z)\nu(dz)\Big)\Big\} dr \\
&- \int_s^T \hat{Z}(r) dW(r) - \int_s^T \int_{\mathbb{R}} \hat{U}(r,z)\tilde{N}(dr, dz), \quad t \le s \le t+h. \qquad (4.23)
\end{aligned}
$$

Applying the a priori estimate (3.18) to the BSDE (4.23), we get

$$
\begin{aligned}
&\mathbb{E}\Big[\big|\hat{Y}(s)\big|^2 + \int_s^{t+h} \big|\hat{Z}(r)\big|^2 dr + \int_s^{t+h} \int_{\mathbb{R}} \big|\hat{U}(r,z)\big|^2 \nu(dz)dr \Big] \\
&\le K\mathbb{E}\Big[\int_s^{t+h} \big|\hat{Y}(r)\big| \Big| \Theta\big(r, \mathscr{X}^{t,x}(r)\big) \\
&\quad + f\Big(r, \mathscr{X}^{t,x}(r), \varphi\big(r, \mathscr{X}^{t,x}(r)\big), \varphi_x\big(r, \mathscr{X}^{t,x}(r)\big)\sigma\big(\mathscr{X}^{t,x}(r)\big), \\
&\quad \int_{\mathbb{R}} \Gamma\big(r, \mathscr{X}^{t,x}(r), z\big)\delta(z)\nu(dz)\Big)\Big| dr \Big] \\
&\le K\mathbb{E}\Big[\int_s^{t+h} \big|\hat{Y}(r)\big| \big(1 + \big|\mathscr{X}^{t,x}(r)\big|^2\big) dr \Big] \\
&\le K\mathbb{E}\Big[\int_s^{t+h} \big|\hat{Y}(r)\big| \big(1 + |x|^2 + \big|\mathscr{X}^{t,x}(r) - x\big|^2\big) dr \Big] \\
&\le K\mathbb{E}\Big[\int_s^{t+h} \big(\big|\hat{Y}(r)\big| + \big|\hat{Y}(r)\big|^2 + \big|\mathscr{X}^{t,x}(r) - x\big|^4\big) dr \Big], \quad t \le s \le t+h, \qquad (4.24)
\end{aligned}
$$

where we use the growth conditions for f, σ, Θ, Γ and the assumption that $\varphi \in \mathscr{C}_b^\infty([0,T] \times \mathbb{R})$. By the classical estimate $|y| \le 1 + |y|^2$ and the moment estimate (4.2) we derive

$$\mathbb{E}\big[|\hat{Y}(s)|^2\big] \le K\mathbb{E}\Big[\int_s^{t+h}(1+|\hat{Y}(r)|^2+h)dr\Big]$$
$$\le K\Big(h+h^2+\mathbb{E}\Big[\int_s^{t+h}|\hat{Y}(r)|^2 dr\Big]\Big), \quad t\le s\le t+h,$$

and applying the Gronwall's inequality, see Lemma 3 in Cohen (2011), for sufficiently small $h>0$ we have

$$\mathbb{E}\big[|\hat{Y}(s)|^2\big] \le K\big(h+h^2\big)e^{Kh} \le Kh, \quad t\le s\le t+h,$$

and

$$\mathbb{E}\big[|\hat{Y}(s)|\big] \le Kh^{1/2}, \quad t\le s\le t+h. \tag{4.25}$$

Combining (4.24), (4.25) and (4.2), for sufficiently small $h>0$ we can now derive

$$\mathbb{E}\Big[\int_t^{t+h}|\hat{Z}(r)|^2 dr + \int_t^{t+h}\int_{\mathbb{R}}|\hat{U}(r,z)|^2 \nu(dz)dr\Big]$$
$$\le K\int_t^{t+h}(h^{1/2}+h+h)dr \le Khh^{1/2}. \tag{4.26}$$

2. The proof of the subsolution property by contradiction. Suppose that for the chosen (t,x) we have

$$-\varphi_t(t,x)-\mathscr{L}\varphi(t,x)-f\big(t,x,u(t,x),\varphi_x(t,x)\sigma(x),\mathscr{J}\varphi(t,x)\big)>0. \tag{4.27}$$

Consequently, there exist $\varepsilon>0$ and $h_0>0, h_0'>0$ such that for $t-h_0\le r\le t+h_0, x-h_0'\le y\le x+h_0'$ we have

$$-\varphi_t(r,y)-\mathscr{L}\varphi(r,y)-f\big(r,y,\varphi(r,y),\varphi_x(r,y)\sigma(y),\mathscr{J}\varphi(r,y)\big)\ge\varepsilon. \tag{4.28}$$

Choose small $h>0$ such that $h<h_0$. Define

$$\zeta_h=\frac{1}{h}\mathbb{E}\Big[\int_t^{t+h}V\big(r,\mathscr{X}^{t,x}(r)\big)dr\Big], \tag{4.29}$$

where

$$V(r,y)=\varphi_t(r,y)+\mathscr{L}\varphi(r,y)+f\big(r,y,\varphi(r,y),\varphi_x(r,y)\sigma(y),\mathscr{J}\varphi(r,y)\big).$$

From (B1)–(B5) and (4.21) we can deduce the growth condition

$$\big|V(r,y)\big|\le K\big(1+|y|^2\big), \quad (r,y)\in[0,T]\times\mathbb{R}. \tag{4.30}$$

We introduce a stopping time

$$\tau=\inf\big\{s\ge t:\big|\mathscr{X}^{t,x}(s)-x\big|>h_0'\big\}.$$

By the moment estimate (4.2) and the Chebyshev inequality we establish

$$\begin{aligned}\mathbb{P}(\tau \le h) &= \mathbb{P}\Big(\sup_{t\le s\le t+h} \big|\mathscr{X}^{t,x}(s) - x\big| > h_0'\Big) \\ &\le \frac{\mathbb{E}[\sup_{t\le s\le t+h} |\mathscr{X}^{t,x}(s) - x|^2]}{|h_0'|^2} \le Kh. \end{aligned}\tag{4.31}$$

By (4.28), (4.30), (4.31) and the Cauchy-Schwarz inequality we derive

$$\begin{aligned}\zeta_h &= \frac{1}{h}\mathbb{E}\Big[\int_t^{t+h} V\big(r, \mathscr{X}^{t,x}(r)\big)dr\mathbf{1}\{\tau > h\}\Big] + \frac{1}{h}\mathbb{E}\Big[\int_t^{t+h} V\big(r, \mathscr{X}^{t,x}(r)\big)dr\mathbf{1}\{\tau \le h\}\Big] \\ &\le -\varepsilon\mathbb{P}(\tau > h) + \frac{1}{h}\sqrt{\mathbb{P}(\tau\le h)}\sqrt{h}\sqrt{\mathbb{E}\Big[\int_t^{t+h} \big|V\big(r, \mathscr{X}^{t,x}(r)\big)\big|^2 dr\Big]} \\ &\le -\varepsilon(1-Kh) + K\sqrt{h}\sqrt{\mathbb{E}\Big[1 + \sup_{t\le s\le t+h}\big|\mathscr{X}^{t,x}(s)\big|^4\Big]},\end{aligned}\tag{4.32}$$

which shows that if (4.27) holds then there exist $\varepsilon_0 > 0$ and $h_0'' > 0$ such that for all $h < h_0''$ we have $\zeta_h \le -\varepsilon_0$. From (4.20) we now conclude that $\hat{Y}(t) = \bar{Y}(t) - \varphi(t, \mathscr{X}^{t,x}(t)) = \bar{Y}(t) - \varphi(t,x) = \bar{Y}(t) - u(t,x) = \bar{Y}(t) - Y^{t,x}(t) \ge 0$, and from (4.23) we obtain

$$\begin{aligned}0 &\le \frac{1}{h}\hat{Y}(t) \\ &= \frac{1}{h}\mathbb{E}\Big[\int_t^{t+h} \Theta\big(r, \mathscr{X}^{t,x}(r)\big)dr + \int_t^{t+h} f\Big(r, \mathscr{X}^{t,x}(r), \varphi\big(r, \mathscr{X}^{t,x}(r)\big) \\ &\quad + \hat{Y}(r), \varphi_x\big(r, \mathscr{X}^{t,x}(r)\big)\sigma\big(\mathscr{X}^{t,x}(r)\big) + \hat{Z}(r), \\ &\quad \int_{\mathbb{R}}\big(\Gamma\big(r, \mathscr{X}^{t,x}(r), z\big) + \hat{U}(r,z)\big)\delta(z)\nu(dz)\Big)dr\Big].\end{aligned}\tag{4.33}$$

Finally, from (4.29), (4.32) and (4.33) we conclude that for sufficiently small $h > 0$ we must have

$$\begin{aligned}\varepsilon_0 &\le \Big|\frac{1}{h}\hat{Y}(t) - \zeta_h\Big| \\ &= \frac{1}{h}\Big|\mathbb{E}\Big[\int_t^{t+h} f\Big(r, \mathscr{X}^{t,x}(r), \varphi\big(r, \mathscr{X}^{t,x}(r)\big) + \hat{Y}(r), \\ &\quad \varphi_x\big(r, \mathscr{X}^{t,x}(r)\big)\sigma\big(\mathscr{X}^{t,x}(r)\big) + \hat{Z}(r), \\ &\quad \int_{\mathbb{R}}\big(\Gamma\big(r, \mathscr{X}^{t,x}(r), z\big) + \hat{U}(r,z)\big)\delta(z)\nu(dz)\Big)dr\end{aligned}$$

$$-\int_t^{t+h} f\Big(r, \mathscr{X}^{t,x}(r), \varphi\big(r, \mathscr{X}^{t,x}(r)\big), \varphi_x\big(r, \mathscr{X}^{t,x}(r)\big)\sigma\big(\mathscr{X}^{t,x}(r)\big),$$

$$\int_{\mathbb{R}} \Gamma\big(r, \mathscr{X}^{t,x}(r), z\big)\delta(z)\nu(dz)\Big)dr\Big]\Big|$$

$$\leq K\mathbb{E}\Big[\frac{1}{h}\int_t^{t+h}|\hat{Y}(r)|dr + \frac{1}{h}\int_t^{t+h}|\hat{Z}(r)|dr + \frac{1}{h}\int_t^{t+h}\int_{\mathbb{R}}|\hat{U}(r,z)\delta(z)|\nu(dz)dr\Big]$$

$$\leq K\Big(\sup_{r\in[t,t+h]}\mathbb{E}\big[|\hat{Y}(r)|\big]$$

$$+\Big(\frac{1}{h}\mathbb{E}\Big[\int_t^{t+h}|\hat{Z}(r)|^2 dr\Big]\Big)^{1/2} + \Big(\frac{1}{h}\mathbb{E}\Big[\int_t^{t+h}\int_{\mathbb{R}}|\hat{U}(r,z)|^2\nu(dz)dr\Big]\Big)^{1/2}\Big)$$

$$\leq K\big(h^{1/2}+h^{1/4}+h^{1/4}\big) \leq Kh^{1/4},$$

where we use the Lipschitz property of f and estimates (4.25) and (4.26). We have obtain a contradiction for sufficiently small $h > 0$. Hence, inequality (4.27) cannot hold and u is a subsolution to (4.14).

The supersolution property of u is proved in an analogous way. For the uniqueness we refer to Theorem 3.5 in Barles et al. (1997). □

Theorem 4.2.2 characterizes the unique solution Y to the BSDE (4.5) by the unique viscosity solution to the PIDE (4.14). This characterization does not require smoothness assumptions. However, in applications we have to derive the control processes Z and U which cannot be characterized in the framework of Theorem 4.2.2. We first use Theorem 4.2.2 to find the unique viscosity solution to the PIDE (4.14). Next, under some additional smoothness assumptions we apply Theorem 4.2.1 or 4.1.4 to derive the control processes.

4.3 Coupled FBSDEs

The FBSDE (4.1), (4.5) is called a decoupled forward-backward stochastic differential equation. In this book we focus on decoupled FBSDEs of the form (4.1), (4.5), which arise in most applications, but it is also worth introducing coupled FBSDEs. We point out that coupled FBSDEs extend the range of possible applications of BSDEs. In Sect. 14.1 we give a financial application of a coupled FBSDE.

We consider the coupled forward-backward stochastic differential equation

$$\mathscr{X}(t) = x + \int_0^t \mu\Big(\mathscr{X}(s-), Y(s-), Z(s), \int_{\mathbb{R}} U(s,z)\delta(z)\nu(dz)\Big)ds$$

$$+\int_0^t \sigma\big(\mathscr{X}(s-), Y(s-)\big)dW(s)$$

$$+\int_0^t \int_{\mathbb{R}} \gamma\big(\mathscr{X}(s-), Y(s-), z\big)\tilde{N}(ds, dz), \quad 0 \le t \le T,$$

$$Y(t) = g\big(\mathscr{X}(T)\big) + \int_t^T f\Big(s, \mathscr{X}(s-), Y(s-), Z(s), \int_{\mathbb{R}} U(s,z)\delta(z)\nu(dz)\Big)ds$$
$$- \int_t^T Z(s)dW(s) - \int_t^T U(s,z)\tilde{N}(ds, dz), \quad 0 \le t \le T. \tag{4.34}$$

Now, not only do the terminal condition and the generator of the backward SDE depend on a solution to the forward SDE, but a solution to the backward SDE also affects the coefficients of the forward SDE. We are interested in finding a solution $(\mathscr{X}, Y, Z, U) \in \mathbb{S}^2(\mathbb{R}) \times \mathbb{S}^2(\mathbb{R}) \times \mathbb{H}^2(\mathbb{R}) \times \mathbb{H}^2_N(\mathbb{R})$ to the system (4.34). This is much more difficult, compared to the decoupled case, due to the interactions between the backward and forward components. We still assume that N is a Poisson random measure.

As in Sects. 4.1 and 4.2, we aim to derive a PIDE for a function which could characterize the solution to the FBSDE. We present heuristic reasoning. Assume that there exists a solution $(\mathscr{X}, Y, Z, U)$ to (4.34). We expect that there exists a measurable function u such that $Y(t) = u(t, \mathscr{X}(t))$. Let $u \in \mathscr{C}^{1,2}([0,T] \times \mathbb{R})$. We apply the Itô's formula and we derive

$$\begin{aligned}
&dY(t)\\
&= du\big(t, \mathscr{X}(t)\big)\\
&= \Big\{u_t\big(t, \mathscr{X}(t-)\big)\\
&+ u_x\big(t, \mathscr{X}(t-)\big)\mu\Big(\mathscr{X}(t-), u\big(t, \mathscr{X}(t-)\big), Z(t), \int_{\mathbb{R}} U(t,z)\delta(z)\nu(dz)\Big)\\
&+ \frac{1}{2}u_{xx}\big(t, \mathscr{X}(t-)\big)\sigma^2\big(\mathscr{X}(t-), u\big(t, \mathscr{X}(t-)\big)\big)\\
&+ \int_{\mathbb{R}} \Big(u\big(t, \mathscr{X}(t-) + \gamma\big(\mathscr{X}(t-), u\big(t, \mathscr{X}(t-)\big), z\big)\big) - u\big(t, \mathscr{X}(t-)\big)\\
&- u_x\big(t, \mathscr{X}(t)\big)\gamma\big(\mathscr{X}(t), u\big(t, \mathscr{X}(t)\big), z\big)\Big)\nu(dz)\Big\}dt\\
&+ u_x\big(t, \mathscr{X}(t-)\big)\sigma\big(\mathscr{X}(t-), u\big(t, \mathscr{X}(t-)\big)\big)dW(t)\\
&+ \int_{\mathbb{R}} \Big(u\big(t, \mathscr{X}(t-) + \gamma\big(\mathscr{X}(t-), u\big(t, \mathscr{X}(t-)\big), z\big)\big)\\
&- u\big(t, \mathscr{X}(t-)\big)\Big)\tilde{N}(dt, dz).
\end{aligned} \tag{4.35}$$

Comparing the coefficients in (4.34) and (4.35), we deduce

$$-f\Big(t, \mathscr{X}(t-), u\big(t, \mathscr{X}(t-)\big), Z(t), \int_{\mathbb{R}} U(t,z)\delta(z)\nu(dz)\Big)$$

$$
\begin{aligned}
&= u_t\big(t, \mathscr{X}(t-)\big) \\
&\quad + u_x\big(t, \mathscr{X}(t-)\big)\mu\left(\mathscr{X}(t-), u\big(t, \mathscr{X}(t-)\big), Z(t), \int_{\mathbb{R}} U(t,z)\delta(z)\nu(dz)\right) \\
&\quad + \frac{1}{2}u_{xx}\big(t, \mathscr{X}(t-)\big)\sigma^2\big(\mathscr{X}(t-), u\big(t, \mathscr{X}(t-)\big)\big) \\
&\quad + \int_{\mathbb{R}} \Big(u\big(t, \mathscr{X}(t-) + \gamma\big(\mathscr{X}(t-), u\big(t, \mathscr{X}(t-)\big), z\big)\big) - u\big(t, \mathscr{X}(t-)\big) \\
&\quad - u_x\big(t, \mathscr{X}(t-)\big)\gamma\big(\mathscr{X}(t-), u\big(t, \mathscr{X}(t-)\big), z\big)\Big)\nu(dz), \quad 0 \le t \le T, \\
&u\big(T, \mathscr{X}(T)\big) = g\big(\mathscr{X}(T)\big),
\end{aligned}
\tag{4.36}
$$

and

$$
\begin{aligned}
Z(t) &= u_x\big(t, \mathscr{X}(t-)\big)\sigma\big(\mathscr{X}(t-), u\big(t, \mathscr{X}(t-)\big)\big), \quad 0 \le t \le T, \\
U(t,z) &= u\big(t, \mathscr{X}(t-) + \gamma\big(\mathscr{X}(t-), u\big(t, \mathscr{X}(t-)\big), z\big)\big) \\
&\quad - u\big(t, \mathscr{X}(t-)\big), \quad 0 \le t \le T,\ z \in \mathbb{R}.
\end{aligned}
\tag{4.37}
$$

We can define the scheme

(S1) Solve the PIDE

$$
\begin{aligned}
&-u_t(t,x) - u_x(t,x)\mu\Big(x, u(t,x), u_x(t,x)\sigma\big(x, u(t,x)\big), \\
&\quad \int_{\mathbb{R}} \big(u\big(t, x + \gamma\big(x, u(t,x), z\big)\big) - u(t,x)\big)\delta(z)\nu(dz)\Big) \\
&\quad - \frac{1}{2}u_{xx}(t,x)\sigma^2\big(x, u(t,x)\big) \\
&\quad - \int_{\mathbb{R}} \big(u\big(t, x + \gamma\big(x, u(t,x), z\big)\big) - u(t,x) - u_x(t,x)\gamma\big(x, u(t,x), z\big)\big)\nu(dz), \\
&\quad - f\Big(t, x, u(t,x), u_x(t,x)\sigma\big(x, u(t,x)\big), \\
&\quad \int_{\mathbb{R}} \big(u\big(t, x + \gamma\big(x, u(t,x), z\big)\big) - u(t,x)\big)\delta(z)\nu(dz)\Big) = 0, \\
&\quad (t,x) \in [0,T) \times \mathbb{R}, \quad u(T,x) = g(x), \quad x \in \mathbb{R}.
\end{aligned}
$$

(S2) Using u, solve the forward SDE

$$
\begin{aligned}
&\mathscr{X}(t) \\
&\quad = x + \int_0^t \mu\Big(\mathscr{X}(s-), u\big(s, \mathscr{X}(s-)\big),
\end{aligned}
$$

$$
\begin{aligned}
&u_x\big(s,\mathscr{X}(s-)\big)\sigma\big(\mathscr{X}(s-),u\big(s,\mathscr{X}(s-)\big)\big),\\
&\int_{\mathbb{R}}\Big(u\big(s,\mathscr{X}(s-)+\gamma\big(\mathscr{X}(s-),u\big(s,\mathscr{X}(s-)\big),z\big)\big)\\
&-u\big(s,\mathscr{X}(s-)\big)\big)\delta(z)\nu(dz)\Big)ds\\
&+\int_0^t\sigma\big(\mathscr{X}(s-),u\big(s,\mathscr{X}(s-)\big)\big)dW(s)\\
&+\int_0^t\int_{\mathbb{R}}\gamma\big(\mathscr{X}(s-),u\big(s,\mathscr{X}(s-)\big),z\big)\tilde{N}(ds,dz),\quad 0\le t\le T,
\end{aligned}
$$

(S3) Set

$$
\begin{aligned}
Y(t)&=u\big(t,\mathscr{X}(t)\big),\quad 0\le t\le T,\\
Z(t)&=u_x\big(t,\mathscr{X}(t-)\big)\sigma\big(\mathscr{X}(t-),u\big(t,\mathscr{X}(t-)\big)\big),\quad 0\le t\le T,\\
U(t,z)&=u\big(t,\mathscr{X}(t-)+\gamma\big(\mathscr{X}(t-),u\big(t,\mathscr{X}(t-)\big),z\big)\big)-u\big(t,\mathscr{X}(t-)\big),\\
&\quad 0\le t\le T, z\in\mathbb{R}.
\end{aligned}
$$

The scheme (S1)–(S3) is closely related to the four step scheme which was originally developed by Ma et al. (1994) for BSDEs driven by Brownian motions. We present the scheme by Ma et al. (2010) which is designed for BSDEs driven by a Brownian motion and a Poisson random measure. However, the scheme does not allow for dependence of σ and γ on (Z,U). In the case when σ and γ are independent of (Z,U), the control processes can be directly determined by (4.37) and only three steps are needed to solve (4.34). Otherwise, a preliminary step is needed to determine the control processes (Z,U). To the best of our knowledge, the general case with σ and γ dependent on (Z,U) has not been considered for fully coupled FBSDEs driven by Brownian motions and random measures.

Theorem 4.3.1 *Consider the scheme* (S1)–(S3). *Assume that*

(i) *the random measure N is a Poisson random measure generated by a Lévy process with a Lévy measure ν,*
(ii) *μ,σ,γ and f are smooth functions with first order bounded derivatives,*
(iii) *$|\mu(x,0,0,0)|+|f(t,x,0,z,u)|\le K$ for all $(t,x,z,u)\in[0,T]\times\mathbb{R}\times\mathbb{R}\times\mathbb{R}$,*
(iv) *$0<\varepsilon\le\sigma^2(x,y)\le K$ for all $(x,y)\in\mathbb{R}\times\mathbb{R}$,*
(v) *$\gamma(0,0,z)\le K(1\wedge|z|^2)$ and $|\delta(z)|\le K(1\wedge|z|)$ for all $z\in\mathbb{R}$,*
(vi) *$g\in\mathscr{C}^{2+\alpha}(\mathbb{R})$ for some $\alpha\in(0,1)$ and g is bounded in $\mathscr{C}^{2+\alpha}$.*

There exists a unique solution $u\in\mathscr{C}^{1,2}([0,T]\times\mathbb{R})$ to the PIDE defined in (S1). *The quadruple $(\mathscr{X},Y,Z,U)$ defined in* (S2)–(S3) *is the unique solution to the coupled FBSDE* (4.34), *and $(\mathscr{X},Y,Z,U)\in\mathbb{S}^2(\mathbb{R})\times\mathbb{S}^2(\mathbb{R})\times\mathbb{H}^2(\mathbb{R})\times\mathbb{H}^2_N(\mathbb{R})$.*

For the proof we refer to Theorem 2 in Ma et al. (2010). We remark that a function is called smooth if it posses partial derivatives of all necessary orders, see Ma et al. (2010). The space $\mathscr{C}^{2+\alpha}$ contains twice differentiable functions with α-Hölder continuous second derivative. In particular, Theorem 4.3.1 gives assumptions under which Theorem 4.2.1 can be applied.

The application of Theorem 4.3.1 relies on the existence of a smooth solution to the PIDE. It is known that we should deal with a non-degenerate PIDE to have a smooth solution, see Chap. 12 in Cont and Tankov (2004). In our framework, $\sigma^2(x, y) \geq \varepsilon > 0$ is the key assumption which guarantees that there exists a unique smooth solution to the PIDE. In fact, the non-degeneracy of σ is responsible for solvability of coupled FBSDEs. As pointed out by Ma and Yong (1995), in order to make a fully coupled FBSDE solvable, it is necessary that the forward diffusion is "random enough". This is an important difference compared to a decoupled FBSDE which has a unique solution even for a degenerate volatility coefficient σ. Notice that in the case of a coupled FBSDE we used a PIDE to establish the existence of a unique solution to the FBSDE, whereas in the case of a decoupled FBSDE we used the method of contraction (see Theorem 3.1.1) to prove the existence of a unique solution. If we used the method of contraction for a coupled FBSDE, then we could only conclude that there exists a unique solution for sufficiently small time horizon or sufficiently small Lipschitz constant of the generator. It turns out that assumptions (B1)–(B5) are not enough to guarantee the existence of a unique solution to a coupled FBSDE and some additional assumptions have to be introduced. Theorem 4.3.1 shows that if we consider smooth coefficients and a non-degenerate volatility coefficient, then we can construct the unique solution by the scheme (S1)–(S3). Let us point out that the four step scheme by Ma et al. (1994) was the first method for solving coupled FBSDEs which removed the restriction on the time horizon or the Lipschitz constant. Instead, strong regularity of the coefficients is required. In general, a coupled FBSDE may have multiple solutions or may not have a solution. Multiple solutions and non-existence of a solution to a coupled FBSDE are discussed in Sect. 14.1.

Bibliographical Notes The proofs of Lemma 4.1.1, Theorems 4.1.4 and 4.2.1 are inspired by El Karoui et al. (1997b) and extended to cover the case of a Poisson random measure, see also Barles et al. (1997). The proof of Theorem 4.2.2 is taken from Barles et al. (1997). We note that the derivative u_x can be characterized as a solution to a FBSDE. Differentiability of a solution to a BSDE with respect to the initial condition of a forward SDE is studied in El Karoui et al. (1997b), Ankirchner et al. (2007) and Bouchard and Elie (2008). Fully coupled FBSDEs driven by Brownian motions are deeply studied in the monograph by Ma and Yong (2000). More recent methods of solving coupled FBSDEs are discussed by Ma et al. (2011).

Chapter 5
Numerical Methods for FBSDEs

Abstract We investigate numerical methods for forward-backward stochastic differential equations driven by a Brownian motion and a compensated Poisson random measure. We consider three approaches to solving FBSDEs. We apply discrete-time approximations and we derive recursive representations of the solution involving conditional expected values. In order to estimate the conditional expected values, we use Least Squares Monte Carlo which overcomes nested Monte Carlo simulations. In the case of a FBSDE driven by a Brownian motion and a compensated Poisson process we replace the original driving noises by discrete-space martingales. We also use the connection with partial integro-differential equations and we present an explicit-implicit finite difference method for solving a PIDE.

We continue the study of (decoupled) forward-backward stochastic differential equations driven by a Brownian motion and a compensated Poisson random measure, which we introduced in the previous chapter. In most cases we cannot derive the solution to a FBSDE in an explicit form and we have to apply a numerical method to solve a FBSDE. In this chapter we investigate three approaches to solving FBSDEs numerically.

5.1 Discrete-Time Approximation and Least Squares Monte Carlo

Let assumptions (B1)–(B5) from Chap. 4 hold. We deal with the forward-backward stochastic differential equation

$$\begin{aligned}
\mathscr{X}(t) &= x + \int_0^t \mu\big(\mathscr{X}(s-)\big)ds + \int_0^t \sigma\big(\mathscr{X}(s-)\big)dW(s) \\
&\quad + \int_0^t \int_{\mathbb{R}} \gamma\big(\mathscr{X}(s-), z\big)\tilde{N}(ds, dz), \quad 0 \le t \le T, \\
Y(t) &= g\big(\mathscr{X}(T)\big) + \int_t^T f\Big(s, \mathscr{X}(s-), Y(s-), Z(s), \int_{\mathbb{R}} U(s, z)\delta(z)\nu(dz)\Big)ds \\
&\quad - \int_t^T Z(s)dW(s) - \int_t^T \int_{\mathbb{R}} U(s, z)\tilde{N}(ds, dz), \quad 0 \le t \le T.
\end{aligned} \tag{5.1}$$

Ł. Delong, *Backward Stochastic Differential Equations with Jumps and Their Actuarial and Financial Applications*, EAA Series, DOI 10.1007/978-1-4471-5331-3_5,

We denote

$$\Psi(t) = \int_{\mathbb{R}} U(t,z)\delta(z)\nu(dz), \quad 0 \le t \le T.$$

First, we consider the case of the random measure N generated by a compound Poisson process.

We aim to solve the FBSDE (5.1). An intuitive idea is to discretize the forward and the backward equation in the spirit of the Euler method. We choose a regular time grid $\pi = \{t_i = ih, i = 0, 1, \dots, n\}$ with step $h = \frac{T}{n}$. The solution to the forward equation (5.1) is approximated by

$$\begin{aligned} &\mathscr{X}^n(0) = x, \\ &\mathscr{X}^n(t_{i+1}) = \mathscr{X}^n(t_i) + \mu\big(\mathscr{X}^n(t_i)\big)h + \sigma\big(\mathscr{X}^n(t_i)\big)\Delta W(i+1) \\ &\qquad + \int_{\mathbb{R}} \gamma\big(\mathscr{X}^n(t_i), z\big)\tilde{N}\big((t_i, t_{i+1}], dz\big), \quad i = 0, \dots, n-1. \end{aligned} \tag{5.2}$$

where $\Delta W(i+1) = W(t_{i+1}) - W(t_i)$ denotes the increment of the Brownian motion. Clearly, there exists a unique $\mathscr{F}$-adapted, square integrable solution $\mathscr{X}^n$ to (5.2). We set $\mathscr{X}^n(t) = \mathscr{X}^n(t_i)$, $t_i \le t < t_{i+1}$. If we apply the Euler-type discretization to the backward equation (5.1), we obtain

$$\begin{aligned} &Y^n(T) = g\big(\mathscr{X}^n(T)\big), \\ &Y^n(t_i) = Y^n(t_{i+1}) + f\left(t_i, \mathscr{X}^n(t_i), Y^n(t_i), Z^n(t_i), \int_{\mathbb{R}} U^n(t_i, z)\delta(z)\nu(dz)\right)h \\ &\qquad - Z^n(t_i)\Delta W(i+1) - \int_{\mathbb{R}} U^n(t_i, z)\tilde{N}\big((t_i, t_{i+1}], dz\big), \quad i = n-1, \dots, 0. \end{aligned} \tag{5.3}$$

Unfortunately, the discrete-time equation (5.3) does not have a solution since the time-discretized Brownian motion and compound Poisson process do not have the predictable representation property, see Briand et al. (2002). However, we use the following backward recursion

$$\begin{aligned} &Y^n(T) = g\big(\mathscr{X}^n(T)\big), \\ &Z^n(t_i) = \frac{1}{h}\mathbb{E}\big[Y^n(t_{i+1})\Delta W(i+1)|\mathscr{F}_{t_i}\big], \quad i = n-1, \dots, 0, \\ &\Psi^n(t_i) = \frac{1}{h}\mathbb{E}\left[Y^n(t_{i+1})\int_{\mathbb{R}} \delta(z)\tilde{N}\big((t_i, t_{i+1}], dz\big)|\mathscr{F}_{t_i}\right], \quad i = n-1, \dots, 0, \\ &Y^n(t_i) = \mathbb{E}\big[Y^n(t_{i+1})|\mathscr{F}_{t_i}\big] \\ &\qquad + f\big(t_i, \mathscr{X}^n(t_i), Y^n(t_i), Z^n(t_i), \Psi^n(t_i)\big)h, \quad i = n-1, \dots, 0. \end{aligned} \tag{5.4}$$

We set $Y^n(t) = Y^n(t_i)$, $Z^n(t) = Z^n(t_i)$, $\Psi^n(t) = \Psi^n(t_i)$, $t_i \le t < t_{i+1}$. The backward recursion (5.4) can be derived by a heuristic reasoning. Let us first recall that for square integrable martingales M_1 and M_2 we have

$$\mathbb{E}\big[M_1(T)M_2(T)\big] = \mathbb{E}\big[[M_1, M_2](T)\big], \tag{5.5}$$

see Corollary II.27.3 in Protter (2004). If we multiply (5.3) by $\Delta W(i+1)$ and $\int_{\mathbb{R}} \delta(z)\tilde{N}((t_i, t_{i+1}], dz)$, take the conditional expected value and use (5.5), then we obtain

$$\mathbb{E}\big[Y^n(t_{i+1})\Delta W(i+1)|\mathscr{F}_{t_i}\big] = \mathbb{E}\big[Z^n(t_i)|\Delta W(i+1)|^2|\mathscr{F}_{t_i}\big] = Z^n(t_i)h,$$

$$\begin{aligned}
&\mathbb{E}\Big[Y^n(t_{i+1})\int_{\mathbb{R}} \delta(z)\tilde{N}\big((t_i, t_{i+1}], dz\big)|\mathscr{F}_{t_i}\Big] \\
&\quad = \mathbb{E}\Big[\int_{\mathbb{R}} U^n(t_i, z)\tilde{N}\big((t_i, t_{i+1}], dz\big)\int_{\mathbb{R}} \delta(z)\tilde{N}\big((t_i, t_{i+1}], dz\big)|\mathscr{F}_{t_i}\Big] \\
&\quad = \mathbb{E}\Big[\Big[\int_{t_i}^{\cdot}\int_{\mathbb{R}} U^n(t_i, z)\tilde{N}(dt, dz), \int_{t_i}^{\cdot}\int_{\mathbb{R}} \delta(z)\tilde{N}(dt, dz)\Big](t_{i+1})|\mathscr{F}_{t_i}\Big] \\
&\quad = \mathbb{E}\Big[\int_{\mathbb{R}} U^n(t_i, z)\delta(z)N\big((t_i, t_{i+1}], dz\big)|\mathscr{F}_{t_i}\Big] \\
&\quad = \int_{\mathbb{R}} U^n(t_i, z)\delta(z)\nu(dz)h = \Psi^n(t_i)h,
\end{aligned}$$

and the formulas for Z^n, Ψ^n can be established. If we take the condition expected value on both sides of (5.3), then the formula for Y^n can be established.

The next theorem justifies the approximations (5.2) and (5.4), see Theorem 2.1, Corollary 2.1 and Remark 2.7 in Bouchard and Elie (2008).

Theorem 5.1.1 *Consider the FBSDE* (5.1) *and the random measure N generated by a compound Poisson process. Assume that* (B2)–(B5) *from Sect.* 4.1 *hold and let the generator f be* 1/2*-Hölder continuous in t. We deal with the approximations* (5.2) *and* (5.4) *of the solution to the FBSDE* (5.1). *We have*

$$\begin{aligned}
&\max_{i=0,1,\ldots,n-1}\mathbb{E}\Big[\sup_{t\in[t_i, t_{i+1}]}\big|\mathscr{X}(t) - \mathscr{X}^n(t_i)\big|^2\Big] \\
&\quad + \max_{i=0,1,\ldots,n-1}\mathbb{E}\Big[\sup_{t\in[t_i, t_{i+1}]}\big|Y(t) - Y^n(t_i)\big|^2\Big] \\
&\quad + \mathbb{E}\Big[\sum_{i=0}^{n-1}\int_{t_i}^{t_{i+1}}\big|Z(t) - Z^n(t_i)\big|^2 dt\Big] \le K\frac{1}{n},
\end{aligned}$$

$$\mathbb{E}\Big[\sum_{i=0}^{n-1}\int_{t_i}^{t_{i+1}}\big|\Psi(t) - \Psi^n(t_i)\big|^2 dt\Big] \le K\Big(\frac{1}{n}\Big)^{1-\varepsilon}, \quad \varepsilon > 0.$$

In addition, if for each $z \in \mathbb{R}$ the mapping $x \mapsto \gamma(x, z)$ is differentiable and

$$\big|\gamma_x(x, z) + 1\big| \ge K > 0, \quad (x, z) \in \mathbb{R}\times\mathbb{R},$$

then

$$\max_{i=0,1,\dots,n-1}\mathbb{E}\Big[\sup_{t\in[t_i,t_{i+1}]}\big|\mathscr{X}(t)-\mathscr{X}^n(t_i)\big|^2\Big]$$
$$+\max_{i=0,1,\dots,n-1}\mathbb{E}\Big[\sup_{t\in[t_i,t_{i+1}]}\big|Y(t)-Y^n(t_i)\big|^2\Big]$$
$$+\mathbb{E}\Bigg[\sum_{i=0}^{n-1}\int_{t_i}^{t_{i+1}}\big|Z(t)-Z^n(t_i)\big|^2dt\Bigg]$$
$$+\mathbb{E}\Bigg[\sum_{i=0}^{n-1}\int_{t_i}^{t_{i+1}}\big|\Psi(t)-\Psi^n(t_i)\big|^2dt\Bigg]\le K\frac{1}{n}.$$

We remark that deriving Y^n from (5.4) involves solving a fixed point equation. For a Lipschitz continuous generator the fixed point procedure of solving (5.4) has a convergence rate of $1/n$. Hence, numerical cost is small. To overcome the fixed point procedure, we can use the following scheme

$$Y^n(T)=g\big(\mathscr{X}^n(T)\big),$$
$$Z^n(t_i)=\frac{1}{h}\mathbb{E}\big[Y^n(t_{i+1})\Delta W(i+1)|\mathscr{F}_{t_i}\big],\quad i=n-1,\dots,0,$$
$$\Psi^n(t_i)=\frac{1}{h}\mathbb{E}\bigg[Y^n(t_{i+1})\int_{\mathbb{R}}\delta(z)\tilde{N}\big((t_i,t_{i+1}],dz\big)|\mathscr{F}_{t_i}\bigg],\quad i=n-1,\dots,0,$$
$$Y^n(t_i)=\mathbb{E}\big[Y^n(t_{i+1})$$
$$+f\big(t_i,\mathscr{X}^n(t_i),Y^n(t_{i+1}),Z^n(t_i),\Psi^n(t_i)\big)h|\mathscr{F}_{t_i}\big],\quad i=n-1,\dots,0,$$

but more complicated conditional expected values have to be calculated instead.

The algorithm (5.4) is still not an implementable scheme since the conditional expectations have to be estimated. Performing Monte Carlo simulations at each point t_i would lead to so-called nested simulations and an enormous numerical cost. Least Squares Monte Carlo can overcome nested simulations.

By the Markov property the conditional expected values in (5.4) can be represented as functions of the state process $\mathscr{X}$. The idea is to approximate the unknown functions by their projections on finite-dimensional function bases. At each point t_i we choose 3 function bases $(b_{l,i}(\cdot))_{l=0,1,2}$ and we approximate each conditional expected value in a vector space spanned by the basis. Each basis $b_{l,i}(\cdot)$ is a $d_{l,i}$-dimensional vector of scalar functions. The vector space spanned by $b_{l,i}$ is denoted by $\alpha b_{l,i}(\cdot)=\sum_{k=1}^{d_{l,i}}\alpha_k b_{l,i}^k(\cdot)$ where $\alpha\in\mathbb{R}^{d_{l,i}}$.

We can use the Least Squares Monte Carlo algorithm:

1. Simulate L independent copies of $(\Delta W^m(i+1),\ i=0,1,\dots,n-1,\ m=1,\dots,L)$ and $(\tilde{N}^m((t_i,t_{i+1}],.),\ i=0,1,\dots,n-1,\ m=1,\dots,L)$,
2. Simulate L independent copies of $(\mathscr{X}^{n,m}(t_i),\ i=1,\dots,n,\ m=1,\dots,L)$,

3. Set $\hat{Y}(T, \mathscr{X}^{n,m}(T)) = g(\mathscr{X}^{n,m}(T))$ for $m = 1, \ldots, L$,
4. Choose function bases $(b_{l,i}(.))_{l=0,1,2,i=0,1,\ldots,n-1}$,
5. Going backwards, for $i = n-1, \ldots, 0$ solve the least squares regression problems

$$\alpha_{1,i} = \arg\inf_{\alpha}\left\{\frac{1}{L}\sum_{i=1}^{L}\left|\frac{1}{h}\hat{Y}\big(t_{i+1}, \mathscr{X}^{n,m}(t_{i+1})\big)\Delta W^m(i+1) - \alpha b_{1,i}\big(\mathscr{X}^{n,m}(t_i)\big)\right|^2\right\},$$

$$\alpha_{2,i} = \arg\inf_{\alpha}\left\{\frac{1}{L}\sum_{i=1}^{L}\left|\frac{1}{h}\hat{Y}\big(t_{i+1}, \mathscr{X}^{n,m}(t_{i+1})\big)\int_{\mathbb{R}}\delta(z)\tilde{N}^m\big((t_i, t_{i+1}], dz\big)\right.\right.$$
$$\left.\left. - \alpha b_{2,i}\big(\mathscr{X}^{n,m}(t_i)\big)\right|^2\right\},$$

6. Set $\hat{Z}(t_i, \mathscr{X}^{n,m}(t_i)) = \alpha_{1,i} b_{1,i}(\mathscr{X}^{n,m}(t_i))$ and $\hat{\Psi}(t_i, \mathscr{X}^{n,m}(t_i)) = \alpha_{2,i} \times b_{2,i}(\mathscr{X}^{n,m}(t_i))$,
7. Solve the least squares regression problem

$$\alpha_{0,i} = \arg\inf_{\alpha}\left\{\frac{1}{L}\sum_{i=1}^{L}\left|\hat{Y}\big(t_{i+1}, \mathscr{X}^{n,m}(t_{i+1})\big) + f\big(t_i, \mathscr{X}^{n,m}(t_i),\right.\right.$$
$$\hat{Y}\big(t_{i+1}, \mathscr{X}^{n,m}(t_{i+1})\big), \hat{Z}\big(t_i, \mathscr{X}^{n,m}(t_i)\big), \hat{\Psi}\big(t_i, \mathscr{X}^{n,m}(t_i)\big)\big)h$$
$$\left.\left. - \alpha b_{0,i}\big(\mathscr{X}^{n,m}(t_i)\big)\right|^2\right\},$$

8. Set $\hat{Y}(t_i, \mathscr{X}^{h,m}(t_i)) = \alpha_{0,i} b_{0,i}(\mathscr{X}^{n,m}(t_i))$,
9. Continue till $t_0 = 0$.

Polynomials, hypercubes and Voronoi partitions are usually used as basis functions, see Gobet et al. (2005). Notice that when we apply the Least Squares Monte Carlo algorithm we additionally face the error of approximating the conditional expectations by estimated regression functions. The total error of the Least Squares Monte Carlo algorithm depends on the number of time steps n, the number of simulations L and the number of basis functions d. The total error is studied in Bouchard and Touzi (2004), Gobet et al. (2005), Gobet et al. (2006), Gobet and Lemor (2006). We also point out that some truncation procedures can be useful in the final application of the algorithm, see Gobet et al. (2005).

We comment on one modification of the algorithm presented. It is shown by Bouchard and Touzi (2004), in the case of BSDEs driven by Brownian motions, that the error of approximating the conditional expectation by an estimator explodes when the mesh of time partition goes to zero, given the accuracy of the estimator. In order to control this approximation error one is forced to simulate more paths as the time partition becomes finer. This significantly increases computational cost. The idea of Bender and Denk (2007), who also investigate BSDEs driven by Brownian motions, is first to approximate the solution (Y, Z) by the Picard iterations

$$Y^I(t) = g\big(\mathscr{X}(T)\big) + \int_t^T f\big(s, \mathscr{X}(s), Y^{I-1}(s), Z^{I-1}(s)\big)ds \\ - \int_t^T Z^I(s)dW(s), \quad 0 \le t \le T, \tag{5.6}$$

and to apply the algorithm (5.4) to derive (Y^I, Z^I). The Picard procedure clearly introduces an additional error, which converges to zero at geometric rate. The advantage of the scheme proposed by Bender and Denk (2007) is that the error of approximating of the conditional expectation by an estimator is reduced and this error does not explode when the mesh of time partition tends to zero and the number of the Picard iteration goes to infinity.

Let us now comment on the case when a FBSDE is driven by a general compensated Poisson random measure. It is known that we cannot simulate small jumps of a Lévy process with an infinite Lévy measure, see Chap. 6 in Cont and Tankov (2004). The usual procedure is to cut off small jumps of a Lévy process and approximate them by an independent Brownian motion. After cutting off small jumps, we can investigate a FBSDE driven by a compensated compound Poisson process and we can apply the algorithm presented in this chapter, see Aazizi (2011) for details.

Finally, we remark that in many applications we end up with a BSDE with zero generator or with a BSDE with generator independent of (Y, Z, U) for which we can derive representations of the solution Y and the control processes (Z, U) in the form of conditional expectations of the state process, see Proposition 4.1.2 and Chap. 8. In those cases the Monte Carlo algorithm is much simpler.

5.2 Discrete-Time and Discrete-Space Martingale Approximation

We deal with the forward-backward stochastic differential equation

$$\begin{aligned} \mathscr{X}(t) &= x + \int_0^t \mu\big(\mathscr{X}(s-)\big)ds + \int_0^t \sigma\big(\mathscr{X}(s-)\big)dW(s) \\ &\quad + \int_0^t \gamma\big(\mathscr{X}(s-)\big)\tilde{N}(ds), \quad 0 \le t \le T, \\ Y(t) &= g\big(\mathscr{X}(T)\big) + \int_t^T f\big(s, \mathscr{X}(s-), Y(s-), Z(s), U(s)\big)ds \\ &\quad - \int_t^T Z(s)dW(s) - \int_t^T U(s)\tilde{N}(ds), \quad 0 \le t \le T, \end{aligned} \tag{5.7}$$

where the measure N is the jump measure of a Poisson process with intensity λ. We consider a discrete-time approximation to (5.7) and we approximate the Brownian motion and the compensated random measure by two discrete-space martingales.

We choose a regular time grid $\pi = \{t_i = ih, i = 0, 1, \ldots, n\}$ with step $h = \frac{T}{n}$. We define two random walks $W^n := (W^n(k), k = 0, 1, \ldots, n)$ and $\tilde{N}^n := (\tilde{N}^n(k),\ k = 0, 1, \ldots, n)$ by

$$\begin{aligned} W^n(0) &= 0, \qquad W^n(k) = \sqrt{h}\sum_{i=1}^{k} \xi_i^n, \quad k = 1, 2, \ldots, n, \\ \tilde{N}^n(0) &= 0, \qquad \tilde{N}^n(k) = \sum_{i=1}^{k} \zeta_i^n, \quad k = 1, 2, \ldots, n, \end{aligned} \tag{5.8}$$

where $\xi_1^n, \ldots, \xi_n^n$ are independent Bernoulli random variables with probabilities

$$\mathbb{P}\big(\xi_k^n = 1\big) = \mathbb{P}\big(\xi_k^n = -1\big) = \frac{1}{2},$$

and $\zeta_1^n, \ldots, \zeta_n^n$ are independent Bernoulli random variables with probabilities

$$\mathbb{P}\big(\zeta_k^n = e^{-\lambda h} - 1\big) = 1 - \mathbb{P}\big(\zeta_k^n = e^{-\lambda h}\big) = e^{-\lambda h}.$$

We also introduce the filtration $\mathscr{F}_k^n = \sigma(\xi_1^n, \ldots, \xi_k^n, \zeta_1^n, \ldots, \zeta_k^n)$, $k = 1, \ldots, n$. The random walks W^n and $\tilde{N}^n$ are $\mathscr{F}^n$-discrete-time-space-martingales.

The first result shows that the random walks are good approximations of the Brownian motion and the compensated Poisson process, see Lemma 3 in Lejay et al. (2010).

Proposition 5.2.1 *The processes $(W^n([\frac{t}{h}]), \tilde{N}^n([\frac{t}{h}]), 0 \le t \le T)$ converge in the J_1-Skorokhod topology in probability to $(W(t), \tilde{N}(t), 0 \le t \le T)$ as $n \to \infty$.*

It is intuitive to approximate the solution to the forward equation (5.7) in the following way

$$\begin{aligned} \mathscr{X}^n(0) &= x, \\ \mathscr{X}^n(t_{i+1}) &= \mathscr{X}^n(t_i) + \mu\big(X^n(t_i)\big)h + \sqrt{h}\sigma\big(\mathscr{X}^n(t_i)\big)\xi_{i+1}^n \\ &\quad + \gamma\big(\mathscr{X}^n(t_i)\big)\zeta_{i+1}^n, \quad i = 0, 1, \ldots, n-1. \end{aligned} \tag{5.9}$$

Clearly, there exists a unique $\mathscr{F}^n$-adapted, square integrable solution $\mathscr{X}^n$ to (5.9). We set $\mathscr{X}^n(t) = \mathscr{X}^n(t_i)$, $t_i \le t < t_{i+1}$. In Sect. 5.1 we claim that the time-discretized Brownian motion and compound Poisson process do not have the predictable representation property. However, in some cases an orthogonal martingale term can be added to recover the predictable representation property, see Briand et al. (2002) and Lejay et al. (2010). We approximate the solution to the backward stochastic differential equation (5.7) by solving the backward stochastic difference equation

$$\begin{aligned} Y^n(T) &= g\big(\mathcal{X}^n(T)\big), \\ Y^n(t_i) &= Y^n(t_{i+1}) + f\big(t_i, Y^n(t_i), Z^n(t_i), U^n(t_i)\big)h \\ &\quad - \sqrt{h} Z^n(t_i)\xi^n_{i+1} - U^n(t_i)\zeta^n_{i+1} - V^n(t_i)\varsigma^n_{i+1}, \quad i = 0, 1, \ldots, n-1, \end{aligned} \tag{5.10}$$

where $(\varsigma^n_i, i = 1, \ldots, n)$ denotes the increments of a third orthogonal discrete-time-space martingale. By the predictable representation property, for an $\mathcal{F}^n$-measurable $\mathcal{X}^n(T)$ there exists a unique $\mathcal{F}^n$-adapted solution (Y^n, Z^n, U^n, V^n) to the backward equation (5.10). We can also derive that solution. Multiplying both sides of (5.10) by ξ^n_{i+1} or ζ^n_{i+1} and taking the conditional expected values, we obtain the representations

$$\begin{aligned} Y^n(T) &= g\big(\mathcal{X}^n(T)\big), \\ Z^n(t_i) &= \frac{1}{\sqrt{h}} \mathbb{E}\big[Y^n(t_{i+1})\xi^n_{i+1} | \mathcal{F}^n_i\big], \quad i = n-1, \ldots, 0, \\ U^n(t_i) &= \frac{1}{e^{-\lambda h}(1 - e^{-\lambda h})} \mathbb{E}\big[Y^n(t_{i+1})\zeta^n_{i+1} | \mathcal{F}^n_i\big], \quad i = n-1, \ldots, 0, \\ Y^n(t_i) &= \mathbb{E}\big[Y^n(t_{i+1}) | \mathcal{F}^n_i\big] \\ &\quad + f\big(t_i, \mathcal{X}^n(t_i), Y^n(t_i), Z^n(t_i), U^n(t_i)\big)h, \quad i = n-1, \ldots, 0. \end{aligned} \tag{5.11}$$

We set $Y^n(t) = Y^n(t_i)$, $Z^n(t) = Z^n(t_i)$, $U^n(t) = U^n(t_i)$, $t_i \le t < t_{i+1}$. The process V^n can also be derived from (5.10) but it is not needed for the approximation of the solution to (5.7). Again, a fixed point procedure has to be applied to derive Y^n from (5.11).

We state the main result of this chapter, see Theorem 1 and Proposition 5 in Lejay et al. (2010).

Theorem 5.2.1 *Consider the FBSDE* (5.7) *and the random measure N generated by a Poisson process. Assume that* (B2)–(B5) *from Sect.* 4.1 *hold and let the generator f satisfy*

$$\begin{aligned} &\big|f(t, x, y, z, u) - f\big(t', x', y', z', u'\big)\big| \\ &\quad \le \varphi\big(t' - t\big) + K\big(\big|x - x'\big| + \big|y - y'\big| + \big|z - z'\big| + \big|u - u'\big|\big), \end{aligned}$$

for all $(t, x, y, z, u), (t', x', y', z', u') \in [0, T] \times \mathbb{R} \times \mathbb{R} \times \mathbb{R} \times \mathbb{R}$, where φ is a bounded, non-decreasing, continuous function such that $\varphi(0) = 0$. We deal with the approximations (5.9) *and* (5.11) *of the solution to the FBSDE* (5.7).

(a) *The process $\mathcal{X}^n$ converges in the J_1-Skorokhod topology in probability to $\mathcal{X}$ as $n \to \infty$.*

(b) *The processes* $(Y^n, \int_0^\cdot Z^n(s)ds, \int_0^\cdot U^n(s)ds)$ *converge in the* J_1*-Skorokhod topology in probability to* $(Y, \int_0^\cdot Z(s)ds, \int_0^\cdot U(s)ds)$ *as* $n \to \infty$.

The efficiency of the algorithm is studied numerically in Lejay et al. (2010).

In the discrete filtration $\mathscr{F}^n$ it is straightforward to calculate the conditional expected values in (5.11). This is the key advantage of the approximation by discrete-space martingales compared to the Least Squares Monte Carlo method. We can use the formula

$$\begin{aligned}
&\mathbb{E}\big[F(\xi_1^n, \ldots, \xi_k^n, \xi_{k+1}^n, \zeta_1^n, \ldots, \zeta_k^n, \zeta_{k+1}^n)|\mathscr{F}_k^n\big] \\
&\quad = F(\xi_1^n, \ldots, \xi_k^n, 1, \zeta_1^n, \ldots, \zeta_k^n, e^{-\lambda h} - 1)\frac{e^{-\lambda h}}{2} \\
&\qquad + F(\xi_1^n, \ldots, \xi_k^n, -1, \zeta_1^n, \ldots, \zeta_k^n, e^{-\lambda h} - 1)\frac{e^{-\lambda h}}{2} \\
&\qquad + F(\xi_1^n, \ldots, \xi_k^n, 1, \zeta_1^n, \ldots, \zeta_k^n, e^{-\lambda h})\frac{1 - e^{-\lambda h}}{2} \\
&\qquad + F(\xi_1^n, \ldots, \xi_k^n, -1, \zeta_1^n, \ldots, \zeta_k^n, e^{-\lambda h})\frac{1 - e^{-\lambda h}}{2}.
\end{aligned}$$

In a low dimension the random walk approximation can provide a numerically efficient alternative to the Monte Carlo simulation. However, complexity becomes very large in multidimensional problems.

5.3 Finite Difference Method

In Sect. 4.2 we establish the connection between the solution to a FBSDE and the solution to a PIDE. The results of that chapter show that in order to derive the solution to the BSDE (4.5) or (5.1) we can solve the PIDE

$$\begin{aligned}
&-u_t(t,x) - \mathscr{L}u(t,x) \\
&\quad - f\big(t, x, u(t,x), u_x(t,x)\sigma(x), \mathscr{J}u(t,x)\big) = 0, \quad (t,x) \in [0,T) \times \mathbb{R}, \qquad (5.12) \\
&u(T,x) = g(x), \quad x \in \mathbb{R}.
\end{aligned}$$

We can apply a finite difference method to solve (5.12).

Let the random measure N be generated by a compound Poisson process. In order to construct a finite difference scheme for (5.12), we have to consider the following steps:

- Localization: the PIDE (5.12) is given on the unbounded domain $\mathbb{R}$. We reduce the original domain to a bounded domain $[-A, A]$ and we impose boundary conditions. The domain of the integral in the operator $\mathscr{J}$ is localized to $[-B, B]$.

- Discretization in space and time: we choose discrete grids $t_k = \frac{T}{n}k, k = 0, 1, \ldots, n$, and $x_i = -A + \frac{2A}{m}i, i = 0, 1, \ldots, m$.
- Approximation of the derivatives: we use finite differences.
- Approximation of the integral in the operator $\mathcal{J}$: we use the trapezoidal quadrature rule.

If we deal with a Poisson random measure N with an infinite Lévy measure, then an additional step is needed to approximate small jumps of a Lévy process by an independent Brownian motion. Consequently, we end up with a random measure N generated by a compound Poisson process.

Using the results from Sect. 12.4 in Cont and Tankov (2004), we can state the following explicit-implicit scheme for solving the PIDE (5.12):

1. Choose n and m which define the spatial and time grid steps: $\Delta t = \frac{T}{n}$ and $\Delta x = \frac{2A}{m}$,
2. Set $u^{n,m}(t_n, x_i) = g(x_i)$ and extend the grid values to all $x \in [-A, A]$ by linear interpolation,
3. Going backward, for $k = n-1, \ldots, 0$ determine the grid values $u(t_k, x_i)$ by solving the difference equation

$$
\begin{aligned}
0 = {} & \frac{u^{n,m}(t_{k+1}, x_i) - u^{n,m}(t_k, x_i)}{\Delta t} + \Bigg(\mu(x_i) \\
& - \sum_{j=0}^{m} \gamma(x_i, z_j)\nu\big((z_j - 1/2, z_j + 1/2]\big)\Bigg) \frac{u^{n,m}(t_k, x_{i+1}) - u^{n,m}(t_k, x_i)}{\Delta x} \\
& + \frac{1}{2}\sigma^2(x_i) \frac{u^{n,m}(t_k, x_{i+1}) - 2u^{n,m}(t_k, x_i) + u^{n,m}(t_k, x_{i-1})}{|\Delta x|^2} \\
& + \sum_{j=0}^{m} \big(u^{n,m}\big(t_{k+1}, x_i + \gamma(x_i, z_j)\big) - u^{n,m}(t_{k+1}, x_i)\big)\nu\big((z_j - 1/2, z_j + 1/2]\big) \\
& + f\Bigg(t_{k+1}, x_i, u^{n,m}(t_{k+1}, x_i), \sigma(x_i)\frac{u^{n,m}(t_{k+1}, x_{i+1}) - u^{n,m}(t_{k+1}, x_i)}{\Delta x}, \\
& \sum_{j=0}^{m} \big(u^{n,m}\big(t_{k+1}, x_i + \gamma(x_i, z_j)\big) - u^{n,m}(t_{k+1}, x_i)\big) \\
& \cdot \delta(z_j)\nu\big((z_j - 1/2, z_j + 1/2]\big)\Bigg),
\end{aligned}
\tag{5.13}
$$

where $z_j = -B + \frac{2B}{m}j, j = 0, 1, \ldots, m$, and extend the grid values to all $x \in [-A, A]$ by linear interpolation.

The implicit scheme is used for the differential operator and the explicit scheme is used for the integral operator. Convergence of explicit-implicit schemes for PIDEs is discussed in Sect. 12.4 in Cont and Tankov (2004).

We point out that solving a PIDE by a finite difference method is efficient in low dimensions (when we deal with few state processes). Least Squares Monte Carlo algorithms perform much better than finite difference methods in high dimensions.

Since in actuarial and financial applications we deal with many risk factors and we consider multidimensional state processes, we can conclude that BSDEs and Monte Carlo methods are more efficient than PDEs (HJB equations) and finite difference methods in solving applied problems. It should be noticed that in many cases the solution to a problem does not involve all control processes of the BSDE and we do not have to estimate all expected values in the algorithms (5.4), (5.11), which simplifies numerical implementations of the algorithms.

Bibliographical Notes The Malliavin calculus plays an important role in proving convergence results for discrete-time approximations of BSDEs. Zhang (2004) was the first who applied the Malliavin calculus to prove path regularity of the solution and convergence of a discrete-time approximation under a deterministic regular time mesh. Bouchard and Elie (2008) followed the arguments from Zhang (2004) and showed path regularity and convergence for BSDEs with Poisson jumps. Various modifications of the Least Squares Monte-Carlo algorithm can be found in Gobet et al. (2006), Gobet and Lemor (2006). An alternative to the Least Square Monte Carlo is to apply Malliavin weights, see Bouchard et al. (2004). A comparison of the regression based approach, the Malliavin weights and the random walk approximation can be found in Bouchard and Warin (2010). Convergence results for discrete-time and martingale approximations to BSDEs driven by Brownian motions are investigated in Briand et al. (2002). In the case of a fully coupled BSDE driven by a Brownian motion, Douglas et al. (1996) provide a modification of a finite difference method from Sect. 5.3. Douglas et al. (1996) also prove convergence of an approximation of the derivative of the value function, which is needed to obtain the control process of a BSDE. Results on numerics for quadratic decoupled FBSDE can be found in Imkeller et al. (2010).

Chapter 6
Nonlinear Expectations and g-Expectations

Abstract We investigate nonlinear expectations. We briefly discuss Choquet expectations and we focus on g-expectations defined by BSDEs. The connection between filtration-consistent nonlinear expectations and g-expectations is presented. We study the properties of translation invariance, positive homogeneity, convexity and sub-linearity of g-expectations and we show that these properties are determined by the generator of the BSDE defining the g-expectation.

The original motivation for studying nonlinear expectations comes from the theory of decision making. The Allais paradox proved that the linear expectation (the expected value operator) might fail in an attempt to describe choices made by decision makers and the Ellsberg's paradox disqualified the notion of linear probability in representing beliefs of decision makers. It was shown that decisions made in the real world contradicted optimal decisions based on additive probabilities and the expected utility theory. Consequently, economists and mathematicians begun to look for a new notion of expectation.

A nonlinear expectation is an operator which preserves all essential properties of the standard expected value operator except linearity. In this chapter we focus on nonlinear expectations called g-expectations which are defined by BSDEs. In Chap. 13 we use g-expectations to define dynamic risk measures which can be used for actuarial and financial valuation.

6.1 Choquet Expectations

Before we study g-expectations, we briefly discuss Choquet expectations. It is well-known that the expected value can be calculated by the formula

$$\mathbb{E}[\xi] = \int_{-\infty}^{0} \big(\Pr(\xi \geq x) - 1\big)dx + \int_{0}^{\infty} \Pr(\xi \geq x)dx. \tag{6.1}$$

The idea by Choquet (1953) was to replace an additive probability measure $\Pr(\cdot)$ with a non-additive capacity measure $V(\cdot)$. We can define a nonlinear operator in

Ł. Delong, *Backward Stochastic Differential Equations with Jumps and Their Actuarial and Financial Applications*, EAA Series, DOI 10.1007/978-1-4471-5331-3_6,

the following way

$$C[\xi] = \int_{-\infty}^{0} \big(V(\xi \geq x) - 1\big)dx + \int_{0}^{\infty} V(\xi \geq x)dx. \tag{6.2}$$

The nonlinear operator (6.2) is called the Choquet expectation or the Choquet integral.

The key example of a non-additive capacity measure, often applied in insurance and finance, is a distorted probability. We can define a non-additive capacity measure by distorting the original probability

$$V(\xi \geq x) = \Psi\big(\Pr(\xi \geq x)\big), \tag{6.3}$$

where we choose a nonlinear function $\Psi : [0,1] \to [0,1]$ such that $\Psi(0) = 0$, $\Psi(1) = 1$ and $x \mapsto \Psi(x)$ is non-decreasing. The function Ψ is called a distortion. The Wang transform is an important distortion function used for actuarial and financial applications, see Wang (2000). We remark that Value-at-Risk and Tail-Value-at-Risk are examples of the Choquet expectations derived under distorted probabilities.

The idea behind the Choquet expectation and distorted probabilities is clear. Consequently, Choquet expectations have found numerous applications in insurance and financial mathematics. Unfortunately, it is very difficult to define a dynamic version of a Choquet expectation. It turns out that BSDEs can be very useful for defining dynamic nonlinear expectations.

6.2 Filtration-Consistent Nonlinear Expectations and g-Expectations

We define a nonlinear expectation and a filtration-consistent nonlinear expectation, see Coquet et al. (2002).

Definition 6.2.1 A functional $\mathscr{E} : \mathbb{L}^2(\Omega, \mathscr{F}_T, \mathbb{P}; \mathbb{R}) \to \mathbb{R}$ is called a nonlinear expectation if it satisfies

(i) the property of strict monotonicity:

$$\xi_1 \geq \xi_2 \quad \Rightarrow \quad \mathscr{E}[\xi_1] \geq \mathscr{E}[\xi_2],$$
$$\xi_1 \geq \xi_2 \quad \text{and} \quad \mathscr{E}[\xi_1] = \mathscr{E}[\xi_2] \quad \Leftrightarrow \quad \xi_1 = \xi_2,$$

for all $\xi_1, \xi_2 \in \mathbb{L}^2(\mathbb{R})$,

(ii) the invariance property:

$$\mathscr{E}[c] = c, \quad \text{for all } c \in \mathbb{R}.$$

Definition 6.2.2 A nonlinear expectation $\mathscr{E}$ is called an $\mathscr{F}$-consistent nonlinear expectation if for each $\xi \in \mathbb{L}^2(\Omega, \mathscr{F}_T, \mathbb{P}; \mathbb{R})$ and $t \in [0, T]$ there exists a random variable $\zeta \in \mathbb{L}^2(\Omega, \mathscr{F}_t, \mathbb{P}; \mathbb{R})$ such that

$$\mathscr{E}[\xi \mathbf{1}_A] = \mathscr{E}[\zeta \mathbf{1}_A], \quad A \in \mathscr{F}_t.$$

We remark that ζ is uniquely determined, see Lemma 3.1 in Coquet et al. (2002). The random variable ζ is denoted by $\mathscr{E}[\xi|\mathscr{F}_t]$. Notice that the definition of an $\mathscr{F}$-consistent nonlinear expectation is analogous to the definition of the standard linear conditional expectation but in Definition 6.2.2 the expectation is taken under a nonlinear operator.

We state properties of $\mathscr{F}$-consistent nonlinear expectations, see Coquet et al. (2002).

Proposition 6.2.1 *Let $\mathscr{E}$ be an $\mathscr{F}$-consistent nonlinear expectation. The following properties hold*:

(a) $\mathscr{E}[\mathscr{E}[\xi|\mathscr{F}_t]|\mathscr{F}_s] = \mathscr{E}[\xi|\mathscr{F}_s]$ *for all* $0 \le s \le t \le T$.
(b) $\mathscr{E}[\xi \mathbf{1}_A|\mathscr{F}_t] = \mathbf{1}_A \mathscr{E}[\xi|\mathscr{F}_t]$ *for all* $0 \le t \le T$ *and* $A \in \mathscr{F}_t$.
(c) *If* $\xi_1 \ge \xi_2$, *then* $\mathscr{E}[\xi_1|\mathscr{F}_t] \ge \mathscr{E}[\xi_2|\mathscr{F}_t]$ *for all* $0 \le t \le T$. *In addition, if* $\mathscr{E}[\xi_1|\mathscr{F}_t] = \mathscr{E}[\xi_2|\mathscr{F}_t]$ *a.s. for some* $t \in [0, T]$, *then* $\xi_1 = \xi_2$.

Proposition 6.2.1 shows that all essential properties of the standard linear conditional expectation, except linearity, are preserved under the notion of an $\mathscr{F}$-consistent nonlinear expectation.

From the modelling point of view, we should be able to generate $\mathscr{F}$-consistent nonlinear expectations in a feasible way. The next example shows one possible way of generating $\mathscr{F}$-consistent nonlinear expectations.

Example 6.1 Choose a continuous, strictly increasing function $\varphi : \mathbb{R} \to \mathbb{R}$. The operator

$$\mathscr{E}[\xi|\mathscr{F}_t] = \varphi^{-1}\big(\mathbb{E}[\varphi(\xi)|\mathscr{F}_t]\big), \quad 0 \le t \le T, \tag{6.4}$$

is an $\mathscr{F}$-consistent nonlinear expectation. The nonlinear expectation (6.4) can be interpreted as the indifference price of ξ determined by an agent with utility φ, see Royer (2006).

It turns out that $\mathscr{F}$-consistent nonlinear expectations can be defined by nonlinear BSDEs. In this chapter we study the BSDEs

$$\begin{aligned} Y(t) = \xi + \int_t^T g\big(s, Y(s), Z(s), U(s,.)\big)ds \\ - \int_t^T Z(s)dW(s) - \int_t^T \int_{\mathbb{R}} U(s,z)\tilde{N}(ds,dz), \quad 0 \le t \le T. \end{aligned} \tag{6.5}$$

By a nonlinear BSDE we mean a BSDE with a nonlinear generator g.

Definition 6.2.3 Consider $g : \Omega \times [0,T] \times \mathbb{R} \times \mathbb{R} \times L^2_Q(\mathbb{R}) \to \mathbb{R}$ such that

(i) g satisfies (A2) from Chap. 3,
(ii) g satisfies the inequality

$$g(\omega,t,y,z,u) - g\big(\omega,t,y,z,u'\big) \leq \int_{\mathbb{R}} \delta^{y,z,u,u'}(t,x)\big(u(x) - u'(x)\big)Q(t,dx)\eta(t),$$

a.s., a.e. $(\omega,t) \in \Omega \times [0,T]$, for all $(y,z,u),(y,z,u') \in \mathbb{R} \times \mathbb{R} \times L^2_Q(\mathbb{R})$, where $\delta^{y,z,u,u'} : \Omega \times [0,T] \times \mathbb{R} \to (-1,\infty)$ is a predictable process such that the mapping $t \mapsto \int_{\mathbb{R}} |\delta^{y,z,u,u'}(t,x)|^2 Q(t,dx)\eta(t)$ is uniformly bounded in (y,z,u,u'),
(iii) $g(t,y,0,0) = 0$ for all $(t,y) \in [0,T] \times \mathbb{R}$.

(a) We define the g-expectation $\mathscr{E}_g : \mathbb{L}^2(\Omega,\mathscr{F}_T,\mathbb{P};\mathbb{R}) \mapsto \mathbb{R}$ by

$$\mathscr{E}_g[\xi] = Y(0),$$

where $Y(0)$ is the unique solution to the BSDE (6.5) with the generator g satisfying (i)–(iii) and the terminal condition $\xi \in \mathbb{L}^2(\Omega,\mathscr{F}_T,\mathbb{P};\mathbb{R})$.
(b) We define the conditional g-expectation $\mathscr{E}_g : \mathbb{L}^2(\Omega,\mathscr{F}_T,\mathbb{P};\mathbb{R}) \to \mathbb{L}^2(\Omega,\mathscr{F}_t,\mathbb{P};\mathbb{R})$ by

$$\mathscr{E}_g[\xi|\mathscr{F}_t] = Y(t), \quad 0 \leq t \leq T,$$

where $Y(t)$ is the unique solution to the BSDE (6.5) with the generator g satisfying (i)–(iii) and the terminal condition $\xi \in \mathbb{L}^2(\Omega,\mathscr{F}_T,\mathbb{P};\mathbb{R})$.

Notice that for the g-expectation its dynamic version is naturally defined.
We state the first key result of this chapter.

Theorem 6.2.1 *The g-expectation $\mathscr{E}_g$ is an $\mathscr{F}$-consistent nonlinear expectation.*

Proof The strict monotonicity of $\mathscr{E}_g$ follows from the comparison principle established in Theorem 3.2.2. Since $g(t,y,0,0) = 0$, we can choose $Y = c$, $Z = U = 0$ as the unique solution to the BSDE (6.5) with $\xi = c$. Hence, the invariance property of $\mathscr{E}_g$ holds. We now prove the $\mathscr{F}$-consistency of $\mathscr{E}_g$. Choose $t \in [0,T]$ and $A \in \mathscr{F}_t$. We investigate

$$Y(0) = \mathscr{E}_g[\xi \mathbf{1}_A], \qquad Y'(0) = \mathscr{E}_g\big[Y(t)\mathbf{1}_A\big],$$

where Y and Y' denote the unique solutions to the BSDEs

$$\begin{aligned} Y(u) = \xi \mathbf{1}_A &+ \int_u^T g\big(s,Y(s),Z(s),U(s)\big)ds \\ &- \int_u^T Z(s)dW(s) - \int_u^T \int_{\mathbb{R}} U(s,z)\tilde{N}(ds,dz), \quad 0 \leq u \leq T, \end{aligned}$$

$$
\begin{aligned}
Y'(u) = Y(t)\mathbf{1}_A &+ \int_u^T g\big(s, Y'(s), Z'(s), U'(s)\big)ds \\
&- \int_u^T Z'(s)dW(s) - \int_u^T \int_{\mathbb{R}} U'(s,z)\tilde{N}(ds,dz), \quad 0 \le u \le T.
\end{aligned}
$$

Since $g(t, y, 0, 0) = 0$, we can put $Y'(s) = Y(t)\mathbf{1}_A$, $Z'(s) = U'(s,z) = 0$, $(s,z) \in [t,T] \times \mathbb{R}$. Consequently, we obtain the equations

$$
\begin{aligned}
Y(u) = Y(t) &+ \int_u^t g\big(s, Y(s), Z(s), U(s)\big)ds \\
&- \int_u^t Z(s)dW(s) - \int_u^t \int_{\mathbb{R}} U(s,z)\tilde{N}(ds,dz), \quad 0 \le u \le t, \\
Y'(u) = Y(t)\mathbf{1}_A &+ \int_u^t g\big(s, Y'(s), Z'(s), U'(s)\big)ds \\
&- \int_u^t Z'(s)dW(s) - \int_u^t \int_{\mathbb{R}} U'(s,z)\tilde{N}(ds,dz), \quad 0 \le u \le t.
\end{aligned}
$$

Consider the BSDE

$$
\begin{aligned}
Y''(u) = \xi &+ \int_u^T g\big(s, Y''(s), Z''(s), U''(s)\big)ds \\
&- \int_u^T Z''(s)dW(s) - \int_u^T \int_{\mathbb{R}} U''(s,z)\tilde{N}(ds,dz), \quad 0 \le u \le T.
\end{aligned}
$$

Since $g(t, y, 0, 0) = 0$ we can also put $Y(s) = Y''(s)\mathbf{1}_A$, $Z(s) = Z''(s)\mathbf{1}_A$, $U(s,z) = U''(s,z)\mathbf{1}_A$ $(s,z) \in [t,T] \times \mathbb{R}$. Hence, we end up with the equations

$$
\begin{aligned}
Y(u) = Y''(t)\mathbf{1}_A &+ \int_u^t g\big(s, Y(s), Z(s), U(s)\big)ds \\
&- \int_u^t Z(s)dW(s) - \int_u^t \int_{\mathbb{R}} U(s,z)\tilde{N}(ds,dz), \quad 0 \le u \le t, \\
Y'(u) = Y''(t)\mathbf{1}_A &+ \int_u^t g\big(s, Y'(s), Z'(s), U'(s)\big)ds \\
&- \int_u^t Z'(s)dW(s) - \int_u^t \int_{\mathbb{R}} U'(s,z)\tilde{N}(ds,dz), \quad 0 \le u \le t.
\end{aligned}
$$

By uniqueness of solutions we finally conclude that $Y(s) = Y'(s)$, $Z(s) = Z'(s)$, $U(s,z) = U'(s,z)$, $(s,z) \in [0,t] \times \mathbb{R}$. Hence, $\mathscr{E}_g$ is a filtration-consistent expectation with the conditional expectation $\mathscr{E}_g[\xi|\mathscr{F}_t] = Y(t)$. □

Any g-expectation clearly satisfies the properties from Proposition 6.2.1, which can now be derived from properties of BSDEs.

Example 6.2 If we consider a BSDE with zero generator, then the g-expectation coincides with the linear conditional expectation. If we consider the BSDE from Proposition 3.3.2 or 3.4.2, then the g-expectation is a filtration-consistent nonlinear expectation.

We are now interested in a converse of Theorem 6.2.1. The first results in this field were proved by Coquet et al. (2002) and Rosazza Gianin (2006) for the Brownian filtration. We present the result proved by Royer (2006) for the filtration generated by a Lévy process. First, we introduce two particular types of g-expectations.

Proposition 6.2.2 *Consider the natural filtration $\mathscr{F}$ generated by a Lévy process with a Lévy measure ν. For $\alpha > 0$ and $-1 < \beta \le 0$ we define the generators*

$$g^{*}_{\alpha,\beta}(t,z,u) = \alpha|z| + \alpha\int_{\mathbb{R}}\big(1\wedge|x|\big)u^{+}(x)\nu(dx) - \beta\int_{\mathbb{R}}\big(1\wedge|x|\big)u^{-}(x)\nu(dx),$$

$$g^{**}_{\alpha,\beta}(t,z,u) = -\alpha|z| - \alpha\int_{\mathbb{R}}\big(1\wedge|x|\big)u^{-}(x)\nu(dx) + \beta\int_{\mathbb{R}}\big(1\wedge|x|\big)u^{+}(x)\nu(dx).$$

The corresponding g-expectations have the representations

$$\mathscr{E}_{g^{*}_{\alpha,\beta}}[\xi|\mathscr{F}_t] = \sup_{\mathbb{Q}\in\mathscr{Q}} \mathbb{E}^{\mathbb{Q}}[\xi|\mathscr{F}_t], \quad 0\le t\le T,$$

$$\mathscr{E}_{g^{**}_{\alpha,\beta}}[\xi|\mathscr{F}_t] = \inf_{\mathbb{Q}\in\mathscr{Q}} \mathbb{E}^{\mathbb{Q}}[\xi|\mathscr{F}_t], \quad 0\le t\le T,$$

where

$$\mathscr{Q} = \left\{\mathbb{Q}\sim\mathbb{P}, \frac{d\mathbb{Q}}{d\mathbb{P}}\Big|\mathscr{F}_t = M^{\phi,\kappa}(t),\ 0\le t\le T\right\},$$

$$\frac{dM^{\phi,\kappa}(t)}{M^{\phi,\kappa}(t-)} = \phi(t)dW(t) + \int_{\mathbb{R}}\kappa(t,x)\tilde{N}(dx,dt), \quad M^{\phi,\kappa}(0) = 1,$$

and (ϕ,κ) are $\mathscr{F}$-predictable processes satisfying

$$\big|\phi(t)\big| \le \alpha, \quad \kappa(t,x) > -1,$$

$$\kappa^{+}(t,x) \le \alpha\big(1\wedge|x|\big), \quad \kappa^{-}(t,x) \le -\beta\big(1\wedge|x|\big), \quad (t,x)\in[0,T]\times\mathbb{R}.$$

Proof The result can be derived by following the arguments from Propositions 3.3.2 and 3.4.2, see also Proposition 3.6 in Royer (2006). □

We now state the second key result of this chapter, see Theorem 4.6 in Royer (2006).

Theorem 6.2.2 *Consider the natural filtration $\mathscr{F}$ generated by a Lévy process with a Lévy measure ν. Let $\mathscr{E}$ be an $\mathscr{F}$-consistent nonlinear expectation such that*

(i) *for all* $\xi_1, \xi_2 \in \mathbb{L}^2(\Omega, \mathscr{F}_T, \mathbb{P}; \mathbb{R})$

$$\mathscr{E}[\xi_1 + \xi_2] - \mathscr{E}[\xi_1] \leq \mathscr{E}_{g^*_{\alpha,\beta}}[\xi_2], \quad \textit{with some } \alpha \geq 0, \ -1 < \beta \leq 0,$$

where $\mathscr{E}_{g^*_{\alpha,\beta}}$ *is the* g*-expectation defined in Proposition* 6.2.2,

(ii) *for all* $\xi_1 \in \mathbb{L}^2(\Omega, \mathscr{F}_T, \mathbb{P}; \mathbb{R})$ *and* $\xi_2 \in \mathbb{L}^2(\Omega, \mathscr{F}_t, \mathbb{P}; \mathbb{R})$

$$\mathscr{E}[\xi_1 + \xi_2 | \mathscr{F}_t] = \mathscr{E}[\xi_1 | \mathscr{F}_t] + \xi_2, \quad 0 \leq t \leq T.$$

Then, there exists a function $g : \Omega \times [0, T] \times \mathbb{R} \times L^2_Q \to \mathbb{R}$ *and the* g*-expectation* $\mathscr{E}_g$ *such that*

$$\mathscr{E}[\xi | \mathscr{F}_t] = \mathscr{E}_g[\xi | \mathscr{F}_t], \quad \xi \in \mathbb{L}^2(\Omega, \mathscr{F}_T, \mathbb{P}; \mathbb{R}), \quad 0 \leq t \leq T.$$

Moreover, the following properties hold:

(i) g *satisfies* (A2) *from Chap.* 3,

(ii) g *satisfies the inequality*

$$g(\omega, t, z, u) - g(\omega, t, z, u') \leq \int_{\mathbb{R}} \delta^{z,u,u'}(t, x)\big(u(x) - u'(x)\big)\nu(dx),$$

a.s., a.e. $(\omega, t) \in \Omega \times [0, T]$, *for all* $(z, u), (z, u') \in \mathbb{R} \times L^2_Q(\mathbb{R})$, *where* $\delta^{z,u,u'}$: $\Omega \times [0, T] \times \mathbb{R} \to (-1, \infty)$ *is a predictable process such that* $\delta^{z,u,u'}(t, x) > -1$ *and* $|\delta^{z,u,u'}(t, x)| \leq K(1 \wedge |x|)$ *for all* $(t, x, z, u, u') \in [0, T] \times \mathbb{R} \times \mathbb{R} \times L^2_Q \times L^2_Q$,

(iii) $g(t, 0, 0) = 0$ *for all* $t \in [0, T]$,

(iv) g *satisfies the growth conditions*

$$g^{**}_{\alpha,\beta}(t, z, u) \leq g(t, z, u) \leq g^*_{\alpha,\beta}(t, z, u),$$

for $(t, z, u) \in [0, T] \times \mathbb{R} \times L^2_Q(\mathbb{R})$.

The first condition of Theorem 6.2.2 is called the domination condition. We remark that a large class of nonlinear expectations satisfies the domination condition, see Rosazza Gianin (2006) and Royer (2006). The second condition requires translation invariance of the nonlinear expectation with respect to "known" pay-offs, which is a reasonable assumption provided that discounting of pay-offs is not allowed in the valuation, see Sect. 13.1.

The importance of Theorem 6.2.2 is obvious. Theorem 6.2.2 shows that all filtration-consistent nonlinear expectations which satisfy the domination condition and the translation invariance property can be derived from BSDEs. Consequently, when we study "regular" filtration-consistent nonlinear expectations we can focus on g-expectations. Notice that the generator derived under the assumptions of Theorem 6.2.2 depends only on the control processes (Z, U) and is independent of Y. This is the consequence of the assumed translation invariance property for the nonlinear expectation.

It is clear that the generator g of a BSDE plays a crucial role in defining a g-expectation. Some important properties of g-expectations can be related to properties of generators g.

Proposition 6.2.3 *Let $\mathcal{E}_g$ be a g-expectation.*

(a) *If g is independent of y, then $\mathcal{E}_g$ is translation invariant*

$$\mathcal{E}_g[\xi + c|\mathcal{F}_t] = \mathcal{E}_g[\xi|\mathcal{F}_t] + c, \quad c \in \mathbb{R},\ 0 \le t \le T.$$

(b) *If g is positively homogenous, then $\mathcal{E}_g$ is positively homogenous*

$$\mathcal{E}_g[c\xi|\mathcal{F}_t] = c\mathcal{E}_g[\xi|\mathcal{F}_t], \quad c > 0,\ 0 \le t \le T.$$

(c) *If g is convex*

$$\begin{aligned} &g\big(t, cy_1 + (1-c)y_2, cz_1 + (1-c)z_2, cu_1 + (1-c)u_2\big) \\ &\quad \le cg(t, y_1, z_1, u_1) + (1-c)g(t, y_2, z_2, u_2), \\ &\qquad c \in (0,1),\ (t, y_1, z_1, u_1), (t, y_2, z_2, u_2) \in [0,T] \times \mathbb{R} \times \mathbb{R} \times L^2_Q, \end{aligned}$$

then $\mathcal{E}_g$ is convex

$$\begin{aligned} &\mathcal{E}_g\big[c\xi_1 + (1-c)\xi_2|\mathcal{F}_t\big] \\ &\quad \le c\mathcal{E}_g[\xi_1|\mathcal{F}_t] + (1-c)\mathcal{E}_g[\xi_2|\mathcal{F}_t], \quad c \in (0,1),\ 0 \le t \le T. \end{aligned}$$

(d) *If g is sub-linear: sub-additive*

$$\begin{aligned} &g(t, y_1 + y_2, z_1 + z_2, u_1 + u_2) \\ &\quad \le g(t, y_1, z_1, u_1) + g(t, y_2, z_2, u_2), \\ &\qquad (t, y_1, z_1, u_1), (t, y_2, z_2, u_2) \in [0,T] \times \mathbb{R} \times \mathbb{R} \times L^2_Q, \end{aligned}$$

and positively homogenous, then $\mathcal{E}_g$ is sub-linear: sub-additive

$$\mathcal{E}_g[\xi_1 + \xi_2|\mathcal{F}_t] \le \mathcal{E}_g[\xi_1|\mathcal{F}_t] + \mathcal{E}_g[\xi_2|\mathcal{F}_t], \quad 0 \le t \le T,$$

positively homogenous.

Proof (a) We deal with two BSDEs

$$\begin{aligned} Y^{\xi+c}(t) &= \xi + c + \int_t^T g\big(s, Z^{\xi+c}(s), U^{\xi+c}(s)\big)ds \\ &\quad - \int_t^T Z^{\xi+c}(s)dW(s) - \int_t^T \int_{\mathbb{R}} U^{\xi+c}(s,z)\tilde{N}(ds,dz), \quad 0 \le t \le T, \end{aligned}$$

$$\begin{aligned}Y^{\xi}(t) = \xi &+ \int_t^T g\big(s, Z^{\xi}(s), U^{\xi}(s)\big)ds \\ &- \int_t^T Z^{\xi}(s)dW(s) - \int_t^T\int_{\mathbb{R}} U^{\xi}(s,z)\tilde{N}(ds,dz), \quad 0 \le t \le T.\end{aligned}$$

We can easily conclude that $Y^{\xi+c}(t) = Y^{\xi}(t) + c$, $Z^{\xi+c}(t) = Z^{\xi}(t)$, $U^{\xi+c}(t,z) = U^{\xi}(t,z)$, $(t,z) \in [0,T] \times \mathbb{R}$.

(b) We deal with two BSDEs

$$\begin{aligned}Y^{c\xi}(t) = c\xi &+ \int_t^T g\big(s, Y^{c\xi}(s), Z^{c\xi}(s), U^{c\xi}(s)\big)ds \\ &- \int_t^T Z^{c\xi}(s)dW(s) - \int_t^T\int_{\mathbb{R}} U^{c\xi}(s,z)\tilde{N}(ds,dz), \quad 0 \le t \le T, \\ Y^{\xi}(t) = \xi &+ \int_t^T g\big(s, Y^{\xi}(s), Z^{\xi}(s), U^{\xi}(s)\big)ds \\ &- \int_t^T Z^{\xi}(s)dW(s) - \int_t^T\int_{\mathbb{R}} U^{\xi}(s,z)\tilde{N}(ds,dz), \quad 0 \le t \le T.\end{aligned}$$

We can easily conclude that $Y^{c\xi}(t) = cY^{\xi}(t)$, $Z^{c\xi}(t) = cZ^{\xi}(t)$, $U^{c\xi}(t,z) = cU^{\xi}(t,z)$, $(t,z) \in [0,T] \times \mathbb{R}$.

(c) We deal with three BSDEs

$$\begin{aligned}Y^{c\xi_1+(1-c)\xi_2}(t) &= c\xi_1 + (1-c)\xi_2 \\ &\quad + \int_t^T g\big(s, Y^{c\xi_1+(1-c)\xi_2}(s), Z^{c\xi_1+(1-c)\xi_2}(s), U^{c\xi_1+(1-c)\xi_2}(s)\big)ds \\ &\quad - \int_t^T Z^{c\xi_1+(1-c)\xi_2}(s)dW(s) \\ &\quad - \int_t^T\int_{\mathbb{R}} U^{c\xi_1+(1-c)\xi_2}(s,z)\tilde{N}(ds,dz), \quad 0 \le t \le T, \\ Y^{\xi_i}(t) = \xi_i &+ \int_t^T g\big(s, Y^{\xi_i}(s), Z^{\xi_i}(s), U^{\xi_i}(s)\big)ds \\ &- \int_t^T Z^{\xi_i}(s)dW(s) - \int_t^T\int_{\mathbb{R}} U^{\xi_i}(s,z)\tilde{N}(ds,dz), \quad i = 1,2,\ 0 \le t \le T.\end{aligned}$$

We introduce the processes $Y(t) = cY^{\xi_1}(t) + (1-c)Y^{\xi_2}(t)$, $Z(t) = cZ^{\xi_1}(t) + (1-c)Z^{\xi_2}(t)$, $U(t,z) = U^{\xi_1}(t,z) + (1-c)U^{\xi_2}(t,z)$. It is straightforward to notice that (Y,Z,U) satisfies the BSDE

$$\begin{aligned}Y(t) &= c\xi_1 + (1-c)\xi_2 \\ &\quad + \int_t^T \Big(cg\big(s, Y^{\xi_1}(s), Z^{\xi_1}(s), U^{\xi_1}(s)\big)\end{aligned}$$

$$+(1-c)g\big(s,Y^{\xi_2}(s),Z^{\xi_2}(s),U^{\xi_2}(s)\big)\big)ds$$
$$-\int_t^T Z(s)dW(s)-\int_t^T\int_{\mathbb{R}} U(s,z)\tilde{N}(ds,dz),\quad 0\le t\le T. \tag{6.6}$$

Since g satisfies

$$cg\big(s,Y^{\xi_1}(s),Z^{\xi_1}(s),U^{\xi_1}(s)\big)+(1-c)g\big(s,Y^{\xi_2}(s),Z^{\xi_2}(s),U^{\xi_2}(s)\big)$$
$$\ge g\big(s,Y(s),Z(s),U(s)\big),\quad 0\le s\le T,$$

the BSDE (6.6) can be written as

$$\begin{aligned}Y(t)&=c\xi_1+(1-c)\xi_2\\&\quad+\int_t^T \big(g\big(s,Y(s),Z(s),U(s)\big)\\&\quad+h\big(s,Y^{\xi_1}(s),Z^{\xi_1}(s),U^{\xi_1}(s),Y^{\xi_2}(s),Z^{\xi_2}(s),U^{\xi_2}(s)\big)\big)ds\\&\quad-\int_t^T Z(s)dW(s)-\int_t^T\int_{\mathbb{R}} U(s,z)\tilde{N}(ds,dz),\quad 0\le t\le T,\end{aligned}$$

with a nonnegative function h. By the comparison principle we get $Y^{c\xi_1+(1-c)x_2}(t)\le Y(t)=cY^{\xi_1}(t)+(1-c)Y^{\xi_2}(t), 0\le t\le T$.

(d) Adapting the arguments from (b) and (c), we can prove the assertion. □

The properties from Proposition 6.2.3 are used in Chap. 13 where we deal with dynamic risk measures.

Bibliographical Notes The Choquet expectation was introduced by Choquet (1953). Properties of Choquet expectations and the Wang transform together with their failures in non-Gaussian financial models are discussed by Nguyen et al. (2012). The g-expectations was introduced by Peng (1997). For the connection between the Choquet expectation and the g-expectation we refer to Chen et al. (2005) and Chen and Kulperger (2006). In the proof of Proposition 6.2.3 we follow the arguments from Rosazza Gianin (2006) and Jiang (2008). We refer to Rosazza Gianin (2006) and Jiang (2008) for stronger relations between static and dynamic properties of g-expectations and generators of BSDEs defining the g-expectations. For a representation of a filtration-consistent nonlinear expectation in a general separable space we refer to Cohen (2011). We remark that g-expectations allow for introducing nonlinear versions of some well-known probabilistic results, see Coquet et al. (2002), Peng (1997), Rosazza Gianin (2006) and Royer (2006). For g-martingales, g-submartingales, g-supermartingales and nonlinear Doob-Meyer decomposition we refer to Coquet et al. (2002) and Royer (2006).

Part II
Backward Stochastic Differential Equations—The Applications

Chapter 7
Combined Financial and Insurance Model

Abstract A combined financial and insurance model is introduced. We consider a Black-Scholes financial model with stochastic coefficients. We use a step process with a stochastic intensity and a random transition kernel to model claims. We investigate a stream of liabilities which consists of annuity, death and survival benefits. We define a set of admissible investment strategies for an insurer (an investor) who trades in the financial market and aims to replicate the stream of liabilities.

We still consider a probability space $(\Omega, \mathscr{F}, \mathbb{P})$ with a filtration $\mathscr{F} = (\mathscr{F}_t)_{0\leq t\leq T}$ satisfying the usual hypotheses of completeness and right continuity and a finite time horizon $T < \infty$. On the space $(\Omega, \mathscr{F}, \mathbb{P})$ we define two independent $\mathscr{F}$-adapted Brownian motions (W, B) and a random measure N generated by an $\mathscr{F}$-adapted step process.

The financial market, the insurance payment process and the set of admissible investment strategies are introduced in this chapter. In next chapters we investigate pricing and hedging problems in our combined financial and insurance model.

7.1 The Financial Market

We consider a Black-Scholes financial model with stochastic coefficients. The financial market consists of a bank account and a stock. The dynamics of the bank account $S_0 := (S_0(t), 0 \leq t \leq T)$ is described by the equation

$$\frac{dS_0(t)}{S_0(t)} = r(t)dt, \quad S_0(0) = 1, \tag{7.1}$$

where $r := (r(t), 0 \leq t \leq T)$ denotes the risk-free rate. The dynamics of the stock price $S := (S(t), 0 \leq t \leq T)$ is given by the forward stochastic differential equation

$$\frac{dS(t)}{S(t)} = \mu(t)dt + \sigma(t)dW(t), \quad S(0) = s > 0, \tag{7.2}$$

where $\mu := (\mu(t), 0 \leq t \leq T)$ denotes the expected return on the stock and $\sigma := (\sigma(t), 0 \leq t \leq T)$ denotes the stock volatility. We denote $\theta(t) = \frac{\mu(t)-r(t)}{\sigma(t)}$. We assume that

Ł. Delong, *Backward Stochastic Differential Equations with Jumps and Their Actuarial and Financial Applications*, EAA Series, DOI 10.1007/978-1-4471-5331-3_7,

(C1) the processes $r : \Omega \times [0, T] \to [0, \infty)$, $\mu : \Omega \times [0, T] \to [0, \infty)$ and $\sigma : \Omega \times [0, T] \to (0, \infty)$ are $\mathscr{F}^W$-predictable and they satisfy

$$\big|r(t)\big| + \big|\mu(t)\big| + \big|\sigma(t)\big| + \big|\sigma(t)\big|^{-1} + \left|\frac{\mu(t) - r(t)}{\sigma(t)}\right| \leq K, \quad 0 \leq t \leq T,$$

$$\mu(t) \geq r(t), \quad 0 \leq t \leq T,$$

(C2) there exists a unique process S which solves (7.2) such that $\sup_{t\in[0,T]} \mathbb{E}[|S(t)|^2] < \infty$.

These are standard assumptions in financial mathematics. We may relax the boundedness assumptions, but other integrability conditions would have to be imposed instead so that we can solve our optimization problems. If r, μ and σ are exogenously given and independent of S, then (C2) is satisfied, see Theorem II.37 in Protter (2004) and Theorem 4.1.1. Since the coefficients may depend on S, assumption (C2) is added.

Continuous-time models (7.1)–(7.2) have become standard in financial applications, see Filipovic (2009), Fouque et al. (2000) and Shreve (2004). We give two important examples of the financial model considered.

Example 7.1 Set r, μ and σ as constants. We can investigate the classical Black-Scholes model with normally distributed log-price, see Shreve (2004).

Example 7.2 Let $r : [0, T] \times (0, \infty) \to [0, \infty)$, $\mu : [0, T] \times (0, \infty) \to [0, \infty)$ and $\sigma : [0, T] \times (0, \infty) \to (0, \infty)$ be measurable functions. We can investigate a local volatility model of the form

$$\frac{dS_0(t)}{S_0(t)} = r\big(t, S(t)\big)dt, \quad S_0(0) = 1,$$

$$\frac{dS(t)}{S(t)} = \mu\big(t, S(t)\big)dt + \sigma\big(t, S(t)\big)dW(t), \quad S(0) = s > 0.$$

Local volatility models provide a better fit to quoted option prices and yield more skewed distributions of asset returns, see Dupire (1997) and Fouque et al. (2000).

By a straightforward generalization we can also investigate stochastic economic factor models such as stochastic interest rate models and stochastic volatility models. We have to introduce more driving noises into the model.

Example 7.3 Let $\mathscr{X} : \Omega \times [0, T] \to \mathbb{R}$ denote an economic factor. We should use a two-dimensional Brownian motion $W = (W_1, W_2)$ to model the stock price S and the economic factor $\mathscr{X}$. Let $r : [0, T] \times [0, \infty) \to [0, \infty)$, $\mu : [0, T] \times [0, \infty) \to [0, \infty)$ and $\sigma : [0, T] \times [0, \infty) \to (0, \infty)$ be measurable functions and assume that $\mathscr{X}$ is an $\mathscr{F}^{W_1} \otimes \mathscr{F}^{W_2}$-adapted process. We define the dynamics

$$\frac{dS_0(t)}{S_0(t)} = r\big(t, \mathscr{X}(t)\big)dt, \quad S_0(0) = 1,$$

$$\frac{dS(t)}{S(t)} = \mu\big(t, \mathscr{X}(t)\big)dt + \sigma\big(t, \mathscr{X}(t)\big)dW_1(t), \quad S(0) = s > 0.$$

We can now consider the Cox-Ingersoll-Ross interest rate model or the stochastic volatility Heston model, see Filipovic (2009) and Fouque et al. (2000).

We can also assume that the stock price S is driven by a Brownian motion and a pure jump Lévy process. Such a dynamics could be desirable in applications as Lévy processes have proved to be very useful for financial modelling, see Cont and Tankov (2004). Since the theory of BSDEs with jumps covers Lévy processes, the extension of the model in this direction is possible. We decide to use a jump process only for claim modelling.

7.2 The Insurance Payment Process

Insurance claims are modelled by a step process J. Let N denote the jump measure of J. We assume that

(C3) the integer-valued random measure N has the $\mathscr{F}$-predictable compensator

$$\vartheta(dz, dt) = Q(t, dz)\eta(t)dt,$$

where $\eta : \Omega \times [0, T] \times \mathbb{R} \to [0, \infty)$ is an $\mathscr{F}$-predictable process, $Q(t, \cdot)$ is a probability measure on $\mathscr{B}(\mathbb{R})$ for $(\omega, t) \in \Omega \times [0, T]$, $Q(\cdot, A) : \Omega \times [0, T] \to [0, 1]$ is an $\mathscr{F}$-predictable process for $A \in \mathscr{B}(\mathbb{R})$, and

$$N\big([0, t], \{0\}\big) = Q\big(t, \{0\}\big) = 0, \quad 0 \le t \le T,$$

$$\int_0^T \eta(t)dt < \infty, \qquad \int_0^T \int_{\mathbb{R}} z^2 Q(t, dz)\eta(t)dt < \infty.$$

Assumption (C3) is in line with assumption (RM) from Sect. 2.1. Since we consider a step process, from the general representation of the compensator of the jump measure, see (2.3), we conclude that Q is a probability transition kernel and η is an integrable intensity of the underlying point process, see Definitions II.D7 and VIII.D5 in Brémaud (1981) and Definition 2.1.6. Notice that both the intensity η and the jump distribution Q are $\mathscr{F}$-predictable processes which may depend on the financial market, the number of claims paid and other sources of uncertainty captured by the filtration $\mathscr{F}$.

We remark that step processes are very often used for claim modelling in actuarial mathematics, see Mikosch (2009) and Rolski et al. (1999). In a more theoretical setting we could also use a Lévy process with infinitely many jumps.

We investigate the insurance payment process

$$P(t) = \int_0^t H(s)ds + \int_0^t \int_{\mathbb{R}} G(s, z)N(ds, dz) + F\mathbf{1}_{t=T}, \quad 0 \le t \le T. \quad (7.3)$$

We will see that the payment process (7.3) can also be used for modelling financial claims. We assume that

(C4) the processes $H:\Omega\times[0,T]\to[0,\infty)$ and $G:\Omega\times[0,T]\times\mathbb{R}\to[0,\infty)$ are $\mathscr{F}$-predictable and the random variable $F:\Omega\to[0,\infty)$ is $\mathscr{F}_T$-measurable. The claims H, G and F satisfy

$$\mathbb{E}\Big[\int_0^T |H(s)|^2 ds\Big]<\infty,\quad \mathbb{E}\Big[\int_0^T\int_{\mathbb{R}}|G(s,z)|^2\eta(s)Q(s,dz)ds\Big]<\infty,$$
$$\mathbb{E}\Big[\int_0^T\int_{\mathbb{R}}|G(s,z)\eta(s)|^2 Q(s,dz)ds\Big]<\infty,\quad \mathbb{E}\big[|F|^2\big]<\infty.$$

Square integrability assumptions are standard in financial mathematics. It is straightforward to conclude that under (C4) we have $\mathbb{E}[|P(T)|^2]<\infty$. We should notice that the stochastic integral with respect to the random measure in (7.3) is a.s. well-defined since the step process J generates a finite number of jumps on $[0,T]$. Consequently, the payment process can be written in the following form

$$P(t)=\int_0^t H(s)ds+\sum_{s\in(0,t]}G\big(s,\Delta J(s)\big)\mathbf{1}_{\{\Delta J(s)\neq 0\}}(s)+F\mathbf{1}_{t=T},\quad 0\le t\le T.$$

The process P represents a very general stream of liabilities. It contains payments H which occur continuously during the term of the contract (annuities), it contains claims G which occur at random times triggered by the jumps of the step process J (death benefits), and finally it contains the liability F which is settled at the end of the contract (a survival benefit). The pay-offs H, G and F may depend on the financial market, the number of claims paid and other sources of uncertainty modelled by $\mathscr{F}$. We point out that we can model unsystematic and systematic insurance risk. By the unsystematic insurance risk we mean the risk of an uncertain number of claims (which is here modelled by the step process J), and by the systematic insurance risk we mean the risk of unpredictable changes in the claim intensity (which is here modelled by a stochastic intensity of the step process J).

To enrich the liability model and extend the area of its applications, we introduce a second $\mathscr{F}$-adapted Brownian motion B independent of the Brownian motion W. The Brownian motion B models a third risk factor (a background source of uncertainty), next to the equity risk modelled by W and the claims risk modelled by N. The Brownian motion B can affect both the claims' pay-offs and the claim intensity. The role of B is clarified in the next examples.

Example 7.4 Let $\hat{F}:(0,\infty)\to[0,\infty)$ be a measurable function. Set $H=G=0$, $F=\hat{F}(S(T))$ and $\eta=0$. We end up with the classical financial setting with a terminal claim contingent on the traded asset S, see Shreve (2004).

Example 7.5 Consider a process $\hat{S}$ which satisfies the forward SDE

$$\frac{d\hat{S}(t)}{\hat{S}(t)} = \hat{\mu}(t)dt + \hat{\sigma}_1(t)dW(t) + \hat{\sigma}_2(t)dB(t), \quad \hat{S}(0) = \hat{s} > 0,$$

where $\hat{\mu}$, $\hat{\sigma_1}$, $\hat{\sigma_2}$ are $\mathscr{F}^W \otimes \mathscr{F}^B$-predictable processes. Let $\hat{F}$ be a measurable functional defined on the space of continuous functions. Set $H = G = 0$, $F = \hat{F}(\hat{S}(t), 0 \leq t \leq T)$ and $\eta = 0$. We can investigate a terminal claim the value of which depends on the path of the non-tradeable index $\hat{S}$ correlated with the traded stock S. The Brownian motion W which is used for modelling the index $\hat{S}$ guarantees that $\hat{S}$ is correlated with the traded stock S, and the Brownian motion B introduces an independent, non-tradeable source of risk which guarantees that $\hat{S}$ is not perfectly correlated with S.

Example 7.6 Consider a predictable process $\lambda : \Omega \times [0, T] \to (0, \infty)$. We define the point process

$$J(t) = \sum_{i=1}^{n} \mathbf{1}\{\tau_i \leq t\}, \quad 0 \leq t \leq T, \tag{7.4}$$

where $(\tau_i, i = 1, \ldots, n)$ is a sequence of random variables which are, conditional on the filtration $\mathscr{F}^\lambda$, independent and exponentially distributed

$$\mathbb{P}(\tau_i > t | \mathscr{F}_t^\lambda) = e^{-\int_0^t \lambda(s)ds}, \quad i = 1, \ldots, n.$$

In actuarial and financial applications the sequence $(\tau_i, i = 1, \ldots, n)$ can model defaults of securities, deceases of persons insured or surrenders of policies. The corresponding characteristics of the point process J take the form

$$Q(t, \{1\}) = 1, \quad \eta(t) = (n - J(t-))\lambda(t), \quad 0 \leq t \leq T. \tag{7.5}$$

Properties of the point process (7.4) are studied by Jeanblanc and Rutkowski (2000) in a credit risk context and Dahl and Møller (2006) in a life insurance context.

Example 7.7 Let $\hat{H} : [0, T] \times (0, \infty) \to [0, \infty)$, $\hat{G} : [0, T] \times (0, \infty) \to [0, \infty)$ and $\hat{F} : (0, \infty) \to [0, \infty)$ be measurable functions and let $\lambda : \Omega \times [0, T] \to (0, \infty)$ be an $\mathscr{F}^B$-predictable process. Set

$$\begin{gathered} H(t) = (n - J(t-))\hat{H}(t, S(t)), \quad G(t, z) = \hat{G}(t, S(t)), \\ F = (n - J(T))\hat{F}(T, S(T)), \\ Q(t, \{1\}) = 1, \quad \eta(t) = (n - J(t-))\lambda(t), \quad 0 \leq t \leq T. \end{gathered} \tag{7.6}$$

We can consider a portfolio consisting of n persons insured and we can investigate life insurance equity-linked claims under longevity risk, see Dahl and Møller (2006) and Dahl et al. (2008). By the longevity risk we mean the risk of uncertain future mortality rates (which are likely to decrease in an unpredictable fashion), which is here modelled by a stochastic process λ independent of the financial market. Both the financial risk of issued guarantees (equity-linked claims) and the longevity risk are the main risk factors faced by life insurers.

If we still use characteristics (7.6) but we assume that λ is an $\mathscr{F}^W$-predictable processes, then we can investigate equity-linked life insurance claims under irrational lapse behavior of policyholders. By the irrational lapse behavior we mean the decision to surrender the policy which is made by a policyholder in a non-optimal way, but after taking into account alternative investment opportunities in the market. Hence, we should model the lapse intensity by a stochastic process which is linked to the financial market. The irrational lapse behavior represents an important risk factor for life insurers, see TP.2.105-111 European Commission QIS5 (2010).

Example 7.8 Let $\lambda : \Omega \times [0, T] \to (0, \infty)$ be an $\mathscr{F}^B$-predictable process. Set

$$Q\big(t, \{1\}\big) = 1, \quad \eta(t) = \big(n - J(t-)\big)\lambda(t), \quad 0 \le t \le T,$$

and assume that the claims H, G, F are contingent on the number of deaths J or the mortality intensity λ in a population. We can investigate mortality derivatives, which are gaining popularity as securitization instruments. For example, a survivor swap contingent on a population consisting of n individuals can be studied by setting $G = F = 0$ and $H(t) = (n - J(t)) - n\hat{p}(t)$ where $\hat{p}(t)$ denotes a survival rate agreed by the parties of the contract, see Dahl et al. (2008).

Example 7.9 Let q be a probability measure supported on $(0, \infty)$, and let $\lambda > 0$. Set $H = F = 0$ and

$$G(t, z) = z, \quad Q(t, dz) = q(dz), \quad \eta(t) = \lambda, \quad 0 \le t \le T, \ z \in (0, \infty). \tag{7.7}$$

We end up with a compound Poisson aggregate claims process, which plays a fundamental role in actuarial mathematics, see Mikosch (2009) and Rolski et al. (1999). If we let $\lambda : \Omega \times [0, T] \to (0, \infty)$ be an $\mathscr{F}^B$-predictable process, then we can consider a compound Cox aggregate claims process. The Cox process with an independent stochastic intensity is a useful generalization of the Poisson process for actuarial applications. It can be used for modelling catastrophic claims since catastrophic claims are triggered by random natural disasters, see Dassios and Jang (2003). It can also be used for modelling seasonal variations of the claim intensity, see Bening and Korolev (2002).

Let $\hat{G} : [0, T] \times (0, \infty) \times (0, \infty) \to [0, \infty)$ and $\lambda : [0, T] \times (0, \infty) \to (0, \infty)$ be measurable functions, and let q be a probability transition kernel such that $q(., A) : [0, T] \times (0, \infty) \to [0, 1]$ is a measurable function for $A \in \mathscr{B}((0, \infty))$ and $q(t, s, .)$ is a probability measure on $\mathscr{B}((0, \infty))$ for $(t, s) \in [0, T] \times (0, \infty)$. Set $H = F = 0$ and

$$\begin{aligned} &G(t, z) = \hat{G}\big(t, \hat{S}(t), z\big), \quad Q(t, z) = q\big(t, \hat{S}(t), dz\big), \\ &\eta(t) = \lambda\big(t, \hat{S}(t)\big), \quad 0 \le t \le T, \ z \in (0, \infty). \end{aligned} \tag{7.8}$$

In order to model aggregate payments we now use a step process with the claim intensity λ and the claim severity q which depend on a non-tradeable asset $\hat{S}$ introduced in Example 7.5. The pay-off $\hat{G}$ is contingent on the claim and also depends on a non-tradeable index $\hat{S}$. The payment process (7.8) may arise when we investigate weather derivatives, see Ankirchner and Imkeller (2008). It is known that the intensity and the severity of weather events (such as earthquakes or hurricanes) is influenced by climate factors. Climate factors are not traded in the market but investors can find traded assets that are correlated with these climate factors. Hence, we should use a non-tradeable index $\hat{S}$ to model a climate risk factor which effects the intensity and the severity of weather claims. We shall remark that weather derivatives are gaining importance in financial markets.

Example 7.10 Let $\lambda : \Omega \times [0,T] \to (0,\infty)$, $\hat{H} : \Omega \times [0,T] \to [0,\infty)$ and $\hat{G} : \Omega \times [0,T] \to [0,\infty)$ be $\mathscr{F}^W$-predictable processes and let $\hat{F} : \Omega \to [0,\infty)$ be an $\mathscr{F}^W$-measurable random variable. Set

$$\begin{aligned} H(t) &= \big(n - J(t-)\big)\hat{H}(t), \quad G(t,z) = \hat{G}(t), \quad F = \big(n - J(T)\big)\hat{F}, \\ Q\big(t,\{1\}\big) &= 1, \quad \eta(t) = \big(n - J(t-)\big)\lambda(t), \quad 0 \le t \le T. \end{aligned} \tag{7.9}$$

We can investigate claims from a portfolio consisting of n defaultable securities, see Bielecki et al. (2004) and Jeanblanc and Rutkowski (2000). In such a framework, $\hat{H}$ denotes a dividend, $\hat{G}$ denotes a recovery rate paid at default and $\hat{F}$ denotes a promised pay-off. The pay-offs and the default intensity are linked to the financial market. We can also introduce an independent, non-tradeable source of risk for the pay-offs and the default intensity and we can assume that the claims and the intensity are modelled by $\mathscr{F}^W \otimes \mathscr{F}^B$-predictable processes. A classical credit default swap (CDS) can be studied if we choose $n = 1$, $F = 0$, $\hat{H} = -\hat{h}$ and $\hat{G} = \hat{g}$, see Bielecki et al. (2008).

We can also investigate a collective credit risk model. Set $H = F = 0$ and

$$G\big(t,\{1\}\big) = \hat{G}(t), \quad Q\big(t,\{1\}\big) = 1, \quad \eta(t) = \lambda(t), \quad 0 \le t \le T.$$

The payment $\hat{G}$ represents a credit loss of a obligor given default. The corresponding payment process models the aggregated credit losses from a portfolio of obligors, see Gundlach and Lehrbass (2004). We can also assume that the credit loss $\hat{G} : \Omega \times [0,T] \times \mathbb{R} \to [0,\infty)$ depends on the financial market and an exogenous factor distributed with $q(dz)$. Then, we set $Q(t,dz) = q(dz)$.

As the examples show, the model (7.1)–(7.3) allows us to consider many desirable extensions of the classical Black-Scholes financial model and the compound Poisson loss model. It is obvious to conclude that the payment process (7.3) can also be used for modelling financial liabilities. The model (7.1)–(7.3) is mathematically tractable and can be used for pricing, hedging and risk management. Various useful financial and insurance models with multiple risk factors can be developed based on (7.1)–(7.3). A further generalization is possible by introducing a regime switching process, see Crépey and Matoussi (2008) and Crépey (2011).

Recalling the discussion in Sect. 2.4, we assume that the weak property of predictable representation holds in our combined financial and insurance model, that is any $\mathscr{F}$-local martingale M has the representation

$$M(t) = M(0) + \int_0^t Z_1(s)dW(s) + \int_0^t Z_2(s)dB(s) + \int_0^t \int_{\mathbb{R}} U(s,z)\tilde{N}(ds,dz), \quad 0 \le t \le T.$$

We point out that the predictable representation may fail under progressive enlargement of the Brownian filtration, which is usually considered in credit risk models, see Jeanblanc and Le Cam (2009) and Jiao et al. (2013). Fortunately, if we construct the default time by the standard method, by an exponentially distributed random variable (see Example 7.6), then the so-called H-hypothesis is satisfied and the predictable representation holds under the progressive enlargement of the Brownian filtration with the information generated by a default process. Consequently, we can assume that the predictable representation holds for a Brownian motion and a compensated default process, see Proposition 3.2 in Jeanblanc and Rutkowski (2000) and Theorem 2.3 in Blanchet-Scalliet et al. (2008). By the construction, the H-hypothesis is also fulfilled by the jumps of the Cox process. Hence, the predictable representation for a Brownian motion and a compensated Cox process can be assumed as well, see Lim (2005). In other cases, we can follow the arguments presented in Sect. 2.4 to conclude that the predictable representation holds. We remark that the assumption of the predictable representation is not controversial for most applications but the reader should be aware that some sophisticated models are excluded from the study, see Jeanblanc and Le Cam (2009) and Jiao et al. (2013).

7.3 Admissible Investment Strategies

We consider an insurer (an investor) who faces the stream of liabilities (7.3) and invests in the bank account (7.1) and the stock (7.2). The insurer's goal is to replicate the liabilities by investing in the assets and to earn a profit from the investing activities. We are interested in constructing investment strategies and investment portfolios which would fulfill these two goals. Such constructions are discussed in next chapters.

Our standing assumption is that

(AF) the combined financial and insurance market is arbitrage-free.

This is the key assumption in pricing and hedging models.

Let X^{π} denote the insurer's investment portfolio (the hedging or replicating portfolio) under an investment strategy π. We assume that X^{π} is self-financing in the sense that the value of the portfolio results from investment gains and claims paid. The dynamics of the investment portfolio $X^{\pi} := (X^{\pi}(t), 0 \le t \le T)$ is given by the forward SDE

$$
\begin{aligned}
dX^{\pi}(t) &= \pi(t)\frac{dS(t)}{S(t)} + \big(X(t)-\pi(t)\big)\frac{dS_0(t)}{S_0(t)} - dP(t) \\
&= \pi(t)\big(\mu(t)dt + \sigma(t)dW(t)\big) + \big(X^{\pi}(t-)-\pi(t)\big)r(t)dt - dP(t), \quad (7.10) \\
X(0) &= x > 0,
\end{aligned}
$$

where π denotes the amount of wealth invested in the risky asset S, and x denotes the initial capital including the premium collected at the inception of the contract. In our applications it is more convenient to deal with the following dynamics

$$
\begin{aligned}
dX^{\pi}(t) &= \pi(t)\big(\mu(t)dt + \sigma(t)dW(t)\big) + \big(X^{\pi}(t-)-\pi(t)\big)r(t)dt, \\
&\quad - H(t)dt - \int_{\mathbb{R}} G(t,z)N(dt,dz), \quad (7.11) \\
X(0) &= x > 0,
\end{aligned}
$$

and subtract the claim F from the terminal wealth $X^{\pi}(T)$. We define a class of admissible investment strategies.

Definition 7.3.1 A strategy $\pi := (\pi(t), 0 \le t \le T)$ is called admissible, written $\pi \in \mathscr{A}$, if it satisfies the conditions:

1. $\pi : [0,T] \times \Omega \to \mathbb{R}$ is an $\mathscr{F}$-predictable process,
2. $\mathbb{E}[\int_0^T |\pi(t)\sigma(t)|^2 dt] < \infty$,
3. there exists a unique càdlàg, $\mathscr{F}$-adapted solution X^{π} to (7.10) (or (7.11)) on $[0,T]$.

If the admissible investment strategy π is independent of the investment portfolio X^{π}, then the unique solution to (7.10) is given by

$$
\begin{aligned}
X^{\pi}(t) &= xe^{\int_0^t r(s)ds} + \int_0^t \big(\mu(s)-r(s)\big)e^{\int_s^t r(u)du}\pi(s)ds \\
&\quad + \int_0^t \sigma(s)e^{\int_s^t r(u)du}\pi(s)dW(s) - \int_0^t \int_{\mathbb{R}} e^{\int_s^t r(u)du} dP(s), \quad 0 \le t \le T.
\end{aligned}
$$

Moreover, the investment portfolio X^{π} is square integrable under $\pi \in \mathscr{A}$. Indeed, by (C1)–(C4) and the Burkholder-Davis-Gundy inequality we can derive

$$
\begin{aligned}
\mathbb{E}\Big[\sup_{t\in[0,T]} \big|X^{\pi}(t)\big|^2\Big] &\le K\bigg(1 + \mathbb{E}\Big[\int_0^T \big|\pi(t)\sigma(t)\big|^2 dt\Big] \\
&\quad + \mathbb{E}\Big[\sup_{t\in[0,T]}\Big|\int_0^t \sigma(s)e^{\int_s^t r(u)du}\pi(s)dW(s)\Big|^2\Big] + \mathbb{E}\big[\big|P(T)\big|^2\big]\bigg) \\
&\le K\bigg(1 + \mathbb{E}\Big[\int_0^T \big|\pi(t)\sigma(t)\big|^2 dt\Big]\bigg) < \infty. \quad (7.12)
\end{aligned}
$$

We point out that square integrability assumptions have a strong financial justification. The next example shows that the square integrability of the investment strategy is required in order to exclude arbitrage strategies.

Example 7.11 By Dudley (1977) it is possible to construct a predictable process V such that

$$\int_0^T V(s)dW(s) = 1, \quad \int_0^T |V(s)|^2 ds < \infty, \text{ a.s.}$$

We now define

$$\mathscr{H}(t) = e^{-\int_0^t r(s)ds - \int_0^t \theta(s)dW(s) - \frac{1}{2}\int_0^t \theta^2(s)ds}, \quad 0 \le t \le T,$$

$$\hat{X}(t) = \mathscr{H}^{-1}(t)\int_0^t V(s)dW(s), \quad 0 \le t \le T,$$

$$\hat{\pi}(t) = \frac{1}{\sigma(t)}\big(\mathscr{H}^{-1}(t)V(t) + \hat{X}(t)\theta(t)\big), \quad 0 \le t \le T.$$

Let $P = 0$. It is straightforward to show that $(\hat{X}, \hat{\pi})$ solves (7.10) with the initial condition $x = 0$ and the terminal value $\hat{X}(T) = \mathscr{H}^{-1}(T) > 0$. Hence, the strategy $\hat{\pi}$ is an arbitrage strategy.

Let us recall that in order to derive a square integrable control process of a BSDE, which is next used to define an arbitrage-free, square integrable investment strategy, we have to assume that the terminal condition and the generator of the BSDE are square integrable. Hence, all our square integrability assumptions are justified.

Bibliographical Notes Detailed references are given in the text. The form of the payment process (7.3) is inspired by Møller and Steffensen (2007). Example 7.11 is taken from El Karoui et al. (1997b).

Chapter 8
Linear BSDEs and Predictable Representations of Insurance Payment Processes

Abstract We solve linear BSDEs which may arise in actuarial applications. Since solving a linear BSDE requires to find the predictable representation of a random variable, we show how to derive the predictable representation of an insurance payment process. We consider the case of a life insurance and a non-life insurance payment process under systematic and unsystematic claims risk. We apply both the Itô's formula and the Malliavin calculus to derive the control processes of linear BSDEs. The representations of the control processes involve conditional expectations which can be explicitly calculated or estimated by Monte Carlo methods.

In this chapter we solve linear BSDEs which may arise in actuarial applications. We illustrate two methods based on the Itô's formula and the Malliavin calculus which can be used to derive the control processes of linear BSDEs. Linear BSDEs are important for applications. In Chap. 9 a linear BSDE is used to characterize the replicating strategy for a liability, and in Chap. 10 linear BSDEs are used to characterize the optimal (in the mean-square sense) hedging strategy for a liability. We also remark that nonlinear BSDEs which we face when investigating pricing and hedging of liabilities under model ambiguity and the entropic risk measure may be reduced to linear BSDEs, see Propositions 3.3.3 and 3.4.3 and Example 3.4. Since solving a linear BSDE requires to find the predictable representation of a random variable, in this chapter we show how to derive the predictable representation of an insurance payment process. We consider the case of a life insurance and a non-life insurance payment process under systematic and unsystematic claims risk.

8.1 The Application of the Itô's Formula

Let us investigate the financial model

$$dS_0(t) = S_0(t)rdt, \quad S_0(0) = 1,$$
$$dS(t) = \mu\big(S(t)\big)dt + \sigma\big(S(t)\big)dW(t), \quad S(0) = s > 0,$$

Ł. Delong, *Backward Stochastic Differential Equations with Jumps and Their Actuarial and Financial Applications*, EAA Series, DOI 10.1007/978-1-4471-5331-3_8,

and the equity-linked (life insurance) liabilities

$$P(t) = \int_0^t \big(n - J(s-)\big)\hat{H}\big(S(s)\big)ds + \int_0^t \hat{G}\big(S(s)\big)dJ(s)$$
$$+ \big(n - J(T)\big)\hat{F}\big(S(T)\big)\mathbf{1}\{t = T\}, \quad 0 \le t \le T,$$

where J denotes the deaths counting process for a life insurance portfolio consisting of n policies. We assume that the mortality intensity is given by the dynamics

$$d\lambda(t) = \mu^\lambda\big(\lambda(t)\big)dt + \sigma^\lambda\big(\lambda(t)\big)dB(t), \quad \lambda(0) = \lambda > 0.$$

For an actuarial and financial motivation we refer to Examples 7.2 and 7.7. We model the longevity risk by a stochastic process which solves a forward SDE describing the time evolution of the mortality intensity. Such stochastic models are advocated in the actuarial literature, see Russo et al. (2011) and Schrager (2006). We can also assume that r depends on S and allow for time-dependent dynamics. Then, we can follow the same arguments to derive the control processes of a BSDE. Let us recall that the jump measure N of the point process J has the compensator $\vartheta(dt, \{1\}) = (n - J(t-))\lambda(t)dt$.

The value of the insurance liabilities is given by

$$Y(t) = \mathbb{E}\bigg[e^{-r(T-t)}\big(n - J(T)\big)\hat{F}\big(S(T)\big) + \int_t^T e^{-r(s-t)}\hat{G}\big(S(s)\big)dJ(s)$$
$$+ \int_t^T e^{-r(s-t)}\big(n - J(s-)\big)\hat{H}\big(S(s)\big)ds \big| \mathscr{F}_t\bigg], \quad 0 \le t \le T. \tag{8.1}$$

If we consider the market-consistent value, then the expectation should be taken under an equivalent martingale measure, see Sect. 9.1. From Propositions 3.3.1 and 3.4.1 we conclude that the value process (8.1) satisfies the linear BSDE

$$Y(t) = \big(n - J(T)\big)\hat{F}\big(S(T)\big) - \int_t^T Y(s-)r ds + \int_t^T \hat{G}\big(S(s)\big)dJ(s)$$
$$+ \int_t^T \big(n - J(s-)\big)\hat{H}\big(S(s)\big)ds$$
$$- \int_t^T Z_1(s)dW(s) - \int_t^T Z_2(s)dB(s) - \int_t^T U(s)\tilde{N}(ds), \quad 0 \le t \le T. \tag{8.2}$$

The dynamics (8.2) is formulated under the real-world probability measure $\mathbb{P}$. If we took the expectation (8.1) under an equivalent probability measure $\mathbb{Q}$, then the BSDE (8.2) would be formulated under $\mathbb{Q}$. Introducing the process $V(t) = U(t) - \hat{G}(S(t))$ and recalling the forward dynamics of S and λ, we can investigate the FBSDE

$$
\begin{aligned}
Y(t) &= \big(n - J(T)\big)\hat{F}\big(S(T)\big) + \int_t^T \big(-Y(s-)r + \hat{G}\big(S(s)\big)\big(n - J(s-)\big)\lambda(s) \\
&\quad + \big(n - J(s-)\big)\hat{H}\big(S(s)\big)\big)ds \\
&\quad - \int_t^T Z_1(s)dW(s) - \int_t^T Z_2(s)dB(s) - \int_t^T V(s)\tilde{N}(ds), \quad 0 \le t \le T, \\
S(t) &= S(0) + \int_0^t \mu\big(S(s)\big)ds + \int_0^t \sigma\big(S(s)\big)dW(s), \quad 0 \le t \le T, \\
\lambda(t) &= \lambda(0) + \int_0^t \mu^\lambda\big(\lambda(s)\big)ds + \int_0^t \sigma^\lambda\big(\lambda(s)\big)dB(s), \quad 0 \le t \le T.
\end{aligned}
\tag{8.3}
$$

The goal is to characterize the control processes (Z_1, Z_2, V) of the linear BSDE (8.3). We use the Itô's formula.

Proposition 8.1.1 *Consider the linear BSDE* (8.3). *Assume that*

(i) *the jump measure N of the point process J has the compensator $\vartheta(dt, \{1\}) = (n - J(t-))\lambda(t)dt$,*
(ii) *the processes S and λ are positive,*
(iii) *the functions $\mu : (0,\infty) \to [0,\infty)$, $\sigma : (0,\infty) \to [0,\infty)$ and $\mu^\lambda : (0,\infty) \to [0,\infty)$,$\sigma^\lambda : (0,\infty) \to [0,\infty)$ are Lipschitz continuous,*
(iv) *the functions $\hat{F} : (0,\infty) \to [0,\infty)$, $\hat{H} : (0,\infty) \to [0,\infty)$ and $\hat{G} : (0,\infty) \to [0,\infty)$ are Lipschitz continuous.*

We define measurable functions $\hat{h} : [0,T] \times [0,T] \times (0,\infty) \to [0,\infty)$, $\hat{f} : [0,T] \times [0,T] \times (0,\infty) \to [0,\infty)$, $\hat{g} : [0,T] \times [0,T] \times (0,\infty) \to [0,\infty)$, $p : [0,T] \times [0,T] \times (0,\infty) \to (0,\infty)$ and $\hat{p} : [0,T] \times [0,T] \times (0,\infty) \to [0,\infty)$ such that for $0 \le t \le u \le T$ we set

$$
\begin{aligned}
\hat{h}(t,u,s) &= \mathbb{E}\big[e^{-r(u-t))}\hat{H}\big(S(u)\big)|S(t) = s\big], \\
\hat{f}(t,u,s) &= \mathbb{E}\big[e^{-r(u-t)}\hat{F}\big(S(u)\big)|S(t) = s\big], \\
\hat{g}(t,u,s) &= \mathbb{E}\big[e^{-r(u-t)}\hat{G}\big(S(u)\big)|S(t) = s\big], \\
p(t,u,\lambda) &= \mathbb{E}\big[e^{-\int_t^u \lambda(v)dv}|\lambda(t) = \lambda\big], \\
\hat{p}(t,u,\lambda) &= \mathbb{E}\big[e^{-\int_t^u \lambda(v)dv}\lambda(u)|\lambda(t) = \lambda\big],
\end{aligned}
\tag{8.4}
$$

and for $0 \le u < t \le T$ we set $\hat{h} = \hat{f} = \hat{g} = p = \hat{p} = 0$. We further assume that

(v) *for each $0 \le u \le T$ the functions $\hat{h}(.,u,.)$, $\hat{f}(.,u,.)$, $\hat{g}(.,u,.)$, $p(.,u,.)$ and $\hat{p}(.,u,.)$ are of the class $\mathscr{C}([0,u] \times (0,\infty)) \cap \mathscr{C}^{1,2}([0,u) \times (0,\infty))$.*

The control processes $(Z_1, Z_2, V) \in \mathbb{H}^2(\mathbb{R}) \times \mathbb{H}^2(\mathbb{R}) \times \mathbb{H}^2_N(\mathbb{R})$ of the BSDE (8.3) *take the form*

$$\begin{aligned}
&Z_1(t)\\
&\quad = \big(n-J(t-)\big)\sigma\big(S(t)\big)p\big(t,T,\lambda(t)\big)\hat{f}_s\big(t,T,S(t)\big)\\
&\qquad + \big(n-J(t-)\big)\sigma\big(S(t)\big)\\
&\qquad \cdot \int_t^T \big(p\big(t,u,\lambda(t)\big)\hat{h}_s\big(t,u,S(t)\big) + \hat{p}\big(t,u,\lambda(t)\big)\hat{g}_s\big(t,u,S(t)\big)\big)du, \quad 0\le t\le T,\\
&Z_2(t)\\
&\quad = \big(n-J(t-)\big)\sigma^{\lambda}\big(\lambda(t)\big)p_{\lambda}\big(t,T,\lambda(t)\big)\hat{f}\big(t,T,S(t)\big)\\
&\qquad + \big(n-J(t-)\big)\sigma^{\lambda}\big(\lambda(t)\big)\\
&\qquad \cdot \int_t^T \big(p_{\lambda}\big(t,u,\lambda(t)\big)\hat{h}\big(t,u,S(t)\big) + \hat{p}_{\lambda}\big(t,u,\lambda(t)\big)\hat{g}\big(t,u,S(t)\big)\big)du, \quad 0\le t\le T,\\
&V(t)\\
&\quad = -p\big(t,T,\lambda(t)\big)\hat{f}\big(t,T,S(t)\big)\\
&\qquad - \int_t^T \big(p\big(t,u,\lambda(t)\big)\hat{h}\big(t,u,S(t)\big) + \hat{p}\big(t,u,\lambda(t)\big)\hat{g}\big(t,u,S(t)\big)\big)du, \quad 0\le t\le T,
\end{aligned}$$

where $\hat{f}_s(t,u,s)=\frac{\partial}{\partial s}\hat{f}(t,u,s)$, $\hat{h}_s(t,u,s)=\frac{\partial}{\partial s}\hat{h}(t,u,s)$, $\hat{g}_s(t,u,s)=\frac{\partial}{\partial s}\hat{g}(t,u,s)$, $p_{\lambda}(t,u,\lambda)=\frac{\partial}{\partial \lambda}p(t,u,\lambda)$, $\hat{p}_{\lambda}(t,u,\lambda)=\frac{\partial}{\partial \lambda}\hat{p}(t,u,\lambda)$.

Proof By Theorems 3.1.1 and 4.1.1 there exists a unique solution (Y,Z_1,Z_2,V,S,λ) to (8.3) and the processes S and λ have finite moments. From Propositions 3.3.1 and 3.4.1 we can deduce that the control processes (Z_1,Z_2,V) are obtained from the predictable representation of the martingale

$$\begin{aligned}
\mathscr{M}(t) = \mathbb{E}\bigg[& e^{-rT}\big(n-J(T)\big)\hat{F}\big(S(T)\big) + \int_0^T e^{-ru}\hat{G}\big(S(u)\big)\big(n-J(u)\big)\lambda(u)du\\
&+ \int_0^T e^{-ru}\big(n-J(u)\big)\hat{H}\big(S(u)\big)du|\mathscr{F}_t\bigg], \quad 0\le t\le T.
\end{aligned} \tag{8.5}$$

Let us consider the first martingale

$$\mathscr{M}^1(t) = \mathbb{E}\big[e^{-rT}\big(n-J(T)\big)\hat{F}\big(S(T)\big)|\mathscr{F}_t\big], \quad 0\le t\le T.$$

We can derive

$$\begin{aligned}
&\mathbb{E}\big[e^{-rT}\big(n-J(T)\big)\hat{F}\big(S(T)\big)|\mathscr{F}_t\big]\\
&\quad = \mathbb{E}\bigg[\sum_{i=1}^n \mathbf{1}\{\tau_i > T\}e^{-rT}\hat{F}\big(S(T)\big)|\mathscr{F}_t\bigg]
\end{aligned}$$

$$= \sum_{i=1}^{n} \mathbf{1}\{\tau_i > t\}\mathbb{E}\big[\mathbf{1}\{\tau_i > T, \tau_i > t\}e^{-rT}\hat{F}\big(S(T)\big)|\mathscr{F}_t\big]$$

$$= \sum_{i=1}^{n} \mathbf{1}\{\tau_i > t\}\mathbb{E}\big[e^{-rT}\hat{F}\big(S(T)\big)\mathbb{E}\big[\mathbf{1}\{\tau_i > T, \tau_i > t\}|\mathscr{F}_T^W \vee \mathscr{F}_T^B \vee \mathscr{F}_t\big]|\mathscr{F}_t\big]$$

$$= \big(n - J(t)\big)\mathbb{E}\big[e^{-\int_t^T \lambda(s)ds}|\mathscr{F}_t\big]\mathbb{E}\big[e^{-rT}\hat{F}\big(S(T)\big)|\mathscr{F}_t\big], \quad 0 \le t \le T, \tag{8.6}$$

where we use representation (7.4), the property of conditional expectations, the exponential conditional distribution of τ_i and the independence of W and B. We get

$$\mathscr{M}^1(t) = e^{-rt}\big(n - J(t)\big)\hat{f}\big(t, T, S(t)\big)p\big(t, T, \lambda(t)\big), \quad 0 \le t \le T.$$

The Lipschitz property of $\hat{F}$, the moment estimate (4.2) for S and square integrability of S imply that $\mathscr{M}^1$ is square integrable. Hence, there exists a unique predictable representation of the martingale $\mathscr{M}^1$ in $\mathbb{H}^2(\mathbb{R}) \times \mathbb{H}^2(\mathbb{R}) \times \mathbb{H}^2_N(\mathbb{R})$ and the martingale $\mathscr{M}^1$ can be represented as a sum of three square integrable stochastic integrals driven by W, B and $\tilde{N}$. Since $\hat{f}, p \in \mathscr{C}^{1,2}([0,T) \times (0,\infty))$, we can apply the Itô's formula and we immediately get the dynamics

$$\begin{aligned}
&\mathscr{M}^1(t) \\
&= \mathscr{M}^1(0) + \int_0^t e^{-ru}\big(n - J(u-)\big)\hat{f}_s\big(u, T, S(u)\big)p\big(u, T, \lambda(u)\big)\sigma\big(S(u)\big)dW(u) \\
&\quad + \int_0^t e^{-ru}\big(n - J(u-)\big)\hat{f}\big(u, T, S(u)\big)p_\lambda\big(u, T, \lambda(u)\big)\sigma^\lambda\big(\lambda(u)\big)dB(u) \\
&\quad - \int_0^t e^{-ru}\hat{f}\big(u, T, S(u)\big)p\big(u, T, \lambda(u)\big)\tilde{N}(du), \quad 0 \le t < T.
\end{aligned} \tag{8.7}$$

We remark that the Lebesque integral in the Itô's formula must vanish by the martingale property. Since $\hat{f}, p \in \mathscr{C}([0,T] \times (0,\infty))$ and $\Delta J(T) = 0$ $a.s.$ by quasi-left continuity, then $\lim_{t\to T-}\mathscr{M}^1(t) = \mathscr{M}^1(T)$ $a.s.$, and in $\mathbb{L}^2(\mathbb{R})$ by the dominated convergence theorem. By the growth properties of $\hat{f}_s$, p, $\hat{f}$, p_λ, which can be deduced from (4.2) and (8.4), the three stochastic integrals in (8.7) are well-defined square integrable martingales on $[0,T]$. Hence, we take the limit in $\mathbb{L}^2(\mathbb{R})$ and we derive

$$\begin{aligned}
&\mathscr{M}^1(t) \\
&= \mathscr{M}^1(0) + \int_0^t e^{-ru}\big(n - J(u-)\big)\hat{f}_s\big(u, T, S(u)\big)p\big(u, T, \lambda(u)\big)\sigma\big(S(u)\big)dW(u) \\
&\quad + \int_0^t e^{-ru}\big(n - J(u-)\big)\hat{f}\big(u, T, S(u)\big)p_\lambda\big(u, T, \lambda(u)\big)\sigma^\lambda\big(\lambda(u)\big)dB(u) \\
&\quad - \int_0^t e^{-ru}\hat{f}\big(u, T, S(u)\big)p\big(u, T, \lambda(u)\big)\tilde{N}(du), \quad 0 \le t \le T.
\end{aligned} \tag{8.8}$$

We deal with the second martingale

$$\mathscr{M}^2(t) = \mathbb{E}\Big[\int_0^T e^{-ru}\big(n - J(u)\big)\hat{H}\big(S(u)\big)du|\mathscr{F}_t\Big], \quad 0 \le t \le T.$$

Fix $u \in [0, T]$. We introduce the process

$$\mathscr{M}^{2,u}(t) = \mathbb{E}\big[e^{-ru}\big(n - J(u)\big)\hat{H}\big(S(u)\big)|\mathscr{F}_t\big], \quad 0 \le t \le u.$$

Following the arguments which led to (8.8), we can obtain

$$\mathscr{M}^{2,u}(t) = e^{-rt}\big(n - J(t)\big)\hat{h}\big(t, u, S(t)\big)p\big(t, u, \lambda(t)\big), \quad 0 \le t \le u,$$

together with the representation

$$\begin{aligned}
&\mathscr{M}^{2,u}(t)\\
&= \mathscr{M}^{2,u}(0) + \int_0^t e^{-rv}\big(n - J(v-)\big)\hat{h}_s\big(v, u, S(v)\big)p\big(v, u, \lambda(v)\big)\sigma\big(S(v)\big)dW(v)\\
&\quad + \int_0^t e^{-rv}\big(n - J(v-)\big)\hat{h}\big(v, u, S(v)\big)p_\lambda\big(v, u, \lambda(v)\big)\sigma^\lambda\big(\lambda(v)\big)dB(v)\\
&\quad - \int_0^t e^{-rv}\hat{h}\big(v, u, S(v)\big)p\big(v, u, \lambda(v)\big)\tilde{N}(dv)\\
&= \mathscr{M}^{2,u}(0) + \int_0^t \mathscr{Z}_1^u(v)dW(v)\\
&\quad + \int_0^t \mathscr{Z}_2^u(v)dB(v) + \int_0^t \mathscr{V}^u(v)\tilde{N}(dv), \quad 0 \le t \le u,
\end{aligned} \tag{8.9}$$

where we introduce the processes $\mathscr{Z}_1^u$, $\mathscr{Z}_2^u$ and $\mathscr{V}^u$. If we apply the a priori estimate (3.4) to the BSDE (8.9) and we use the Lipschitz property of $\hat{H}$ and the moment estimate (4.2) for S, then we can derive

$$\begin{aligned}
&\mathbb{E}\Big[\int_0^u \big|\mathscr{Z}_1^u(v)\big|^2 dv + \int_0^u \big|\mathscr{Z}_2^u(v)\big|^2 dv + \int_0^u \big|\mathscr{V}^u(v)\big|^2\big(n - J(v)\big)\lambda(v)dv\Big]\\
&\quad \le K\mathbb{E}\big[\big|e^{-ru}\big(n - J(u)\big)\hat{H}\big(S(u)\big)\big|^2\big] \le K\Big(1 + \mathbb{E}\Big[\sup_{v\in[0,T]}\big|S(v)\big|^2\Big]\Big) \le K,
\end{aligned} \tag{8.10}$$

with K independent of u. Estimate (8.10) also yields

$$\begin{aligned}
&\int_0^T\int_0^T \big|\mathscr{Z}_1^u(v)\big|^2\mathbf{1}\{v \le u\}dvdu + \int_0^T\int_0^T \big|\mathscr{Z}_2^u(v)\big|^2\mathbf{1}\{v \le u\}dvdu\\
&\quad + \int_0^T\int_0^T \big|\mathscr{V}^u(v)\big|^2\mathbf{1}\{v \le u\}\big(n - J(v)\big)\lambda(v)dvdu < \infty.
\end{aligned} \tag{8.11}$$

We now obtain the representation of the second martingale $\mathscr{M}^2$. Using (8.9), we get

$$\begin{aligned}
&\mathscr{M}^2(t) \\
&= \int_0^t e^{-ru}\big(n - J(u)\big)\hat{H}\big(S(u)\big)du + \mathbb{E}\Big[\int_t^T e^{-ru}\big(n - J(u)\big)\hat{H}\big(S(u)\big)du|\mathscr{F}_t\Big] \\
&= \int_0^t \Big(\mathscr{M}^{2,u}(0) + \int_0^t \mathscr{Z}_1^u(v)\mathbf{1}\{v \le u\}dW(v) + \int_0^t \mathscr{Z}_2^u(v)\mathbf{1}\{v \le u\}dB(v) \\
&\quad + \int_0^t \mathscr{V}^u(v)\mathbf{1}\{v \le u\}\tilde{N}(dv)\Big)du \\
&\quad + \int_t^T \Big(\mathscr{M}^{2,u}(0) + \int_0^t \mathscr{Z}_1^u(v)\mathbf{1}\{v \le u\}dW(v) + \int_0^t \mathscr{Z}_2^u(v)\mathbf{1}\{v \le u\}dB(v) \\
&\quad + \int_0^t \mathscr{V}^u(v)\mathbf{1}\{v \le u\}\tilde{N}(dv)\Big)du \\
&= \int_0^T \mathscr{M}^{2,u}(0)du + \int_0^T \Big(\int_0^t \mathscr{Z}_1^u(v)\mathbf{1}\{v \le u\}dW(v)\Big)du \\
&\quad + \int_0^T \Big(\int_0^t \mathscr{Z}_2^u(v)\mathbf{1}\{v \le u\}dB(v)\Big)du \\
&\quad + \int_0^T \Big(\int_0^t \mathscr{V}^u(v)\mathbf{1}\{v \le u\}\tilde{N}(dv)\Big)du, \quad 0 \le t \le T. \qquad (8.12)
\end{aligned}$$

By the measurability assumptions and property (8.11) we can apply the Fubini's theorem for stochastic integrals, see Theorem IV.65 in Protter (2004). We change the order of integration in (8.12) and we derive the representation

$$\begin{aligned}
\mathscr{M}^2(t) &= \mathscr{M}^2(0) + \int_0^t \Big(\int_v^T \mathscr{Z}_1^u(v)du\Big)dW(v) \\
&\quad + \int_0^t \Big(\int_v^T \mathscr{Z}_2^u(v)du\Big)dB(v) + \int_0^t \Big(\int_v^T \mathscr{V}^u(v)du\Big)\tilde{N}(dv), \quad 0 \le t \le T.
\end{aligned} \qquad (8.13)$$

The representation of the third martingale

$$\mathscr{M}^3(t) = \mathbb{E}\Big[\int_0^T e^{-ru}\hat{G}\big(S(u)\big)\big(n - J(u)\big)\lambda(u)du|\mathscr{F}_t\Big], \quad 0 \le t \le T,$$

is obtained analogously to (8.13). From the representation of the martingale $\mathscr{M}$ and Propositions 3.3.1 and 3.4.1 we deduce the formulas for the control processes (Z_1, Z_2, V). □

The assumptions of Proposition 8.1.1 should hold in many cases. We remark that the Lipschitz continuity assumptions of Proposition 8.1.1 can be relaxed, but additional existence and moment assumptions for S and λ would have to be introduced instead. The assumption that the functions $\hat{f}$, $\hat{h}$, $\hat{g}$, $\hat{p}$ and p are of the class $\mathscr{C}([0,T]\times(0,\infty))\cap\mathscr{C}^{1,2}([0,T)\times(0,\infty))$ is mathematically convenient. In fact, this is the key assumption which allows us to apply the Itô's formula and, consequently, to find the predictable representation and the control processes of the linear BSDE. We recall that prices of call options, put options and survival probabilities in classical financial and actuarial models are sufficiently smooth under appropriate conditions, see Sect. 12.1 in Cont and Tankov (2004) and Chap. 5 in Filipovic (2009). However, the smoothness of prices is not guaranteed and it may require very strong assumptions.

The representation of the control processes from Proposition 8.1.1 involve derivatives. If we deal with the classical Black-Scholes model and we assume that the mortality intensity follows the Cox-Ingersoll-Ross process, then $\hat{f}_s$, $\hat{h}_s$, $\hat{g}_s$, $\hat{p}_\lambda$ and p_λ have closed form solutions, see Chap. 5 in Shreve (2004) and Chap. 5 in Filipovic (2009). If we consider a more general model, then we can use Proposition 4.1.2 to calculate (or estimate) the derivatives.

8.2 The Application of the Malliavin Calculus

We investigate the financial model

$$\begin{aligned} dS_0(t) &= S_0(t)rdt, \quad S_0(0)=1,\\ dS(t) &= \mu\big(S(t)\big)dt+\sigma\big(S(t)\big)dW(t), \quad S(0)=s>0, \end{aligned}$$

and the equity-linked (non-life insurance) liabilities

$$P(t)=\int_0^t\int_{\mathbb{R}}\hat{G}\big(S(s),z\big)N(ds,dz), \quad 0\le t\le T,$$

where the random measure N is generated by a compound Poisson process with intensity λ and jump size distribution q. For an actuarial and financial motivation we refer to Examples 7.9 and 7.10.

The value of the insurance liabilities is given by

$$Y(t)=\mathbb{E}\bigg[\int_t^T\int_{\mathbb{R}}e^{-r(s-t)}\hat{G}\big(S(s),z\big)N(ds,dz)|\mathscr{F}_t\bigg], \quad 0\le t\le T.$$

Following the reasoning that led to (8.3), we consider the FBSDE

$$\begin{aligned} Y(t) &= \int_t^T\bigg(-Y(s-)r+\int_{\mathbb{R}}\hat{G}\big(S(s),z\big)\lambda q(dz)\bigg)ds\\ &\quad -\int_t^T Z(s)dW(s)-\int_t^T\int_{\mathbb{R}}V(s,z)\tilde{N}(ds,dz), \quad 0\le t\le T, \quad (8.14)\\ S(t) &= S(0)+\int_0^t\mu\big(S(s)\big)ds+\int_0^t\sigma\big(S(s)\big)dW(s), \quad 0\le t\le T. \end{aligned}$$

Using the Malliavin calculus and the results for FBSDEs from Chap. 4, we show how to derive the control processes (Z, V) without applying the Itô's formula. The Malliavin calculus allows us to relax smoothness assumptions in our models.

Proposition 8.2.1 *Consider the linear BSDE* (8.14). *Assume that*

(i) *the random measure N is generated by a compound Poisson process with intensity λ and jump size distribution q,*
(ii) *the process S is positive,*
(iii) *the functions $\mu : (0, \infty) \to [0, \infty)$ and $\sigma : (0, \infty) \to (0, \infty)$ are twice continuously differentiable with bounded derivatives,*
(iv) *the function $\hat{G} : (0, \infty) \times \mathbb{R} \to [0, \infty)$ is measurable, Lipschitz continuous in the sense that*

$$\big|\hat{G}(s, z) - \hat{G}\big(s', z\big)\big| \leq K\big|s - s'\big||z|, \quad (s, z), \big(s', z\big) \in [0, T] \times \mathbb{R},$$

and satisfies the growth condition

$$\hat{G}(0, z) \leq K|z|, \quad z \in \mathbb{R}.$$

The control processes $(Z, V) \in \mathbb{H}^2(\mathbb{R}) \times \mathbb{H}^2_N(\mathbb{R})$ of the BSDE (8.14) *take the form*

$$Z(t) = \sigma\big(S(t)\big)\mathbb{E}\bigg[\int_t^T \int_{\mathbb{R}} e^{-r(u-t)}\hat{G}_s\big(S(u), z\big)\frac{\mathscr{Y}(u)}{\mathscr{Y}(t)}\lambda q(dz)du|\mathscr{F}_t\bigg], \quad 0 \leq t \leq T,$$

$$V(t, z) = 0, \quad 0 \leq t \leq T, \ z \in \mathbb{R},$$

where

$$\mathscr{Y}(t) = 1 + \int_0^t \mathscr{Y}(u)\mu'\big(S(u)\big)du + \int_0^t \mathscr{Y}(u)\sigma'\big(S(u)\big)dW(u), \quad 0 \leq t \leq T,$$

and $\hat{G}_s(s, z) = \frac{\partial}{\partial s}\hat{G}(s, z)$.

Proof By Theorems 3.1.1 and 4.1.1 there exists a unique solution (Y, Z, V, S) to (8.14). Since we deal with the Markovian dynamics, we have $Y(t) = u(t, S(t))$ where

$$\begin{aligned} u(t, s) &= \mathbb{E}\bigg[\int_t^T \int_{\mathbb{R}} e^{-r(u-t)}\hat{G}\big(S^{t,s}(u), z\big)\lambda q(dz)du\bigg] \\ &= \int_t^T \int_{\mathbb{R}} e^{-r(u-t)}\mathbb{E}\big[\hat{G}\big(S^{t,s}(u), z\big)\big]\lambda q(dz)du, \quad (t, s) \in [0, T] \times (0, \infty). \end{aligned}$$

By Theorem 4.1.2 the law of S is absolutely continuous. From Theorem 4.1.2 we conclude that S is Malliavin differentiable and, next, from Propositions 2.6.4–2.6.5 we deduce that $\int_{\mathbb{R}} \hat{G}(S(s), z)\lambda q(dz)$ is Malliavin differentiable for $0 \leq s \leq T$. Hence, Theorem 4.1.4 yields the control processes

$$Z(t) = u_s\big(t, S(t)\big)\sigma\big(S(t)\big), \quad V(t, z) = 0, \quad 0 \leq t \leq T, \ z \in \mathbb{R}.$$

By the Lipschitz property of $\hat{G}$ and the moment estimates (4.2) we obtain the inequalities

$$\mathbb{E}\left[\left|\frac{\hat{G}(S^{t,s}(u),z)-\hat{G}(S^{t,s'}(u),z)}{s-s'}\right|\right]$$
$$\leq K\frac{\mathbb{E}[|S^{t,s}(u)-S^{t,s'}(u)|]}{|s-s'|}|z|\leq K\frac{\sqrt{\mathbb{E}[|S^{t,s}(u)-S^{t,s'}(u)|^2]}}{|s-s'|}|z|\leq K|z|, \quad (8.15)$$

and

$$\mathbb{E}\left[\left|\frac{\hat{G}(S^{t,s}(u),z)-\hat{G}(S^{t,s'}(u),z)}{s-s'}\right|^2\right]\leq K|z|^2, \quad (8.16)$$

for all $(t,u,s,s',z)\in[0,T]\times[0,t]\times(0,\infty)\times(0,\infty)\times\mathbb{R}$. From (8.16) we deduce that the family

$$A_{s'}=\left|\frac{\hat{G}(S^{t,s}(u),z)-\hat{G}(S^{t,s'}(u),z)}{s-s'}\right|, \quad s'\in(0,\infty),$$

is uniformly integrable for fixed $(t,u,s,z)\in[0,T]\times[0,T]\times(0,\infty)\times\mathbb{R}$, see the proof of Proposition 4.1.2. By (8.15), the dominated convergence theorem, the uniform integrability of $A_{s'}$ and Fubini's theorem we finally derive

$$\begin{aligned}
u_s(t,s)&=\lim_{s'\to s}\frac{u(t,s)-u(t,s')}{s-s'}\\
&=\lim_{s'\to s}\int_t^T\int_{\mathbb{R}}e^{-r(t-u)}\mathbb{E}\left[\frac{\hat{G}(S^{t,s}(u),z)-\hat{G}(S^{t,s'}(u),z)}{s-s'}\right]\lambda q(dz)du\\
&=\mathbb{E}\left[\int_t^T\int_{\mathbb{R}}e^{-r(t-u)}\hat{G}_s\big(S^{t,s}(u),z\big)\mathscr{Y}^{t,s}(u)\lambda q(dz)du\right],\\
&\quad (t,s)\in[0,T]\times(0,\infty).
\end{aligned}$$

where

$$\begin{aligned}
\mathscr{Y}^{t,s}(u)=1&+\int_t^u\mathscr{Y}^{t,s}(v)\mu'\big(S^{t,s}(v)\big)dv\\
&+\int_t^u\mathscr{Y}^{t,s}(v)\sigma'\big(S^{t,s}(v)\big)dW(v), \quad 0\leq t\leq u\leq T,
\end{aligned}$$

and we use the fact that $s\mapsto\hat{G}(S^{t,s}(u),z)$ is a.s. differentiable for $z\in\mathbb{R}$, $u\in(t,T]$. □

The proof of Proposition 8.2.1 relies on Theorem 4.1.4 which we established for a FBSDE driven by a Brownian motion and a compensated Poisson random measure.

It is interesting to note that Theorem 4.1.4 can be applied to BSDEs driven by more general random measures provided that appropriate modifications are introduced.

Let us investigate the financial model

$$dS_0(t) = S_0(t)r dt, \quad S_0(0) = 1,$$
$$dS(t) = \mu\big(S(t)\big)dt + \sigma\big(S(t)\big)dW(t), \quad S(0) = s > 0,$$

and the equity-linked (non-life insurance) liabilities

$$P(t) = \int_0^t \int_{\mathbb{R}} \hat{G}\big(S(s), z\big)N(ds, dz), \quad 0 \le t \le T,$$

where the random measure N is generated by a compound Cox process with compensator $\vartheta(dt, dz) = \lambda(S(t))q(dz)dt$. Since the intensity λ is contingent on the stock, the systematic claims risk is now considered. For an actuarial and financial motivation we again refer to Examples 7.9 and 7.10. We show that we can replace the random measure N generated by a compound Cox process with a Poisson random measure.

Let us assume that the intensity λ is bounded and set $E = \{y : 0 \le y \le \sup_{s>0} \lambda(s)\}$. Let N^p denote a Poisson random measure on $\Omega \times \mathscr{B}([0, T]) \times \mathscr{B}(\mathbb{R}) \times \mathscr{B}(E)$. We assume that the compensator of the Poisson random measure N^p is $\vartheta(dt, dz, dy) = q(dz)dydt$. We introduce the random measure

$$\mathscr{N}(dt, dz) = \int_E \mathbf{1}_{[0,\lambda(S(t))]}(y)N^p(dt, dz, dy). \tag{8.17}$$

We can notice that

$$\int_0^t \int_{\mathbb{R}} \mathbf{1}_A(\omega, s, z)\mathscr{N}(ds, dz) - \int_0^t \int_{\mathbb{R}} \mathbf{1}_A(\omega, s, z)\lambda\big(S(s)\big)q(dz)ds$$
$$= \int_0^t \int_{\mathbb{R}\times E} \mathbf{1}_A(\omega, s, z)\mathbf{1}_{[0,\lambda(S(s))]}(y)\tilde{N}^p(ds, dz, dy), \quad 0 \le t \le T,$$

is a martingale for $A \in \mathscr{P} \otimes \mathscr{B}(\mathbb{R})$ by Theorem 2.3.3. Hence, by Definition 2.1.4 the random measure $\mathscr{N}$ has the compensator $\lambda(S(t))q(dz)dt$. From Theorem 11.5 in He et al. (1992) we conclude that the measures N and $\mathscr{N}$ are indistinguishable. Consequently, we can use the results derived for BSDEs driven by Poisson random measures also in the case when we deal with random measures generated by Cox processes.

The value of the insurance liability is given by

$$Y(t) = \mathbb{E}\bigg[\int_t^T \int_{\mathbb{R}} e^{-r(s-t)}\hat{G}\big(S(s), z\big)N(ds, dz)|\mathscr{F}_t\bigg]$$
$$= \mathbb{E}\bigg[\int_t^T \int_{\mathbb{R}\times E} e^{-r(s-t)}\hat{G}\big(S(s), z\big)\mathbf{1}_{[0,\lambda(S(s))]}(y)N^p(ds, dz, dy)|\mathscr{F}_t\bigg],$$
$$0 \le t \le T,$$

and we investigate the FBSDE

$$\begin{aligned} Y(t) &= \int_t^T \left(-Y(s-)r + \int_{\mathbb{R}\times E} \hat{G}\big(S(s), z\big)\mathbf{1}_{[0,\lambda(S(s))]}(y) q(dz)dy \right) ds \\ &\quad - \int_t^T Z(s)dW(s) - \int_t^T \int_{\mathbb{R}\times E} V(s,z,y)\tilde{N}^P(ds,dz,dy), \quad 0 \le t \le T, \\ S(t) &= S(0) + \int_0^t \mu\big(S(s)\big)ds + \int_0^t \sigma\big(S(s)\big)dW(s), \quad 0 \le t \le T. \end{aligned} \tag{8.18}$$

Proposition 8.2.2 *Consider the linear BSDE* (8.18). *Assume that*

(i) *the process S is positive,*
(ii) *the function $\lambda : (0,\infty) \to E$ is positive and Lipschitz continuous, and the set E is bounded,*
(iii) *the random measure N^P is defined on $\Omega \times \mathscr{B}([0,T]) \times \mathscr{B}(\mathbb{R}) \times \mathscr{B}(E)$ and it is a Poisson random measure with the compensator $\vartheta(dt,dz,dy) = q(dz)dydt$ such that $q(\mathbb{R}) = 1$ and $\int_{\mathbb{R}} |z|^2 q(dz) < \infty$,*
(iv) *the functions $\mu : (0,\infty) \to [0,\infty)$ and $\sigma : (0,\infty) \to (0,\infty)$ are twice continuously differentiable with bounded derivatives,*
(v) *the function $\hat{G} : (0,\infty) \times \mathbb{R} \to [0,\infty)$ is measurable, Lipschitz continuous in the sense that*

$$\big|\hat{G}(s,z) - \hat{G}\big(s',z\big)\big| \le K\big|s-s'\big||z|, \quad (s,z), \big(s',z\big) \in (0,\infty)\times\mathbb{R},$$

and satisfies the growth condition

$$\hat{G}(s,z) \le K|z|, \quad (s,z) \in (0,\infty)\times\mathbb{R}.$$

The control processes $(Z,V) \in \mathbb{H}^2(\mathbb{R}) \times \mathbb{H}^2_{N^P}(\mathbb{R})$ of the BSDE (8.18) *take the form*

$$\begin{aligned} Z(t) &= \sigma\big(S(t)\big)\mathbb{E}\left[\int_t^T \int_{\mathbb{R}} e^{-r(u-t)}\hat{G}_s\big(S(u),z\big)\frac{\mathscr{Y}(u)}{\mathscr{Y}(t)}\lambda\big(S(u)\big)q(dz)du \Big| \mathscr{F}_t\right] \\ &\quad + \sigma\big(S(t)\big)\mathbb{E}\left[\int_t^T \int_{\mathbb{R}} e^{-r(u-t)}\hat{G}\big(S(u),z\big)\frac{\mathscr{Y}(u)}{\mathscr{Y}(t)}\lambda'\big(S(u)\big)q(dz)du \Big| \mathscr{F}_t\right], \\ &\quad 0 \le t \le T, \\ V(t,z,y) &= 0, \quad 0 \le t \le T,\ z \in \mathbb{R},\ y \in E, \end{aligned}$$

where

$$\mathscr{Y}(t) = 1 + \int_0^t \mathscr{Y}(u)\mu'\big(S(u)\big)du + \int_0^t \mathscr{Y}(u)\sigma'\big(S(u)\big)dW(u), \quad 0 \le t \le T,$$

and $\hat{G}_s(s,z) = \frac{\partial}{\partial s}\hat{G}(s,z)$.

Proof We cannot apply Proposition 8.2.1 since the indicator function in the generator of (8.18) is not smooth enough. Let $E=\{y:0\le y\le c\}$ where $c=\sup_{s>0}\lambda(s)$. Choose a smooth function $\varphi\in\mathscr{C}^2(\mathbb{R})$ such that $\varphi(x)=1$, $x\le 0$, and $\varphi(x)=0$, $x\ge 1$. It is easy to show

$$\lim_{n\to\infty}\varphi\big(n\big(y-\lambda(s)\big)\big)=\mathbf{1}_{[0,\lambda(s)]}(y),\quad (y,s)\in E\times(0,\infty). \tag{8.19}$$

By Theorems 3.1.1 and 4.1.1 there exists a unique solution (Y,Z,V,S) to (8.18), and by Theorem 4.1.2 the law of S is absolutely continuous with respect to the Lebesgue measure. Consider the FBSDE

$$\begin{aligned}
Y^n(t)={}&\int_t^T\Big(-Y^n(s-)r+\int_{\mathbb{R}\times E}\hat{G}\big(S(s),z\big)\varphi\big(n\big(y-\lambda\big(S(s)\big)\big)\big)q(dz)dy\Big)ds\\
&-\int_t^T Z^n(s)dW(s)-\int_t^T\int_{\mathbb{R}\times E}V^n(s,z,y)\tilde{N}^P(ds,dz,dy),\\
&0\le t\le T,
\end{aligned} \tag{8.20}$$

$$S(t)=S(0)+\int_0^t\mu\big(S(s)\big)ds+\int_0^t\sigma\big(S(s)\big)dW(s),\quad 0\le t\le T.$$

By Theorems 3.1.1 and 4.1.1 there also exists a unique solution (Y^n,Z^n,V^n,S) to (8.20). The a priori estimates (3.4) and (3.6) together with the Cauchy-Schwarz inequality yield

$$\begin{aligned}
&\|Y-Y^n\|^2_{\mathbb{S}^2}+\|Z-Z^n\|^2_{\mathbb{H}^2}+\|V-V^n\|^2_{\mathbb{H}^2_{NP}}\\
&\le K\mathbb{E}\Big[\int_0^T\Big|\int_{\mathbb{R}\times E}\hat{G}\big(S(u),z\big)\mathbf{1}_{[0,\lambda(S(u))]}(y)q(dz)dy\\
&\quad-\int_{\mathbb{R}\times E}\hat{G}\big(S(u),z\big)\varphi\big(n\big(y-\lambda\big(S(u)\big)\big)\big)q(dz)dy\Big|^2du\Big]\\
&\le K\mathbb{E}\Big[\int_0^T\int_{\mathbb{R}\times E}\big|\hat{G}\big(S(u),z\big)\big|^2\big|\mathbf{1}_{[0,\lambda(S(u))]}(y)-\varphi\big(n\big(y-\lambda\big(S(u)\big)\big)\big)\big|^2q(dz)dydu\Big],
\end{aligned}$$

and by the dominated convergence theorem and (8.19) we get

$$\lim_{n\to\infty}\big\{\|Y-Y^n\|^2_{\mathbb{S}^2}+\|Z-Z^n\|^2_{\mathbb{H}^2}+\|V-V^n\|^2_{\mathbb{H}^2_{NP}}\big\}=0.$$

In order to find the solution (Z,V), we find the solution (Z^n,V^n) to the BSDE (8.20) and we take the limit. We can now apply Proposition 8.2.1 to (8.20). We obtain the control processes

$$\begin{aligned}
&Z^n(t)\\
&\quad=\sigma\big(S(t)\big)\mathbb{E}\bigg[\int_t^T\int_{\mathbb{R}\times E}e^{-r(u-t)}\hat{G}_s\big(S(u),z\big)\\
&\qquad\cdot\frac{\mathcal{Y}(u)}{\mathcal{Y}(t)}\varphi\big(n\big(y-\lambda\big(S(u)\big)\big)\big)q(dz)dydu|\mathcal{F}_t\bigg]\\
&\qquad+\sigma\big(S(t)\big)\mathbb{E}\bigg[\int_t^T\int_{\mathbb{R}\times E}e^{-r(u-t)}\hat{G}\big(S(u),z\big)\\
&\qquad\cdot\big(-\varphi'\big(n\big(y-\lambda\big(S(u)\big)\big)\big)n\lambda'\big(S(u)\big)\big)\frac{\mathcal{Y}(u)}{\mathcal{Y}(t)}q(dz)dydu|\mathcal{F}_t\bigg],\quad 0\leq t\leq T,\\
&V^n(t,z,y)=0,\quad 0\leq t\leq T,\ z\in\mathbb{R},\ y\in E.
\end{aligned}\tag{8.21}$$

Notice that for a.a. $s>0$, for which the derivative $\lambda'(s)$ exists, we have

$$\begin{aligned}
&-\int_E\varphi'\big(n\big(y-\lambda(s)\big)\big)n\lambda'(s)dy\\
&\quad=-\int_{\lambda(s)}^{(\lambda(s)+\frac{1}{n})\wedge c}\varphi'\big(n\big(y-\lambda(s)\big)\big)ndy\lambda'(s)\\
&\quad=-\int_0^{\frac{1}{n}\wedge(c-\lambda(s))}\varphi'(nv)ndv\lambda'(s)=\big(1-\varphi\big(1\wedge n\big(c-\lambda(s)\big)\big)\big)\lambda'(s),
\end{aligned}\tag{8.22}$$

and

$$\lim_{n\to\infty}\big(1-\varphi\big(1\wedge n\big(c-\lambda(s)\big)\big)\big)\lambda'(s)=\lambda'(s),\tag{8.23}$$

where we use properties of φ. Combining (8.21) with (8.22), we get

$$\begin{aligned}
&Z^n(t)\\
&\quad=\sigma\big(S(t)\big)\mathbb{E}\bigg[\int_t^T\int_{\mathbb{R}\times E}e^{-r(u-t)}\hat{G}_s\big(S(u),z\big)\\
&\qquad\cdot\frac{\mathcal{Y}(u)}{\mathcal{Y}(t)}\varphi\big(n\big(y-\lambda\big(S(u)\big)\big)\big)q(dz)dydu|\mathcal{F}_t\bigg]\\
&\qquad+\sigma\big(S(t)\big)\mathbb{E}\bigg[\int_t^T\int_{\mathbb{R}}e^{-r(u-t)}\hat{G}\big(S(u),z\big)\\
&\qquad\cdot\big(1-\varphi\big(1\wedge n\big(c-\lambda\big(S(u)\big)\big)\big)\big)\lambda'\big(S(u)\big)\frac{\mathcal{Y}(u)}{\mathcal{Y}(t)}q(dz)dydu|\mathcal{F}_t\bigg],\quad 0\leq t\leq T,
\end{aligned}$$

and the limit of Z^n can be established by the dominated convergence theorem and properties (8.19), (8.23). □

The results derived so far are only applicable to Markovian dynamics. The power of the Malliavin calculus lies in the fact that it can be applied in a general non-Markovian setting. Let us investigate the financial model

$$\begin{aligned}
dS_0(t) &= S_0(t)rdt, \quad S_0(0) = 1,\\
dS(t) &= S(t)\big(\mu(t)dt + S(t)\sigma(t)dW(t)\big), \quad S(0) = s > 0,
\end{aligned}$$

and the liabilities

$$P(t) = \int_0^t \int_{\mathbb{R}} G(s,z)N(ds,dz), \quad 0 \le t \le T,$$

where the random measure N is generated by a compound Poisson process with intensity λ and jump size distribution q. The drift and volatility of the stock S and the claim G are non-Markov processes. We consider the BSDE

$$\begin{aligned}
Y(t) = {}& \int_t^T \Big(-Y(s-)r + \int_{\mathbb{R}} G(s,z)\lambda q(dz)\Big)ds\\
& - \int_t^T Z(s)dW(s) - \int_t^T \int_{\mathbb{R}} V(s,z)\tilde{N}(ds,dz), \quad 0 \le t \le T. \qquad (8.24)
\end{aligned}$$

The next proposition shows the benefit of the Malliavin calculus in the theory and applications of BSDEs. Recalling the key results established for the Malliavin derivative and FBSDEs, we can immediately derive the control processes (Z, V) of the BSDE (8.24).

Proposition 8.2.3 *Let us consider the natural filtration $\mathscr{F}$ generated by a Brownian motion and a compound Poisson process with intensity λ and jump size distribution q. We investigate the linear BSDE* (8.24). *Assume that*

(i) (C1)–(C4) *from Chap. 7 hold,*
(ii) *υ-a.e. $(s,z) \in [0,T] \times \mathbb{R}$ the random variable $G(s,z)$ is Malliavin differentiable, and the Malliavin derivative satisfies*

$$\mathbb{E}\Big[\int_0^T \int_0^T \int_{\mathbb{R}} \big|D_u G(s,z)\big|^2 \lambda q(dz)dsdu\Big] < \infty,$$

$$\mathbb{E}\Big[\int_0^T \int_{\mathbb{R}} \int_0^T \int_{\mathbb{R}} \big|D_{u,x} G(s,z)\big|^2 \lambda q(dz)dsq(dx)du\Big] < \infty.$$

The control processes $(Z, V) \in \mathbb{H}^2(\mathbb{R}) \times \mathbb{H}^2_N(\mathbb{R})$ of the BSDE (8.24) *take form*

$$Z(t) = \Big(\mathbb{E}\Big[\int_t^T \int_{\mathbb{R}} e^{-r(u-t)} D_t G(u,z)\lambda q(dz)du \big| \mathscr{F}_t\Big]\Big)^{\mathscr{P}}, \quad 0 \le t \le T,$$

$$V(t,z) = z\Big(\mathbb{E}\Big[\int_t^T \int_{\mathbb{R}} e^{-r(u-t)} D_{t,z} G(u,z)\lambda q(dz)du \big| \mathscr{F}_t\Big]\Big)^{\mathscr{P}}, \quad 0 \le t \le T,\ z \in \mathbb{R},$$

where $(\cdot)^{\mathscr{P}}$ denotes the predictable projection of the process.

Proof The result follows from Theorem 3.5.1, Propositions 3.5.2, 2.6.3 and 2.6.5. □

Proposition 8.2.1 arises now as a special case of Proposition 8.2.3.

Let us remark that in applications we may face a linear BSDE with a generator which also have a linear term in the control processes (Z, U). By Propositions 3.3.1 and 3.4.1 such a linear term may be absorbed by the random noises and the change of measure. We end up with a linear BSDE with a generator independent of (Z, U) with the dynamics under a different probability measure. After the change of measure, we can use the techniques presented in this chapter to derive the control processes.

We have discussed two methods which can be applied to derive the predicable representations of random variables and representations of the control processes of linear BSDEs. The representations involve expectations of state processes. Such representations are useful since we can use Monte Carlo simulations in numerical applications. As the control processes of BSDEs determine hedging strategies, the results of this chapter point out methods which can be used to establish implementable formulas for hedging strategies.

Bibliographical Notes In the proof of Proposition 8.1.1 we closely follow the proof from Delong (2010), see also Møller (2001). The idea of replacing a random measure generated by a Cox process by a Poisson random measure is taken from Ankirchner and Imkeller (2008). In the proof of Proposition 8.2.2 we closely follow the arguments from Ankirchner and Imkeller (2008). For applications of the Malliavin calculus to linear BSDEs driven by Brownian motions we refer to El Karoui et al. (1997b).

Chapter 9
Arbitrage-Free Pricing, Perfect Hedging and Superhedging

Abstract We consider arbitrage-free pricing of assets and liabilities. We start with perfect replication of a terminal financial claim in the Black-Scholes model. Next, we study superhedging strategies for the insurance payment process. Finally, we investigate perfect replication of a stream of life insurance liabilities with a mortality bond. We characterize the arbitrage-free prices and the replicating strategies by linear BSDEs. The superhedging price and the superhedging strategy are characterized as a supersolution to a BSDE.

We give a brief introduction to arbitrage-free pricing of assets and liabilities. We show that BSDEs arise naturally when we deal with hedging problems. In this chapter we focus on perfect replication and superhedging of liabilities.

9.1 Arbitrage-Free Pricing and Market-Consistent Valuation

Traditional actuarial pricing is based on the law of large numbers and the idea of diversification. However, if we deal with equity-linked payments, then diversification arguments cannot be applied. The risk of equity-linked claims can only be mitigated if an asset portfolio is found which is strongly correlated with the claims. The cost of setting such an asset portfolio (called a hedging or replicating portfolio) gives the price of the liability. Such an approach to pricing insurance liabilities is called market-consistent valuation and refers to the non-arbitrage pricing theory from financial mathematics. We should remark that market-consistent valuation is advocated by Solvency II Directive and many accounting standards.

The fundamental theorem of financial mathematics states that in a market model that is arbitrage-free (with No Free Lunch with Vanishing Risk) we can construct an equivalent probability measure $\mathbb{Q}$ such that the discounted prices of traded instruments are $\mathbb{Q}$-local martingales, see Delbaen and Schachermayer (1994). Such a measure $\mathbb{Q}$ is called an equivalent martingale measure. The price of a liability is given by the expectation of the discounted pay-off under an equivalent martingale measure.

In order to price liabilities in our combined financial and insurance model (7.1)–(7.3), we have to define the set of equivalent martingale measures $\mathscr{Q}^m$. From

Ł. Delong, *Backward Stochastic Differential Equations with Jumps and Their Actuarial and Financial Applications*, EAA Series, DOI 10.1007/978-1-4471-5331-3_9,

Sect. 2.5 we conclude that the set $\mathcal{Q}^m$ is of the form

$$\mathcal{Q}^m = \Big\{\mathbb{Q} \sim \mathbb{P}, \ \frac{d\mathbb{Q}}{d\mathbb{P}}\Big|\mathcal{F}_t = M^m(t), \ 0 \leq t \leq T, \\ M^m \text{ is a positive } \mathcal{F}\text{-martingale}\Big\}, \tag{9.1}$$

$$\frac{dM^m(t)}{M^m(t-)} = -\theta(t)dW(t) + \phi(t)dB(t) + \int_{\mathbb{R}} \kappa(t,z)\tilde{N}(dt,dz), \quad M(0) = 1,$$

where $\theta(t) = \frac{\mu(t)-r(t)}{\sigma(t)}$, and ϕ and κ are $\mathcal{F}$-predictable processes such that

$$\int_0^T |\phi(t)|^2 dt < \infty, \qquad \int_0^T \int_{\mathbb{R}} |\kappa(t,z)|^2 Q(t,dz)\eta(t)dt < \infty,$$
$$\kappa(t,z) > -1, \quad (t,z) \in [0,T] \times \mathbb{R}.$$

It is easy to show that under (C1) the discounted stock price $e^{-\int_0^t r(s)ds}S(t)$ is a $\mathbb{Q}$-martingale for any $\mathbb{Q} \in \mathcal{Q}^m$, see Example 2.9. The insurance payment process (7.3) should be priced by the principle

$$\text{Price of } P \text{ at time } t = \mathbb{E}^{\mathbb{Q}}\Big[\int_t^T e^{-\int_s^T r(u)du} dP(s)|\mathcal{F}_t\Big], \quad 0 \leq t \leq T, \tag{9.2}$$

with some $\mathbb{Q} \in \mathcal{Q}^m$. The measure $\mathbb{Q}$ is determined by the triple (θ, ϕ, κ). The process θ is called the market price of the financial risk or the risk premium required by investors for taking the financial risk. The processes ϕ and κ are called the market prices of the insurance risk or the risk premiums required by investors for taking, respectively, the systematic and the unsystematic insurance risk. Notice that in our combined financial and insurance model the risk premium for the financial risk can be uniquely derived from the traded stock. The risk premiums for the systematic and unsystematic insurance risks cannot be uniquely derived since "insurance instruments" are not traded in the market. In order to apply the principle (9.2), the insurer has to decide on the insurance risk premiums. We remark that the insurance risk premiums (ϕ, κ) and the corresponding equivalent martingale measure should not be taken out of the blue. The price of a liability should be related to a hedging portfolio. We should first state a hedging objective (a performance criterion) and then we should derive the optimal hedging strategy. The cost of setting the hedging portfolio gives the price of the liability, from which the insurance risk premiums can be deduced. In next chapters of this book we will focus on hedging of liabilities.

9.2 Perfect Hedging in the Financial Market

We consider the financial market (7.1)–(7.2). We start with the simplest hedging problem in which the insurer faces a terminal claim F depending on the performance of the financial market. We investigate perfect hedging (perfect replication) of the claim F. Let $\pi \in \mathscr{A}$ be an admissible investment strategy, see Definition 7.3.1. The dynamics of the investment portfolio X^π under the strategy π is given by the equation

$$dX^\pi(t) = \pi(t)\big(\mu(t)dt + \sigma(t)dW(t)\big) + \big(X^\pi(t) - \pi(t)\big)r(t)dt. \tag{9.3}$$

The goal is to find an initial value of the investment portfolio and an admissible investment strategy which perfectly replicate the liability F, i.e. an initial capital $X^\pi(0)$ and a strategy $\pi \in \mathscr{A}$ such that $X^\pi(T) = F$. We set

$$Z(t) = \pi(t)\sigma(t), \quad Y(t) = X^\pi(t), \quad 0 \le t \le T.$$

We can notice that the problem of finding a replicating strategy for the claim F in the financial market (7.1)–(7.2) is equivalent to the problem of finding processes (Y, Z) which satisfy the equation

$$dY(t) = Y(t)r(t)dt + Z(t)\theta(t)dt + Z(t)dW(t),$$

with the terminal condition $Y(T) = F$. Consequently, the problem of finding a replicating strategy for the claim F in the financial market (7.1)–(7.2) is equivalent to the problem of solving the BSDE

$$\begin{aligned} Y(t) = F &+ \int_t^T \big(-Y(s)r(s) - Z(s)\theta(s)\big)ds \\ &- \int_t^T Z(s)dW(s), \quad 0 \le t \le T. \end{aligned} \tag{9.4}$$

Hence, the fundamental problem in financial mathematics can be described by a solution to a BSDE. Notice that the connection between the perfect replication of F and the BSDE (9.4) arises very naturally.

Recalling Proposition 3.3.1, we can state the following result.

Theorem 9.2.1 *Assume that* (C1)–(C2) *from Chap.* 7 *hold and let F be an $\mathscr{F}^W$-measurable random variable. The replicating portfolio X^* and the admissible replicating strategy $\pi^* \in \mathscr{A}$ for the claim F are given by*

$$X^*(t) = Y(t) = \mathbb{E}^{\mathbb{Q}^*}\big[e^{-\int_t^T r(s)ds}F|\mathscr{F}_t^W\big], \quad 0 \le t \le T,$$

$$\pi^*(t) = \frac{Z(t)}{\sigma(t)}, \quad 0 \le t \le T,$$

where $(Y, Z) \in \mathbb{S}^2(\mathbb{R}) \times \mathbb{H}^2(\mathbb{R})$ is the unique solution to the BSDE (9.4), *and the equivalent martingale measure $\mathbb{Q}^*$ is defined by*

$$\frac{d\mathbb{Q}^*}{d\mathbb{P}}\Big|\mathscr{F}_t^W = e^{-\int_0^t \theta(s)dW(s) - \frac{1}{2}\int_0^t |\theta(s)|^2 ds}, \quad 0 \le t \le T.$$

By Proposition 3.3.1 we have $Z(t)=\mathscr{Z}(t)e^{\int_0^t r(s)ds}$ where $\mathscr{Z}$ is derived from the predictable representation

$$e^{-\int_0^T r(s)ds}F=\mathbb{E}^{\mathbb{Q}^*}\big[e^{-\int_0^T r(s)ds}F\big]+\int_0^T \mathscr{Z}(s)dW^{\mathbb{Q}^*}(s). \qquad (9.5)$$

Theorem 9.2.1 shows that the replicating portfolio and the replicating strategy for the claim F solve the linear BSDE (9.4). The price of the claim F, the value of the replicating portfolio, is arbitrage-free since it is of the form (9.2). Notice that in the case of the financial model (7.1)–(7.2) the set $\mathscr{Q}^m$ consists of one equivalent martingale measure, hence the arbitrage-free price is uniquely defined. The advantage of formulating the BSDE (9.4), compared to referring directly to the fundamental pricing theorem (9.2), is that the BSDE gives the replicating strategy and the price $Y(0)$ can be seen as the cost of setting the replicating portfolio. The square integrability of the claim guarantees that there exists a unique solution to the BSDE (9.4) and, consequently, a unique replicating portfolio for F. We already know from Example 7.11 that the square integrability assumption excludes arbitrage strategies. We remark that uniqueness of a replicating portfolio coincides with the notion of non-arbitrage. Since solving the linear BSDE (9.4) requires to derive the predictable representation (9.5), in Theorem 9.2.1 we prove the well-known result of financial mathematics.

Example 9.1 We consider the classical Black-Scholes model with constant coefficients r,μ,σ. We are interested in pricing and hedging the put option $F=(K-S(T))^+$. In order to find the replicating strategy and the replicating portfolio for the claim F, we have to solve the BSDE

$$Y(t)=\big(K-S(T)\big)^+ +\int_t^T\big(-Y(s)r-Z(s)\theta\big)ds-\int_t^T Z(s)dW(s), \quad 0\le t\le T.$$

By Proposition 3.3.1 we have to find the predictable representation of the random variable $e^{-rT}(K-S(T))^+$ under the equivalent martingale measure $\mathbb{Q}^*$ which is defined in Theorem 9.2.1. Recalling Girsanov's theorem and Example 2.9, we deduce that $dS(t)=S(t)rdt+S(t)\sigma dW^{\mathbb{Q}^*}(t)$, or $S(T)=S(0)e^{rT-\frac{1}{2}\sigma^2T+\sigma W^{\mathbb{Q}^*}(T)}$. Following the steps from Example 3.2 and applying the Clark-Ocone formula, we can derive the representation

$$e^{-rT}\big(K-S(T)\big)^+=\mathbb{E}^{\mathbb{Q}^*}\big[e^{-rT}\big(K-S(T)\big)^+\big]+\int_0^T\varphi\big(t,S(t)\big)dW^{\mathbb{Q}^*}(t),$$

where

$$\varphi\big(t,S(t)\big)=-\sigma e^{-rT}\mathbb{E}^{\mathbb{Q}^*}\big[S(T)\mathbf{1}\big\{S(T)<K\big\}|\mathscr{F}_t\big], \quad 0\le t\le T.$$

By Proposition 3.3.1 and Theorem 9.2.1 the replicating strategy for the put option is given by the formula

$$\pi^*(t)=-e^{-r(T-t)}\mathbb{E}^{\mathbb{Q}^*}\big[S(T)\mathbf{1}\big\{S(T)<K\big\}|\mathscr{F}_t\big]$$

$$= -S(t)\Phi\big(-d\big(t, S(t)\big)\big), \quad 0 \le t \le T, \tag{9.6}$$

and the price of the put option, and the initial value of the replicating portfolio, is equal to

$$\begin{aligned} X^*(0) = Y(0) &= \mathbb{E}^{\mathbb{Q}^*}\big[e^{-rT}\big(K - S(T)\big)^+\big] \\ &= Ke^{-rT}\Phi\big(-d\big(0, S(0)\big) - \sigma\sqrt{T}\big) - S(0)\Phi\big(-d\big(0, S(0)\big)\big), \end{aligned}$$

where Φ denotes the distribution function for the standard normal random variable and $d(t,s) = \frac{\ln(\frac{s}{K}) + (r+\sigma^2/2)(T-t)}{\sigma\sqrt{T-t}}$. The value of the replicating portfolio is determined by the process

$$\begin{aligned} X^*(t) = Y(t) &= \mathbb{E}^{\mathbb{Q}^*}\big[e^{-r(T-t)}\big(K - S(T)\big)^+ | \mathscr{F}_t\big] \\ &= Ke^{-r(T-t)}\Phi\big(-d\big(t, S(t)\big) - \sigma\sqrt{T-t}\big) \\ &\quad - S(t)\Phi\big(-d\big(t, S(t)\big)\big), \quad 0 \le t \le T. \end{aligned} \tag{9.7}$$

Let us now focus on the Markovian dynamics

$$\frac{dS_0(t)}{S_0(t)} = r\big(S(t)\big)dt, \quad S_0(0) = 1, \tag{9.8}$$

$$\frac{dS(t)}{S(t)} = \mu\big(S(t)\big)dt + \sigma\big(S(t)\big)dW(t), \quad S(0) = s > 0, \tag{9.9}$$

and the terminal claim $F = \hat{F}(S(T))$. Clearly, the replicating portfolio and the replicating strategy solve the FBSDE

$$\begin{aligned} S(t) &= s + \int_0^t S(u)\mu\big(S(u)\big)du + \int_0^t S(u)\sigma\big(S(u)\big)dW(u), \quad 0 \le t \le T, \\ Y(t) &= \hat{F}\big(S(T)\big) + \int_t^T \big(-Y(u)r\big(S(u)\big) - Z(u)\theta\big(S(u)\big)\big)du \\ &\quad - \int_t^T Z(u)dW(u), \quad 0 \le t \le T. \end{aligned} \tag{9.10}$$

In the Markovian setting we can go beyond Theorem 9.2.1. We can recall results on FBSDEs, and from Proposition 4.1.1, Theorems 4.2.2 and 4.1.4 we conclude that the value of the replicating portfolio (the price) is given by

$$\begin{aligned} X^*(t) = Y(t) &= u\big(t, S(t)\big) \\ &= \mathbb{E}^{\mathbb{Q}^*}\big[e^{-\int_t^T r(S(u))du}\hat{F}\big(S(T)\big)|\mathscr{F}_t^W\big], \quad 0 \le t \le T, \end{aligned} \tag{9.11}$$

and the replicating strategy takes the form

$$\pi^*(t) = \frac{Z(t)}{\sigma(t)} = u_s\big(t, S(t)\big)S(t), \quad 0 \le t \le T, \tag{9.12}$$

where the function u solves the PDE

$$\begin{aligned} &u_t(t,s)+r(s)su_s(t,s)+\frac{1}{2}s^2\sigma^2(s)u_{ss}(t,s) \\ &\quad -r(s)u(t,s)=0, \quad (t,s)\in[0,T)\times(0,\infty), \\ &u(T,s)=\hat{F}(s), \quad s\in(0,\infty). \end{aligned} \tag{9.13}$$

These are classical results in financial mathematics, see Chaps. 5 and 6 in Shreve (2004), which have been deduced from the properties of FBSDEs. The strategy (9.12) is the delta hedging strategy and the PDE (9.13) is the Black-Scholes equation. The relation between the PDE (9.13) and the expectation (9.11) is the famous Feynman-Kac formula. It should be now clear that there is a strong connection between the theory of BSDEs and financial applications.

From Chap. 4 we recall that in a Markovian model with a state process $\mathscr{X}$ the control process Z of a BSDE can be related to derivative u_x of the function u characterizing the solution Y. If u denotes the arbitrage-free price of a liability in the Markovian model (9.8)–(9.9) with the risk factor S, then the control process Z of the BSDE (9.10) can be related to u_s and the famous delta hedging strategy (9.12) arises. In the problems considered in this book we will characterize the price and the hedging strategy as a unique solution to a BSDE. If the control process Z of a BSDE can be related to the first derivatives of the price of a liability with respect to risk factors driven by the Brownian motion W, then the investment strategy for the stock S given by $\pi(t)=Z(t)/\sigma(t)$ will be called a delta hedging strategy. We point out that there is an advantage of characterizing hedging strategies by BSDEs. Notice that the application of the delta hedging strategy (9.12) is possible only if we deal with a Markovian model and there exists a smooth solution to the PDE (9.13), whereas the control process Z of the BSDE (9.10) can be used both in a Markovian model and a non-Markovian model and the control process Z exists under weak square integrability assumption of the claim. Consequently, we can apply BSDEs to solve hedging problems in more general models, where neither continuity of the pay-off nor a Markovian dynamics is assumed.

We give two more examples of perfect replication.

Example 9.2 We consider the financial market (7.1)–(7.2). It is reasonable to assume that the investor borrows at a rate r^b which is greater than the rate r earned in the bank account. If $r^b>r$, then the investment portfolio process X^π under a strategy π is given by the equation

$$\begin{aligned} dX^\pi(t) &= X^\pi(t)r(t)+\pi(t)\sigma(t)\theta(t)-\big(r^b(t)-r(t)\big)\big(X^\pi(t)-\pi(t)\big)^- dt \\ &\quad +\pi(t)\sigma(t)W(t). \end{aligned}$$

For an $\mathscr{F}^W$-measurable, square integrable claim F we aim to find an initial capital $X^\pi(0)$ and a strategy $\pi\in\mathscr{A}$ such that $X^\pi(T)=F$. As in (9.4), we end up with a BSDE with a Lipschitz (but nonlinear) generator. We can conclude that there exists a unique admissible replicating strategy $\pi^* \in \mathscr{A}$ and a unique replicating portfolio X^{π^*} for F. Since we obtain a nonlinear BSDE, numerical methods have to be applied to derive the solution.

Example 9.3 We consider the financial market (7.1)–(7.2) and an insurer who faces the life insurance payment process introduced in Example 7.7. We may apply diversification arguments and assume that the insurance risk is fully diversified. We replace the random evolution of the point process J, which counts the claims in a life insurance portfolio, with deterministic expectations. Such an approach has a long tradition in life insurance. By diversification arguments we investigate the payment process

$$P^e(t) = \int_0^t \hat{H}\big(s, S(s)\big)np^e(s)ds + \int_0^t \hat{G}\big(s, S(s)\big)np^e(s)\lambda^e(s)ds \\ + \hat{F}\big(S(T)\big)np^e(T)\mathbf{1}_{t=T}, \quad 0 \le t \le T,$$

where $p^e(t) = e^{-\int_0^t \lambda^e(s)ds}$ denotes the survival probability and λ^e is the expected claim intensity. The expectations λ^e and p^e are fixed at time $t = 0$. Notice that we end up with purely financial claims weighted with some deterministic factors. The goal is to find $\pi \in \mathscr{A}$ and X^π which satisfy the equation

$$dX^\pi(t) = \pi(t)\big(\mu(t)dt + \sigma(t)dW(t)\big) + \big(X^\pi(t) - \pi(t)\big)r(t)dt - dP^e(t),$$

with the terminal condition $X^\pi(T) = 0$. Hence, if $\mathbb{E}[|P^e(T)|^2] < \infty$, then the unique replicating portfolio and the admissible replicating strategy for P^e are given by

$$X^*(t) = \mathbb{E}^{\mathbb{Q}^*}\bigg[\int_t^T e^{-\int_t^s r(u)du} dP^e(s)|\mathscr{F}_t^W\bigg], \quad 0 \le t \le T,$$

$$\pi^*(t) = \frac{\mathscr{Z}(t)}{\sigma(t)} e^{\int_0^t r(s)ds}, \quad 0 \le t \le T,$$

where $\mathscr{Z}$ is derived from the predictable representation

$$\int_0^T e^{-\int_0^s r(u)du} dP^e(s) = \mathbb{E}^{\mathbb{Q}^*}\bigg[\int_0^T e^{-\int_0^s r(u)du} dP^e(s)\bigg] + \int_0^T \mathscr{Z}(s)dW^{\mathbb{Q}^*}(s),$$

and the equivalent martingale measure $\mathbb{Q}^*$ is defined in Theorem 9.2.1.

From the point of view of static pricing, the diversification argument leads to a reasonable initial price. However, the diversification argument fails in the context of dynamic pricing and hedging. The replicating strategy derived in Example 9.3 is based on the insurance assumptions made at time $t = 0$ which are not updated over the lifetime of the contract. It is clear that the insurer should adapt the hedging strategy and the hedging portfolio (the reserve) to the claim experience. More importantly, diversification turns the incomplete combined financial and insurance market, in which claims cannot be perfectly hedged, into a complete financial market, in which all claims can be perfectly hedged. This is definitely too strong assumption from the point of risk management. The insurer should quantify unhedgeable risks and develop risk mitigating techniques such as efficient hedging strategies.

9.3 Superhedging in the Financial and Insurance Market

Let us assume that the insurer faces a terminal claim F which now depends on all three sources of uncertainty (W, B, N). The predictable representation of the $\mathscr{F}_T$-measurable, square integrable claim F under any equivalent martingale measure $\mathbb{Q}$ gives

$$\begin{aligned} e^{-\int_0^T r(s)ds} F &= \mathbb{E}^{\mathbb{Q}}\big[e^{-\int_0^T r(s)ds} F\big] + \int_0^T \mathscr{Z}_1(s) dW^{\mathbb{Q}}(s) \\ &\quad + \int_0^T \mathscr{Z}_2(t) dB^{\mathbb{Q}}(s) + \int_0^T \int_{\mathbb{R}} \mathscr{U}(t, z) \tilde{N}^{\mathbb{Q}}(dt, dz), \end{aligned}$$

which cannot be matched with the dynamics of the investment portfolio (9.3). Hence, the reasoning that led to Theorem 9.2.1 fails. We conclude that perfect hedging of a general $\mathscr{F}_T$-measurable claim is not possible in the financial market (7.1)–(7.2).

Let us consider the financial market (7.1)–(7.2) and the insurance payment process (7.3). Let X^π be the insurer's investment portfolio given by (7.10). Since perfect hedging of the payment process is not possible, the insurer could be interested in finding an investment strategy π which would yield $X^\pi(T) \geq 0$. Such an objective is called a superhedging objective.

We have to introduce a new definition of an admissible investment strategy and an investment portfolio. We adapt the definition from El Karoui and Quenez (1995) to our model.

Definition 9.3.1 A strategy (π, C) is called admissible for the superhedging problem, written $(\pi, C) \in \mathscr{A}^{super}$, if it consists of a predictable process π satisfying

$$\int_0^T \left|\pi(t)\sigma(t)\right|^2 dt < \infty,$$

and a càdlàg, adapted, non-decreasing process C such that $C(0) = 0$. An investment portfolio $X^{\pi,C} := (X^{\pi,C}(t), 0 \leq t \leq T)$ under $(\pi, C) \in \mathscr{A}^{super}$ is called admissible if $X^{\pi,C}$ is a càdlàg, adapted, non-negative process which satisfies the dynamics

$$\begin{aligned} dX^{\pi,C}(t) &= \pi(t)\big(\mu(t)dt + \sigma(t)dW(t)\big) \\ &\quad + \big(X^{\pi,C}(t) - \pi(t)\big)r(t)dt - dP(t) - dC(t), \quad 0 \leq t \leq T, \end{aligned}$$

together with the terminal condition $X^{\pi,C}(T)=0$. The lowest admissible investment portfolio is called the superhedging selling price of P and the corresponding $(\pi,C)\in\mathscr{A}^{super}$ is called the superhedging strategy for P.

In order to exclude arbitrage strategies, an admissible investment portfolio is here constrained to take non-negative values, instead of assuming the square integrability of an investment strategy as in Definition 7.3.1. The process π characterizes the amount of wealth invested in the stock and the process C represents the cumulative amount the insurer can withdraw from the investment portfolio. Under the superhedging objective the insurer requires a premium $X(0)$ which suffices to cover the payment process P in all scenarios. As time passes, the capital invested in the portfolio may become too large, if some of the worst scenarios are no longer possible, and the insurer can withdraw the capital from the investment portfolio. The process C represents the profit arising from selling the payments P for the price $X(0)$.

The arbitrage-free price of the insurance payment process is defined by $\mathbb{E}^{\mathbb{Q}}[\int_t^T e^{-\int_t^T r(s)ds}dP(s)|\mathscr{F}_t]$ for some $\mathbb{Q}\in\mathscr{Q}^m$. Notice that the set of equivalent martingale measures $\mathscr{Q}^m$ in the combined financial and insurance model is not a singleton. In the view of the superhedging objective applied, it is reasonable to study the price

$$X^*(t)=\operatorname*{ess\ sup}_{\mathbb{Q}\in\mathscr{Q}^m}\mathbb{E}^{\mathbb{Q}}\left[\int_t^T e^{-\int_t^T r(s)ds}dP(s)|\mathscr{F}_t\right],\quad 0\le t\le T. \tag{9.14}$$

We show that the process (9.14) defines the lowest admissible investment portfolio in the sense of Definition 9.3.1. First, we state two important results from El Karoui and Quenez (1995). By $\mathbb{Q}^0$ we denote the equivalent martingale measure defined by the Radon-Nikodym derivative (9.1) with $\phi=\kappa=0$.

Theorem 9.3.1 *Consider the set of equivalent martingale measures $\mathscr{Q}^m$ defined by (9.1). Let ξ be an $\mathscr{F}$-measurable random variable such that $\xi\ge 0$ and $\sup_{\mathbb{Q}\in\mathscr{Q}^m}\mathbb{E}^{\mathbb{Q}}[\xi]<\infty$. There exists a càdlàg, $\mathscr{F}$-adapted process*

$$Y(t)=\operatorname*{ess\ sup}_{\mathbb{Q}\in\mathscr{Q}^m}\mathbb{E}^{\mathbb{Q}}[\xi|\mathscr{F}_t],\quad 0\le t\le T.$$

The process Y is characterized as the smallest right-continuous $\mathbb{Q}$-supermartingale, for any $\mathbb{Q}\in\mathscr{Q}^m$, which is equal to ξ at time T. Also, $\mathbb{Q}^*\in\mathscr{Q}^m$ is optimal if and only if Y is a $\mathbb{Q}^*$-martingale.

Theorem 9.3.2 *Consider the set of equivalent martingale measures $\mathscr{Q}^m$ defined by (9.1). Let ξ be an $\mathscr{F}$-measurable random variable such that $\xi\ge 0$ and $\sup_{\mathbb{Q}\in\mathscr{Q}^m}\mathbb{E}^{\mathbb{Q}}[\xi]<\infty$. The process Y defined in Theorem 9.3.1 has the representation*

$$Y(t)=Y(0)+\int_0^t Z(s)dW^{\mathbb{Q}^0}(s)-C(t),\quad 0\le t\le T, \tag{9.15}$$

where Z is a predictable process such that

$$\int_0^T |Z(t)|^2 dt < \infty,$$

and C is a càdlàg, adapted, non-decreasing process such that $C(0) = 0$.

We refer to Theorems 2.1.1, 2.1.2 and 2.3.1 in El Karoui and Quenez (1995). The next result gives the solution to the superhedging problem.

Theorem 9.3.3 *Consider the set of equivalent martingale measures $\mathcal{Q}^m$ defined by* (9.1). *Assume that* (C1)–(C4) *from Chap. 7 hold and let* $\sup_{\mathbb{Q}\in\mathcal{Q}^m} \mathbb{E}^{\mathbb{Q}}[P(T)] < \infty$. *Define the process*

$$\mathscr{Y}(t) = \text{ess} \sup_{\mathbb{Q}\in\mathcal{Q}^m} \mathbb{E}^{\mathbb{Q}}\left[\int_0^T e^{-\int_0^s r(u)du} dP(s)|\mathscr{F}_t\right], \quad 0 \le t \le T, \tag{9.16}$$

which has the representation

$$\mathscr{Y}(t) = \mathscr{Y}(0) + \int_0^t \hat{\pi}(s)\sigma(s) dW^{\mathbb{Q}_0}(s) - \hat{C}(t), \quad 0 \le t \le T,$$

where $(\hat{\pi}, \hat{C}) \in \mathscr{A}^{super}$. Then the process

$$X^*(t) = \text{ess} \sup_{\mathbb{Q}\in\mathcal{Q}^m} \mathbb{E}^{\mathbb{Q}}\left[\int_t^T e^{-\int_t^s r(u)du} dP(s)|\mathscr{F}_t\right], \quad 0 \le t \le T, \tag{9.17}$$

is the superhedging selling price of the payment process P, and the investment strategy

$$\pi^*(t) = \hat{\pi}(t) e^{\int_0^t r(s)ds}, \quad 0 \le t \le T,$$

is the superhedging strategy for the payment process P.

Proof It is straightforward to notice that we have

$$X^*(t) = \mathscr{Y}(t) e^{\int_0^t r(s)ds} - e^{\int_0^t r(s)ds} \int_0^t e^{-\int_0^s r(u)du} dP(s), \quad 0 \le t \le T. \tag{9.18}$$

The process X^* is càdlàg, adapted and non-negative by Theorem 9.3.1, representation (9.17) and non-negativity of the claims. We can derive the dynamics

$$\begin{aligned} dX^*(t) &= r(t)X^*(t) + \hat{\pi}(t)\sigma(t) e^{\int_0^t r(s)ds} dW^{\mathbb{Q}^0}(t) - e^{\int_0^t r(s)ds} d\hat{C}(t) - dP(t) \\ &= \pi^*(t)\big(\mu(t)dt + \sigma(t)dW(t)\big) + \big(X^*(t) - \pi^*(t)\big)r(t)dt - dC^*(t) - dP(t), \end{aligned}$$

where we introduce $C^*(t) = \int_0^t e^{\int_0^s r(u)du} d\hat{C}(s)$. From Theorem 9.3.2 we can conclude that (π^*, C^*) is an admissible investment strategy. Hence, the process X^* is an admissible investment portfolio in the sense of Definition 9.3.1. We have to prove that the candidate X^* is the lowest admissible investment portfolio. To do this, we first show that for any admissible control $(\pi, C) \in \mathscr{A}^{super}$ the process

$$\tilde{X}^{\pi,C}(t) = X^{\pi,C}(t)e^{-\int_0^t r(s)ds} + \int_0^t e^{-\int_0^s r(u)du} dP(s), \quad 0 \le t \le T, \quad (9.19)$$

is a $\mathbb{Q}$-supermartingale, for any $\mathbb{Q} \in \mathscr{Q}^m$. Clearly, $\tilde{X}^{\pi,C}$ is càdlàg and adapted. From Definition 9.3.1 of an admissible process $X^{\pi,C}$ we deduce

$$X^{\pi,C}(t)e^{-\int_0^t r(s)ds} = X^{\pi,C}(0) + \int_0^t \pi(s)\sigma(s)e^{-\int_0^s r(u)du} dW^{\mathbb{Q}_0}(s)$$
$$- \int_0^t e^{-\int_0^s r(u)du} dC(s) - \int_0^t e^{-\int_0^s r(u)du} dP(s), \quad 0 \le t \le T.$$

Hence, we get the representation

$$\tilde{X}^{\pi,C}(t) = X^{\pi,C}(0) + \int_0^t \pi(s)\sigma(s)e^{-\int_0^s r(u)du} dW^{\mathbb{Q}_0}(s)$$
$$- \int_0^t e^{-\int_0^s r(u)du} dC(s), \quad 0 \le t \le T. \quad (9.20)$$

Since $(\pi, C) \in \mathscr{A}^{super}$, the process $\int_0^t \pi(s)\sigma(s)e^{-\int_0^s r(u)du} dW^{\mathbb{Q}^0}(s)$ is a $\mathbb{Q}$-local martingale, for any $\mathbb{Q} \in \mathscr{Q}^m$, $X^{\pi,C}(t) \ge 0$ and $\tilde{X}^{\pi,C}(t) \ge 0$ by (9.19). From these properties and representation (9.20) we conclude that the process $X^{\pi,C}(0) + \int_0^t \pi(s)\sigma(s)e^{-\int_0^s r(u)du} dW^{\mathbb{Q}^0}(s)$ is a non-negative $\mathbb{Q}$-local martingale, hence a $\mathbb{Q}$-supermartingale, for any $\mathbb{Q} \in \mathscr{Q}^m$. From (9.19), (9.20) and the terminal value $X^{\pi,C}(T) = 0$ we deduce

$$\int_0^T e^{-\int_0^s r(u)du} dC(s) = X^{\pi,C}(0) - \tilde{X}^{\pi,C}(T) + \int_0^t \pi(s)\sigma(s)e^{-\int_0^s r(u)du} dW^{\mathbb{Q}_0}(s)$$
$$= X^{\pi,C}(0) + \int_0^t \pi(s)\sigma(s)e^{-\int_0^s r(u)du} dW^{\mathbb{Q}_0}(s)$$
$$- \int_0^T e^{-\int_0^s r(u)du} dP(s), \quad 0 \le t \le T,$$

and we conclude that $\int_0^T e^{-\int_0^s r(u)du} dC(s)$ is integrable under any $\mathbb{Q} \in \mathscr{Q}^m$ since the stochastic integral and the aggregated discounted claims are integrable under any $\mathbb{Q} \in \mathscr{Q}^m$. The supermartingale property of $\tilde{X}^{\pi,C}$ now follows from representation (9.20), the supermartingale property of $X^{\pi,C}(0) + \int_0^t \pi(s)\sigma(s)e^{-\int_0^s r(u)du} dW^{\mathbb{Q}^0}(s)$ and non-negativity of the integrable process $\int_0^t e^{-\int_0^s r(u)du} dC(s)$. By Theorem 9.3.1 we get $\mathscr{Y}(t) \le \tilde{X}^{\pi,C}(t), 0 \le t \le T$, and finally $X^*(t) \le X^{\pi,C}(t), 0 \le t \le T$, by (9.18) and (9.19). □

We remark that the strategy $(\mathscr{Y}, \hat{\pi}, \hat{C})$ from Theorem 9.3.3 can be viewed as a supersolution to the BSDE

$$\mathscr{Y}(t) = \int_0^T e^{-\int_0^s r(u)du} dP(s) - \int_t^T \hat{\pi}(s)\sigma(s)dW^{\mathbb{Q}^0}(s) + \big(\hat{C}(T) - \hat{C}(t)\big), \quad 0 \le t \le T. \quad (9.21)$$

For the notion of a supersolution to a BSDE we refer to Sect. 2.3 in El Karoui et al. (1997b) and Peng (1999).

The superhedging price and the superhedging strategy are difficult to compute. Let us show that for the Brownian filtration $\mathscr{F} = \mathscr{F}^W \otimes \mathscr{F}^B$ the process X^* can be approximated with a sequence of solutions to BSDEs which can be easily computed. Since a comparison principle plays a crucial role in the proof, we restrict our study to the Brownian filtration.

We define the set of equivalent martingale measures

$$\mathscr{Q}_n^m = \Big\{ \mathbb{Q} \sim \mathbb{P}, \ \frac{d\mathbb{Q}}{d\mathbb{P}}\Big|\mathscr{F}_t = M_n^m(t), \ 0 \le t \le T, \ M_n^m \text{ is a positive } \mathscr{F}^W \otimes \mathscr{F}^B\text{-martingale} \Big\}, \quad (9.22)$$

$$\frac{dM_n^m(t)}{M_n^m(t)} = -\theta(t)dW(t) + \phi_n(t)dB(t), \quad M_n^m(0) = 1,$$

where ϕ_n is an $\mathscr{F}^W \otimes \mathscr{F}^B$-predictable process such that

$$|\phi_n(t)| \le n, \quad 0 \le t \le T.$$

We investigate the optimization problem

$$Y^n(t) = \operatorname*{ess\ sup}_{\mathbb{Q} \in \mathscr{Q}_n^m} \mathbb{E}^{\mathbb{Q}}[\xi | \mathscr{F}_t], \quad 0 \le t \le T. \quad (9.23)$$

Following the proof of Proposition 3.3.2, we can derive the solution to the problem (9.23).

Proposition 9.3.1 *Let us deal with the Brownian filtration $\mathscr{F} = \mathscr{F}^W \otimes \mathscr{F}^B$, and let us consider the set of equivalent martingale measures $\mathscr{Q}_n^m$ defined by* (9.22). *We assume that* (C1)–(C2) *from Chap. 7 hold and ξ is an $\mathscr{F}$-measurable random variable such that $\mathbb{E}[|\xi|^2] < \infty$.*

(a) *For each $n \in \mathbb{N}$ there exists a unique solution $(Y^n, Z_1^n, Z_2^n) \in \mathbb{S}^2(\mathbb{R}) \times \mathbb{H}^2(\mathbb{R}) \times \mathbb{H}^2(\mathbb{R})$ to the BSDE*

$$Y^n(t) = \xi - \int_t^T Z_1^n(s)\theta(s)ds + \int_t^T n|Z_2^n(s)|ds$$

$$-\int_t^T Z_1^n(s)dW(s) - \int_t^T Z_2^n(s)dB(s), \quad 0 \le t \le T, \quad (9.24)$$

and the process Y^n *has representation* (9.23).

(b) *The optimal control* ϕ_n^* *for the optimization problem* (9.23) *is given by*

$$\phi_n^*(t) = n\frac{|Z_2^n(t)|}{Z_2^n(t)}\mathbf{1}\{Z_2^n(t) \neq 0\}, \quad 0 \le t \le T.$$

Let $Y(t) = \text{ess sup}_{\mathbb{Q}\in\mathcal{Q}^m} \mathbb{E}^{\mathbb{Q}}[\xi|\mathcal{F}_t]$. The superhedging price and the superhedging investment portfolio Y in the diffusion model can be obtained as an increasing limit of the prices Y^n solving the BSDEs (9.24).

Proposition 9.3.2 *Under the assumptions of Theorem* 9.3.1 *and Proposition* 9.3.1 *we have* $\lim_{n\to\infty} Y^n(t) = Y(t)$, $0 \le t \le T$.

Proof It is clear that $0 \le Y^n(t) \le Y(t)$, $0 \le t \le T$. By the comparison principle the sequence $(Y^n)_{n\in\mathbb{N}}$ is non-decreasing. Let us define $Y^\infty(t) = \lim_{n\to\infty} Y^n(t)$, $0 \le t \le T$. We have $0 \le Y^n(t) \le Y^\infty(t) \le Y(t)$, $0 \le t \le T$. By Theorem 9.3.1 the processes Y^n and Y are càdlàg, $\mathcal{F}^W \otimes \mathcal{F}^B$-adapted, $\mathbb{Q}^0$-supermartingales. We can also deduce that Y^∞ is càdlàg and $\mathcal{F}^W \otimes \mathcal{F}^B$-adapted. By the dominated convergence theorem Y^∞ is a $\mathbb{Q}^0$-supermartingale. Since the process Y is the smallest $\mathbb{Q}^0$-supermartingale, we get the inequality $Y(t) \le Y^\infty(t)$, $0 \le t \le T$. The proof is complete. □

Superhedging strategies are not popular in financial markets. Firstly, they are difficult to compute. Secondly, superhedging prices are not arbitrage-free and they are usually too high to be accepted by buyers. Notice that the price process Y^n from Proposition 9.3.1 is arbitrage-free since Y^n can be represented as the expectation under an equivalent martingale measure. However, the limit of the sequence $(Y^n)_{n\in\mathbb{N}}$ and the superhedging price process Y may not have such a representation.

Example 9.4 Consider the claim $\xi = \mathbf{1}\{J(T) = 0\}$ where J is a Poisson process with intensity λ. The claim ξ may represent a survival benefit. Let $r = 0$. From the representation of the superhedging price, see Theorem 9.3.1, we conclude that $Y(0) \le 1$. Let us introduce the subset of equivalent martingale measures $\mathcal{Q}_\kappa^m \subset \mathcal{Q}^m$ which arises from $\mathcal{Q}^m$ by considering $\kappa(t) = \kappa$. An element of $\mathcal{Q}_\kappa^m$ is denoted by $\mathbb{Q}_\kappa$. We have

$$Y(0) \ge \sup_{\kappa\in(-1,\infty)} \mathbb{E}^{\mathbb{Q}_\kappa}\big[\mathbf{1}\{J(T) = 0\}\big] = \sup_{\kappa\in(-1,\infty)} e^{-(1+\kappa)\lambda T} = 1, \quad 0 \le t \le T,$$

where we use the fact under $\mathbb{Q}_\kappa$ the Poisson process has the intensity $(1+\kappa)\lambda$. Hence, the superhedging price $Y(0) = 1$. Clearly, this price is too high and the insurer would never sell the contract for such a price. The price $Y(0)$ yields arbitrage since it guarantees a profit with a positive probability and no shortfall for the insurer.

One can show that the sequence of equivalent probability measures $\mathbb{Q}_\kappa$ converges weakly in distribution, as $\kappa \to -1$, to a probability measure under which the Poisson process equals zero, see El Karoui and Quenez (1995). Hence, the limit is not an equivalent probability measure.

Since superhedging strategies are too high to be of practical use in the markets, we have to look for other strategies which minimize the risk of not fulfilling the obligation. If the insurer has an access to a traded derivative which is perfectly correlated with the insurance liability, then the risk can be eliminated. We consider such a case in the next chapter. However, in most cases the insurer is not able to find assets which perfectly match the insurance liabilities. Then, the goal should be to replicate the liabilities as good as possible by optimizing a hedging error and solving an asset allocation problem. We investigate such optimization problems in Chaps. 10, 11, 12.

9.4 Perfect Hedging in the Financial and Insurance Market Completed with a Mortality Bond

In order to help insurance companies to hedge mortality risk, mortality derivatives have been introduced in global financial markets. In this chapter we assume that the financial market consists of the bank account (7.1), the stock (7.2) and a mortality bond. Since pay-offs from mortality bonds are contingent on the mortality experience in a population, see Blake et al. (2010) and Wills and Sherris (2010), the insurer can use a mortality bond to hedge death and survival benefits from its portfolio.

We investigate a life insurance portfolio consisting of n persons insured whose future lifetimes $(\tau_i, i = 1, \ldots, n)$ are assumed to be independent and exponentially distributed

$$\mathbb{P}(\tau_i > t) = e^{-\int_0^t \lambda(s)ds}, \quad i = 1, \ldots, n,$$

where $\lambda : [0, T] \to (0, \infty)$ denotes a (deterministic) mortality intensity. We deal with the life insurance payment process

$$P(t) = \int_0^t \big(n - J(s)\big)\hat{H}(s)ds + \int_0^t \hat{G}(s)dJ(s) + \big(n - J(T)\big)\hat{F}\mathbf{1}_{t=T}, \quad 0 \le t \le T, \tag{9.25}$$

where J is the deaths counting process for the insurance portfolio. For details we refer to Examples 7.6–7.7. We consider a mortality bond which pays a unit for each insured person who survives till the maturity of the policy. We show that such a mortality bond, which is assumed to be traded in the market, completes the combined financial and insurance model (7.1), (7.2), (9.25) in the sense that the life insurance payment process (9.25) can be perfectly replicated by investing in the bank account, the stock and the mortality bond.

First, we derive the price dynamics of the mortality bond. The combined financial and insurance market with the mortality bond is arbitrage-free. We assume that the mortality bond is already priced by the market, i.e. the equivalent martingale measure $\mathbb{Q} \in \mathcal{Q}^m$ or the process κ, which is used to price the unsystematic mortality risk, is given. We set $\phi = 0$, since the Brownian motion B is not used here. The arbitrage-free price of the mortality bond is given by

$$D(t) = \mathbb{E}^{\mathbb{Q}}\big[e^{-\int_t^T r(s)ds}\big(n - J(T)\big)|\mathcal{F}_t\big], \quad 0 \le t \le T. \tag{9.26}$$

Proposition 9.4.1 *Assume that*

(i) (C1) *from Chap. 7 holds,*
(ii) *the compensator of the point process J is of the form $\vartheta(dt, \{1\}) = (n - J(t-))\lambda(t)dt$ where $\lambda : [0, T] \to (0, \infty)$ is continuous,*
(iii) *the equivalent martingale measure $\mathbb{Q} \in \mathcal{Q}^m$ is given by* (9.1) *with an $\mathcal{F}^W$-predictable, bounded process κ.*

The price of the mortality bond (9.26) *satisfies the dynamics*

$$\begin{aligned} dD(t) = D(t-)\mathbf{1}\big\{n - J(t-) > 0\big\}\bigg(\bigg(r(t) + \big(1 + \kappa(t)\big)\lambda(t) + \frac{Z^E(t)}{E(t)}\theta(t)\bigg)dt \\ - \frac{1}{n - J(t-)}dJ(t) + \frac{Z^E(t)}{E(t)}dW(t)\bigg), \quad 0 \le t \le T, \end{aligned} \tag{9.27}$$

where

$$\begin{aligned} E(t) &= \mathbb{E}^{\mathbb{Q}}\big[e^{-\int_0^T r(s)ds}e^{-\int_0^T (1+\kappa(s))\lambda(s)ds}|\mathcal{F}_t^W\big], \quad 0 \le t \le T, \\ E(t) &= E(0) + \int_0^t Z^E(s)dW^{\mathbb{Q}}(s), \quad 0 \le t \le T. \end{aligned} \tag{9.28}$$

Proof By the Girsanov's theorem the compensator of the point process J under $\mathbb{Q}$ is $(1 + \kappa(t))(n - J(t-))\lambda(t)dt$. We can also deduce that the $\mathcal{F}^W$-conditional distribution of the future lifetime τ of an insured is exponential under $\mathbb{Q}$. Consequently, we can derive

$$\begin{aligned} D(t) &= \big(n - J(t)\big)\mathbb{E}^{\mathbb{Q}}\big[e^{-\int_t^T r(s)ds}e^{-\int_t^T (1+\kappa(s))\lambda(s)ds}|\mathcal{F}_t^W\big] \\ &= \big(n - J(t)\big)e^{\int_0^t r(s)ds + \int_0^t (1+\kappa(s))\lambda(s)ds}E(t), \quad 0 \le t \le T, \end{aligned}$$

see also (8.6). By the predictable representation of E from (9.28) and the Itô's formula we obtain the dynamics

$$\begin{aligned} dD(t) = &-e^{\int_0^t r(s)ds + \int_0^t (1+\kappa(s))\lambda(s)ds}E(t)dJ(t) \\ &+ \big(n - J(t-)\big)e^{\int_0^t r(s)ds + \int_0^t (1+\kappa(s))\lambda(s)ds}E(t)\big(r(t) + \big(1 + \kappa(t)\big)\lambda(t)\big)dt \\ &+ \big(n - J(t-)\big)e^{\int_0^t r(s)ds + \int_0^t (1+\kappa(s))\lambda(s)ds}Z^E(t)dW^{\mathbb{Q}}(t) \end{aligned}$$

$$= D(t-)\mathbf{1}\{n - J(t-) > 0\}\Bigg(\Big(r(t) + \big(1+\kappa(t)\big)\lambda(t)\Big)dt - \frac{1}{n - J(t-)}dJ(t)$$

$$+ \frac{Z^E(t)}{E(t)}dW^{\mathbb{Q}}(t)\Bigg).$$

If we change the measure to $\mathbb{P}$, then (9.27) can be derived. □

We remark that the restriction that the process κ (or the drift in (9.27)) is $\mathscr{F}^W$-predictable is reasonable. In particular, it implies that the future lifetimes are, conditional on $\mathscr{F}^W$, independent and exponentially distributed under $\mathbb{Q}$. Notice that the deterministic intensity λ can be changed into a random intensity if an $\mathscr{F}^W$-predictable process κ is used. A deterministic κ is likely to be chosen in actuarial applications.

We now solve the perfect replication problem for the insurance payment process (9.25). Let $\pi := (\pi(t), 0 \le t \le T)$ and $\chi := (\chi(t), 0 \le t \le T)$ denote the amount of wealth invested in the stock S and in the mortality bond D. The dynamics of the investment portfolio is given by

$$\begin{aligned}
dX^{\pi,\chi}(t) &= \pi(t)\frac{dS(t)}{S(t)} + \chi(t)\frac{dD(t)}{D(t-)} + \big(X^{\pi,\chi}(t-) - \pi(t) - \chi(t)\big)\frac{dS_0(t)}{S_0(t)} - dP(t)\\
&= \pi(t)\big(\mu(t)dt + \sigma(t)dW(t)\big)\\
&\quad + \chi(t)\mathbf{1}\{n - J(t-) > 0\}\Bigg(\bigg(r(t) + \big(1+\kappa(t)\big)\lambda(t) + \frac{Z^E(t)}{E(t)}\theta(t)\bigg)dt\\
&\quad - \frac{1}{n - J(t-)}dJ(t) + \frac{Z^E(t)}{E(t)}dW(t)\Bigg)\\
&\quad + \big(X^{\pi,\chi}(t-) - \pi(t) - \chi(t)\mathbf{1}\{n - J(t-) > 0\}\big)r(t)dt\\
&\quad - dP(t), \quad X^{\pi,\chi}(0) = x.
\end{aligned} \tag{9.29}$$

We define a class of admissible investment strategies.

Definition 9.4.1 A strategy (π, χ) is called admissible, written $(\pi, \chi) \in \mathscr{A}^l$, if it satisfies the conditions:

(i) $\pi : [0, T] \times \Omega \to \mathbb{R}$ and $\chi : [0, T] \times \Omega \to \mathbb{R}$ are predictable processes,
(ii) $\mathbb{E}[\int_0^T |\pi(t)\sigma(t)|^2 dt] < \infty$, $\mathbb{E}[\int_0^T |\chi(t)|^2 \lambda(t)dt] < \infty$ and $\mathbb{E}[\int_0^T |\chi(t) \times Z^E(t)|^2 dt] < \infty$,
(iii) there exists a unique càdlàg, adapted solution $X^{\pi,\chi}$ to (9.29) on $[0, T]$.

The goal is to find an initial value of the investment portfolio $X^{\pi,\chi}(0)$ and an admissible investment strategy $(\pi, \chi) \in \mathscr{A}^l$ such that $X^{\pi,\chi}(T) = 0$. This replication problem is equivalent to the problem of solving the BSDE

$$\begin{aligned}Y(t) &= \big(n - J(T)\big)\hat{F} + \int_t^T \big(-Y(s)r(s) - \theta(s)Z(s) \\ &\quad + \big(n - J(s-)\big)U(s)\kappa(s)\lambda(s) + \hat{H}(s)\big(n - J(s-)\big) \\ &\quad + \hat{G}(s)\big(1 + \kappa(s)\big)\big(n - J(s-)\big)\lambda(s)\big)ds \\ &\quad - \int_t^T Z(s)dW(s) - \int_t^T U(s)\tilde{N}(ds), \quad 0 \le t \le T, \end{aligned} \tag{9.30}$$

which can be immediately derived from the wealth process (9.29) by introducing the variables

$$\begin{aligned}Y(t) &= X^{\pi,\chi}(t), \quad 0 \le t \le T, \\ Z(t) &= \left(\pi(t)\sigma(t) + \chi(t)\frac{Z^E(t)}{E(t)}\mathbf{1}\big\{n - J(t-) > 0\big\}\right), \quad 0 \le t \le T, \\ U(t) &= \frac{-\chi(t)}{n - J(t-)}\mathbf{1}\big\{n - J(t-) > 0\big\} - \hat{G}(t), \quad 0 \le t \le T,\end{aligned} \tag{9.31}$$

and the jump measure N of the point process J.

We state the main result of this chapter.

Theorem 9.4.1 *Consider the payment process* (9.25) *with* $\mathscr{F}^W$*-measurable* $\hat{F}$ *and* $\mathscr{F}^W$*-predictable* $\hat{G}$ *and* $\hat{H}$*. Assume that* (C1)–(C4) *from Chap.* 7 *hold and let the assumptions of Proposition* 9.4.1 *be satisfied.*

(a) *There exists a unique solution* $(Y, Z, U) \in \mathbb{S}^2(\mathbb{R}) \times \mathbb{H}^2(\mathbb{R}) \times \mathbb{H}^2_N(\mathbb{R})$ *to the BSDE* (9.30).

(b) *The replicating portfolio* X^* *for the payment process* P *is given by*

$$X^*(t) = Y(t) = \mathbb{E}^{\mathbb{Q}}\left[\int_t^T e^{-\int_t^s r(u)du} dP(s)|\mathscr{F}_t\right], \quad 0 \le t \le T,$$

where the equivalent martingale measure $\mathbb{Q}$ *is given by* (9.26).

(c) *If*

$$\mathbb{E}\left[\int_0^T \big|U(t)Z^E(t)\big|^2 dt\right] < \infty, \qquad \mathbb{E}\left[\int_0^T \big|\hat{G}(t)Z^E(t)\big|^2 dt\right] < \infty, \tag{9.32}$$

then the admissible replicating strategy $(\pi^*, \chi^*) \in \mathscr{A}^l$ *for the payment process* P *takes the form*

$$\begin{aligned}\pi^*(t) &= \frac{Z(t)}{\sigma(t)} - \chi^*(t)\frac{Z^E(t)}{\sigma(t)E(t)}, \quad 0 \le t \le T, \\ \chi^*(t) &= -\big(n - J(t-)\big)\big(U(t) + \hat{G}(t)\big), \quad 0 \le t \le T.\end{aligned}$$

Proof (a) The existence of a unique solution to the BSDE (9.30) follows from Theorem 3.1.1.

(b) Following the proof of Propositions 3.3.1 and 3.4.1, we get the representation

$$Y(t) = \mathbb{E}^{\mathbb{Q}}\Big[e^{-\int_t^T r(s)ds}\big(n - J(T)\big)\hat{F} + \int_t^T e^{-\int_t^s r(u)du}\big(\hat{H}(s)\big(n - J(s-)\big)$$
$$+ \hat{G}(s)\big(1 + \kappa(s)\big)\big(n - J(s-)\big)\lambda(s)\big)ds|\mathscr{F}_t\Big].$$

Since $(1 + \kappa(t))(n - J(t-))\lambda(t)dt$ is the compensator of the jump measure N under $\mathbb{Q}$, we have

$$\mathbb{E}^{\mathbb{Q}}\Big[\int_0^{\tau_n} e^{-\int_t^s r(u)du}\hat{G}(s)\big(1 + \kappa(s)\big)\big(n - J(s-)\big)\lambda(s)ds\Big]$$
$$= \mathbb{E}^{\mathbb{Q}}\Big[\int_0^{\tau_n} e^{-\int_t^s r(u)du}\hat{G}(s)N(ds)\Big],$$

where $(\tau_n)_{n\geq 1}$ is a sequence of localizing stopping times, and we use Theorem 2.3.2. By the monotone convergence theorem we obtain

$$\mathbb{E}^{\mathbb{Q}}\Big[\int_0^{T} e^{-\int_t^s r(u)du}\hat{G}(s)\big(1 + \kappa(s)\big)\big(n - J(s-)\big)\lambda(s)ds\Big]$$
$$= \mathbb{E}^{\mathbb{Q}}\Big[\int_0^{T} e^{-\int_t^s r(u)du}\hat{G}(s)N(ds)\Big],$$

and the representation of Y follows.

(c) The form of the replicating strategy follows from (9.31). From the square integrability assumption (9.32) we deduce that the strategy (π^*, χ^*) is square integrable. Clearly, there exists a unique, càdlàg, adapted solution X^{π^*,χ^*} to (9.29). The admissibility of (π^*, χ^*) has been proved. □

Notice that if r and κ are deterministic, then $Z^E(t) = 0$ and the square integrability assumptions (9.32) are trivially satisfied. From Propositions 3.3.1 and 3.4.1 we deduce that

$$Z(t) = \mathscr{Z}(t)e^{\int_0^t r(s)ds}, \quad U(t) + \hat{G}(t) = \mathscr{U}(t)e^{\int_0^t r(s)ds}, \quad 0 \leq t \leq T,$$

where $\mathscr{Z}$ and $\mathscr{U}$ are derived from the predictable representation

$$\int_0^T e^{-\int_0^s r(u)du}dP(s) = \mathbb{E}^{\mathbb{Q}}\Big[\int_0^T e^{-\int_0^s r(u)du}dP(s)\Big]$$
$$+ \int_0^T \mathscr{Z}(s)dW^{\mathbb{Q}}(s) + \int_0^T \mathscr{U}(s)\tilde{N}^{\mathbb{Q}}(ds).$$

Theorem 9.4.1 shows that the replicating portfolio and the replicating strategy for the life insurance payment process (9.25) are characterized by the linear BSDE (9.30). The price of the payment process, the initial value of the replicating portfolio, is arbitrage-free. The unsystematic insurance risk of P is priced with the same risk premium κ which is used to price the mortality bond, which agrees with the concept of market-consistent valuation. Let us now focus on the replicating strategy and the control processes of the linear BSDE (9.30). Recalling Theorem 4.2.1, we can deduce that the amount χ^* invested in the mortality bond hedges the change in the price of the liability (including the death benefit) due to the death of a policyholder. Hence, χ^* is a delta hedging strategy for the mortality risk. The strategy π^*, which determines the amount invested in the stock, consists of two terms. From Theorem 4.2.1 we conclude that the first term hedges the change in the price of the liability due to movements in the financial market. Hence, the first term of π^* is a delta hedging strategy for the financial risk. Since the price of the mortality bond depends on the financial market through the stochastic interest rate r and the risk premium κ, the optimal strategy π^*, which hedges the financial risk, should also take into account the position in the mortality bond kept for hedging the mortality risk. Consequently, the second term of π^* hedges the change in the price of the mortality bond due to movements in the financial market.

We now study an example.

Example 9.5 We consider the classical Black-Scholes model with constant coefficients r, μ, σ and an insurer who issues a portfolio of n unit-linked endowment policies with a capital guarantee. We are interested in pricing and hedging the claim $F = (n - J(T))(K - S(T))^+$. We remark that put options on funds, which guarantee a minimal value of the terminal pay-off, are often embedded in unit-linked insurance contracts. Let us consider a constant mortality intensity λ and assume that the market believes that the unsystematic mortality risk can be diversified. Hence, $\kappa = 0$. In order to find the replicating strategy and the replicating portfolio for the claim F, we have to solve the BSDE

$$
\begin{aligned}
Y(t) = {} & \big(n - J(T)\big)\big(K - S(T)\big)^+ + \int_t^T \big(-Y(s)r - \theta Z(s)\big)ds \\
& - \int_t^T Z(s)dW(s) - \int_t^T U(s)\tilde{N}(ds), \quad 0 \le t \le T.
\end{aligned}
$$

If we change the measure, then we end up with

$$
\begin{aligned}
Y(t) = {} & \big(n - J(T)\big)\big(K - S(T)\big)^+ - \int_t^T Y(s)r ds \\
& - \int_t^T Z(s)dW^{\mathbb{Q}^*}(s) - \int_t^T U(s)\tilde{N}^{\mathbb{Q}^*}(ds), \quad 0 \le t \le T, \qquad (9.33)
\end{aligned}
$$

where $\mathbb{Q}^*$ is the equivalent martingale measure defined in Theorem 9.2.1. Let $\hat{f}(t,s) = \mathbb{E}^{\mathbb{Q}^*}[e^{-r(T-t)}(K - S^{t,s}(T))^+]$. Recalling Examples 4.1 and 9.1 and formulas (4.13), (9.6), (9.7), we can derive

$$\begin{aligned}\hat{f}(t,s) &= Ke^{-r(T-t)}\Phi\big(-d(t,s)-\sigma\sqrt{T-t}\big)\\ &\quad - s\Phi\big(-d(t,s)\big), \quad (t,s)\in[0,T]\times(0,\infty),\\ \hat{f}_s(t,s) &= -\Phi\big(-d(t,s)\big), \quad (t,s)\in[0,T)\times(0,\infty).\end{aligned} \tag{9.34}$$

From Proposition 8.1.1 we can now deduce that the solution to the BSDE (9.33) takes the form

$$\begin{aligned}Y(t) &= \big(n-J(t)\big)e^{-\lambda(T-t)}\big(Ke^{-r(T-t)}\Phi\big(-d\big(t,S(t)\big)-\sigma\sqrt{T-t}\big)\\ &\quad - S(t)\Phi\big(-d\big(t,S(t)\big)\big)\big), \quad 0\le t\le T,\\ Z(t) &= -\big(n-J(t-)\big)e^{-\lambda(T-t)}\sigma S(t)\Phi\big(-d\big(t,S(t)\big)\big), \quad 0\le t\le T,\\ U(t) &= -e^{-\lambda(T-t)}\big(Ke^{-r(T-t)}\Phi\big(-d\big(t,S(t)\big)-\sigma\sqrt{T-t}\big)\\ &\quad - S(t)\Phi\big(-d\big(t,S(t)\big)\big)\big), \quad 0\le t\le T.\end{aligned}$$

By Theorem 9.4.1 the replicating strategy is given by the formulas

$$\begin{aligned}\pi^*(t) &= -\big(n-J(t-)\big)e^{-\lambda(T-t)}S(t)\Phi\big(-d\big(t,S(t)\big)\big), \quad 0\le t\le T,\\ \chi^*(t) &= \big(n-J(t-)\big)e^{-\lambda(T-t)}\big(Ke^{-r(T-t)}\Phi\big(-d\big(t,S(t)\big)-\sigma\sqrt{T-t}\big)\\ &\quad - S(t)\Phi\big(-d\big(t,S(t)\big)\big)\big), \quad 0\le t\le T,\end{aligned}$$

and the price of the claim, and the initial value of the replicating portfolio, is equal to

$$X^*(0)=Y(0)=ne^{-\lambda T}\big(Ke^{-rT}\Phi\big(-d\big(0,S(0)\big)-\sigma\sqrt{T}\big)-S(0)\Phi\big(-d\big(0,S(0)\big)\big)\big).$$

The value of the replicating portfolio is determined by the process

$$\begin{aligned}X^*(t)=Y(t) &= \big(n-J(t)\big)e^{-\lambda(T-t)}\big(Ke^{-r(T-t)}\Phi\big(-d\big(t,S(t)\big)-\sigma\sqrt{T-t}\big)\\ &\quad - S(t)\Phi\big(-d\big(t,S(t)\big)\big)\big), \quad 0\le t\le T.\end{aligned}$$

The results of this chapter can also be applied if we aim to hedge defaultable securities with a defaultable bond, see Example 7.10. We remark that in credit risk models it is usually assumed that the default intensity λ is an $\mathscr{F}^W$-predictable process. Since we allow for an $\mathscr{F}^W$-predictable mortality intensity under $\mathbb{Q}$, the replicating strategy for defaultable securities can be found by following the same arguments that led to Theorem 9.4.1, see Blanchet-Scalliet et al. (2008), Blanchet-Scalliet and Jeanblanc (2004) and Kharroubi and Lim (2012).

Bibliographical Notes For arbitrage-free pricing theory we refer to Shreve (2004). Market-consistent valuation of insurance liabilities is discussed in Møller and Steffensen (2007), see also European Commission QIS5 (2010). The results of Sect. 9.2 are classical in financial mathematics, see Shreve (2004) and El Karoui et al. (1997b). Example 9.2 is taken from El Karoui et al. (1997b). In the proof of Theorem 9.3.3 we follow the arguments from El Karoui and Quenez (1995) by adapting them to our setting with the payment process. The proof of Proposition 9.3.2 and Example 9.4 are taken from El Karoui and Quenez (1995). For superhedging under constraints we refer to Bender and Kohlmann (2008). Counterparts of Proposition 9.4.1 and Theorem 9.4.1 for credit risk models can be found in Blanchet-Scalliet et al. (2008) and Blanchet-Scalliet and Jeanblanc (2004).

Chapter 10
Quadratic Pricing and Hedging

Abstract We investigate pricing and hedging of the insurance payment process under quadratic objectives. Four types of quadratic loss functions are considered. First, we deal with a minimal hedging error in a mean-square sense. The hedging error is evaluated both under an equivalent martingale measure and the real-world measure. Next, we investigate locally risk minimizing strategies which lead to non-self-financing investment portfolio processes. Finally, we minimize an instantaneous mean-variance risk measure of the insurer's surplus to derive a hedging strategy. The pricing and hedging strategies are characterized by linear and nonlinear BSDEs.

Since a self-financing investment portfolio cannot perfectly replicate the insurance payment process, we aim to find an investment strategy which hedges the payment process with a minimal replication error. It is natural to measure the replication error with a quadratic function. In this chapter we investigate four types of quadratic objectives for pricing and hedging in incomplete markets. Quadratic objectives have gained their importance in portfolio optimization since Markowitz (1952) solved a one-period portfolio selection problem. We point out that the Markowitz mean-variance portfolio selection is used in practice for investment decision making and asset-liability management, see Zenios and Ziemba (2006) and Adam (2007). Let us also remark that replicating portfolios for insurance liabilities are constructed in practice by means of a mean-square criterion, see Boekel et al. (2009).

10.1 Quadratic Pricing and Hedging Under an Equivalent Martingale Measure

Let an equivalent martingale measure $\mathbb{Q} \in \mathscr{Q}^m$ be given. We formulate our pricing and hedging objective under the measure $\mathbb{Q}$. First, we define a class of admissible strategies under $\mathbb{Q}$.

Definition 10.1.1 A strategy $\pi := (\pi(t), 0 \leq t \leq T)$ is called admissible under the measure $\mathbb{Q}$, written $\pi \in \mathscr{A}^{\mathbb{Q}}$, if it satisfies the conditions:

Ł. Delong, *Backward Stochastic Differential Equations with Jumps and Their Actuarial and Financial Applications*, EAA Series, DOI 10.1007/978-1-4471-5331-3_10,

(i) $\pi : [0, T] \times \Omega \to \mathbb{R}$ is a predictable process,
(ii) $\mathbb{E}^{\mathbb{Q}}[\int_0^T |\pi(t)\sigma(t)|^2 dt] < \infty$,
(iii) there exists a unique càdlàg, adapted solution X^π to (7.11) on $[0, T]$.

If assumptions (C2)–(C4) hold under $\mathbb{Q}$, then the wealth process X^π is square integrable under $\mathbb{Q}$ for any admissible strategy $\pi \in \mathscr{A}^{\mathbb{Q}}$, see (7.12).

The insurer faces the payment process (7.3). We control the wealth of the insurer at time T which may denote the planning horizon in an ALM study or the point of time when the insurance portfolio terminates. The goal is to find an initial capital x and an admissible investment strategy $\pi \in \mathscr{A}^{\mathbb{Q}}$ which minimize the mean-square hedging error

$$\inf_{x, \pi \in \mathscr{A}^{\mathbb{Q}}} \mathbb{E}^{\mathbb{Q}}\big[\big|e^{-\int_0^T r(s)ds} X^{\pi,x}(T) - e^{-\int_0^T r(s)ds} F\big|^2\big], \tag{10.1}$$

where the investment portfolio process $X^{\pi,x}$ is given by (7.11). Hence, we aim to find an investment portfolio which hedges the insurance payment process with a minimal replication error at time T. The use of the mean-square objective for measuring the replication error is very natural. The formulation of the objective under an equivalent martingale measure is mathematically and practically convenient, as we discuss later. The optimal capital, or the initial value of the optimal hedging portfolio, yields the price of the payment process.

Theorem 10.1.1 *Let an equivalent martingale measure $\mathbb{Q} \in \mathscr{Q}^m$ be given and assume that* (C1)–(C4) *from Chap.* 7 *hold under $\mathbb{Q}$. We consider the quadratic pricing and hedging problem* (10.1).

(a) *There exists a unique solution $(Y, Z_1, Z_2, U) \in \mathbb{S}^2(\mathbb{R}) \times \mathbb{H}^2(\mathbb{R}) \times \mathbb{H}^2(\mathbb{R}) \times \mathbb{H}^2_N(\mathbb{R})$ to the BSDE*

$$\begin{aligned} Y(t) = {} & \int_0^T e^{-\int_0^s r(u)du} dP(s) - \int_t^T Z_1(s)dW^{\mathbb{Q}}(s) - \int_t^T Z_2(s)dB^{\mathbb{Q}}(s) \\ & - \int_t^T \int_{\mathbb{R}} U(s,z)\tilde{N}^{\mathbb{Q}}(ds,dz), \quad 0 \le t \le T, \end{aligned} \tag{10.2}$$

where the square integrability conditions for (Y, Z_1, Z_2, U) hold under $\mathbb{Q}$.

(b) *The optimal initial value of the hedging portfolio x^* and the optimal admissible hedging strategy $\pi^* \in \mathscr{A}^{\mathbb{Q}}$ for the payment process P take the form*

$$x^* = Y(0) = \mathbb{E}^{\mathbb{Q}}\left[\int_0^T e^{-\int_0^s r(u)du} dP(s)\right],$$

$$\pi^*(t) = \frac{Z_1(t)}{\sigma(t)} e^{\int_0^t r(s)ds}, \quad 0 \le t \le T.$$

Proof (a) The assertion follows from Theorem 3.1.1.

(b) We recall that the dynamics of $X^{\pi,x}$ is described by (7.11). By the change of measure, the Girsanov's theorem and discounting we obtain

$$\begin{aligned} &e^{-\int_0^T r(s)ds} X^{\pi,x}(T) \\ &\quad = x + \int_0^T e^{-\int_0^s r(u)du} \pi(s)\sigma(s)dW^{\mathbb{Q}}(s) \\ &\qquad - \int_0^T e^{-\int_0^s r(u)du} H(s)ds - \int_0^T \int_{\mathbb{R}} e^{-\int_0^s r(u)du} G(s,z)N(ds,dz). \end{aligned} \tag{10.3}$$

From (10.3) we conclude that the mean-square error (10.1) is given by

$$\mathbb{E}^{\mathbb{Q}}\Big[\Big|x + \int_0^T e^{-\int_0^s r(u)du}\pi(s)\sigma(s)dW^{\mathbb{Q}}(s) - \int_0^T e^{-\int_0^s r(u)du} dP(s)\Big|^2\Big],$$

and by (10.2) we get

$$\begin{aligned} \mathbb{E}^{\mathbb{Q}}\Big[\Big|&x - Y(0) + \int_0^T \big(e^{-\int_0^s r(u)du}\pi(s)\sigma(s) - Z_1(s)\big)dW^{\mathbb{Q}}(s) \\ &- \int_0^T Z_2(s)dB^{\mathbb{Q}}(s) - \int_0^T \int_{\mathbb{R}} U(s,z)\tilde{N}^{\mathbb{Q}}(ds,dz)\Big|^2\Big]. \end{aligned}$$

Taking the squares and using moment properties of stochastic integrals, see Theorem 2.3.3, we derive

$$\begin{aligned} &\big(x - Y(0)\big)^2 + \mathbb{E}^{\mathbb{Q}}\Big[\int_0^T \big|e^{-\int_0^s r(u)du}\pi(s)\sigma(s) - Z_1(s)\big|^2 ds\Big] \\ &\quad + \mathbb{E}^{\mathbb{Q}}\Big[\int_0^T \big|Z_2(s)\big|^2 ds\Big] + \mathbb{E}^{\mathbb{Q}}\Big[\int_0^T \int_{\mathbb{R}} \big|U(s,z)\big|^2 Q(s,dz)\eta(s)ds\Big] \\ &\quad - 2\mathbb{E}^{\mathbb{Q}}\Big[\int_0^T \big(e^{-\int_0^s r(u)du}\pi(s)\sigma(s) - Z_1(s)\big)dW^{\mathbb{Q}}(s) \int_0^T Z_2(s)dB^{\mathbb{Q}}(s)\Big] \\ &\quad - 2\mathbb{E}^{\mathbb{Q}}\Big[\int_0^T \big(e^{-\int_0^s r(u)du}\pi(s)\sigma(s) - Z_1(s)\big)dW^{\mathbb{Q}}(s) \int_0^T \int_{\mathbb{R}} U(s,z)\tilde{N}^{\mathbb{Q}}(ds,dz)\Big] \\ &\quad + 2\mathbb{E}^{\mathbb{Q}}\Big[\int_0^T Z_2(s)dB^{\mathbb{Q}}(s) \int_0^T \int_{\mathbb{R}} U(s,z)\tilde{N}^{\mathbb{Q}}(ds,dz)\Big]. \end{aligned}$$

Finally, property (5.5) yields the hedging error

$$\begin{aligned} &\big(x - Y(0)\big)^2 + \mathbb{E}^{\mathbb{Q}}\Big[\int_0^T \big|e^{-\int_0^s r(u)du}\pi(s)\sigma(s) - Z_1(s)\big|^2 ds\Big] \\ &\quad + \mathbb{E}^{\mathbb{Q}}\Big[\int_0^T \big|Z_2(s)\big|^2 ds\Big] + \mathbb{E}^{\mathbb{Q}}\Big[\int_0^T \int_{\mathbb{R}} \big|U(s,z)\big|^2 Q(s,dz)\eta(s)ds\Big], \end{aligned}$$

and the optimality of (x^*, π^*) follows. The admissability of π^* is clear. □

The price and the hedging strategy for the payment process P are characterized by the predictable representation of the discounted payments (or by the linear BSDE (10.2)). The price x^* is arbitrage-free. Recalling the interpretation given in Sect. 9.2 and Theorem 4.2.1, we can conclude that the hedging strategy π^* is a delta-hedging strategy. The hedging strategy π^* is updated with the current information on the financial and the insurance risk.

Example 10.1 We consider the classical Black-Scholes model with constant coefficients r, μ, σ and an insurer who issues a portfolio of n unit-linked endowment policies with a capital guarantee. We are interested in pricing and hedging the claim $F = (n - J(T))(K - S(T))^+$ where J is the deaths counting process for the insurance portfolio. We consider a constant mortality intensity λ and we choose the equivalent martingale measure $\mathbb{Q}^*$ defined in Theorem 9.2.1. In order to find the price and the optimal hedging strategy for the claim F, we have to solve the BSDE

$$\begin{aligned} Y(t) &= e^{-rT}\big(n - J(T)\big)\big(K - S(T)\big)^+ \\ &\quad - \int_t^T Z(s)dW^{\mathbb{Q}^*}(s) - \int_t^T U(s)\tilde{N}^{\mathbb{Q}^*}(ds), \quad 0 \le t \le T. \end{aligned} \tag{10.4}$$

From Proposition 8.1.1 and formula (9.34) we conclude that the solution to the BSDE (10.4) is given by the triple

$$\begin{aligned} Y(t) &= \big(n - J(t)\big)e^{-\lambda(T-t)}e^{-rt}\big(Ke^{-r(T-t)}\Phi\big(-d\big(t, S(t)\big) - \sigma\sqrt{T-t}\big) \\ &\quad - S(t)\Phi\big(-d\big(t, S(t)\big)\big)\big), \quad 0 \le t \le T, \\ Z(t) &= -\big(n - J(t-)\big)e^{-\lambda(T-t)}\sigma S(t)e^{-rt}\Phi\big(-d\big(t, S(t)\big)\big), \quad 0 \le t \le T, \\ U(t) &= -e^{-\lambda(T-t)}e^{-rt}\big(Ke^{-r(T-t)}\Phi\big(-d\big(t, S(t)\big) - \sigma\sqrt{T-t}\big) \\ &\quad - S(t)\Phi\big(-d\big(t, S(t)\big)\big)\big), \quad 0 \le t \le T. \end{aligned}$$

By Theorem 10.1.1 the optimal hedging strategy is given by the formula

$$\pi^*(t) = -\big(n - J(t-)\big)e^{-\lambda(T-t)}S(t)\Phi\big(-d\big(t, S(t)\big)\big), \quad 0 \le t \le T,$$

and the price of the claim, and the initial value of the optimal hedging portfolio, is equal to

$$x^* = Y(0) = ne^{-\lambda T}\big(Ke^{-rT}\Phi\big(-d\big(0, S(0)\big) - \sigma\sqrt{T}\big) - S(0)\Phi\big(-d\big(0, S(0)\big)\big)\big).$$

The value of the optimal hedging portfolio is determined by the process

$$X^*(t) = x^*e^{rt} + \int_0^t \pi^*(s)e^{r(t-s)}\big(\theta\sigma ds + \sigma dW(s)\big), \quad 0 \le t \le T.$$

The solution to the optimization problem (10.1) is easy to derive but the objective (10.1) has two drawback from the financial point of view. Firstly, we have to choose the equivalent martingale measure $\mathbb{Q}$. This choice determines the price. When we investigate pricing in incomplete markets we expect to find the equivalent martingale measure. Consequently, the equivalent martingale measure used for pricing should be the output of an optimization problem, rather than the input. Secondly, the use of an equivalent martingale measure for evaluating the hedging error is questionable. Profits and losses or the performance of the hedging portfolio should be evaluated under the real-world measure.

10.2 Quadratic Pricing and Hedging Under the Real-World Measure

We aim to find an investment portfolio which hedges the insurance payment process (7.3) with a minimal replication error at time T and we evaluate the performance of the hedging portfolio under the real-world measure. The goal is to find an initial capital x and an admissible investment strategy $\pi \in \mathscr{A}$ which minimize the mean-square hedging error

$$\inf_{x,\pi\in\mathscr{A}} \mathbb{E}\big[\big|X^{\pi,x}(T) - F\big|^2\big], \tag{10.5}$$

where $X^{\pi,x}$ is given by (7.11), and we use the class of admissible strategies $\mathscr{A}$ from Definition 7.3.1. If one prefers to minimize the hedging error (10.5) for the discounted quantities as in (10.1), then obvious modifications should be introduced.

We deal with two backward stochastic differential equations

$$\begin{aligned} Y(t) = 1 &+ \int_t^T \left(2Y(s)r(s) - \frac{|Z(s)|^2}{Y(s)} - \big|\theta(s)\big|^2 Y(s) - 2\theta(s)Z(s)\right)ds \\ &- \int_t^T Z(s)dW(s), \quad 0 \le t \le T, \end{aligned} \tag{10.6}$$

and

$$\begin{aligned} \mathscr{Y}(t) = F &+ \int_t^T \bigg(-\mathscr{Y}(s-)r(s) + H(s) + \int_{\mathbb{R}} G(s,z)Q(s,dz)\eta(s) \\ &- \mathscr{Z}_1(s)\theta(s)\bigg)ds - \int_t^T \mathscr{Z}_1(s)dW(s) \\ &- \int_t^T \mathscr{Z}_2(s)dB(s) - \int_t^T\int_{\mathbb{R}} \mathscr{U}(s,z)\tilde{N}(ds,dz), \quad 0 \le t \le T. \end{aligned} \tag{10.7}$$

Equation (10.6) is called a stochastic Riccati equation. First, we establish existence and uniqueness results for these two BSDEs.

Proposition 10.2.1 *Assume that* (C1)–(C2) *from Chap. 7 hold. There exists a unique solution* $(Y, Z) \in \mathbb{S}^{\infty}(\mathbb{R}) \times \mathbb{H}^2(\mathbb{R})$ *to the BSDE* (10.6) *such that* $Y(t) \geq k > 0$, $0 \leq t \leq T$. *Moreover, the process* $(\int_0^t Z(s)dW(s), 0 \leq t \leq T)$ *is a BMO-martingale.*

Proof Since we look for a process Y which is uniformly bounded away from zero, we can introduce new variables

$$\hat{Y}(t) = \frac{1}{Y(t)}, \quad \hat{Z}(t) = \frac{-Z(t)}{|Y(t)|^2}, \quad 0 \leq t \leq T.$$

The Itô's formula yields

$$\begin{aligned}\hat{Y}(t) = 1 &+ \int_t^T \big(\big(|\theta(s)|^2 - 2r(s)\big)\hat{Y}(s) - 2\theta(s)\hat{Z}(s)\big)ds \\ &- \int_t^T \hat{Z}(s)dW(s), \quad 0 \leq t \leq T.\end{aligned} \tag{10.8}$$

We end up with a linear BSDE. By Proposition 3.3.1 there exists a unique solution $(\hat{Y}, \hat{Z}) \in \mathbb{S}^2(\mathbb{R}) \times \mathbb{H}^2(\mathbb{R})$ to the linear BSDE (10.8). From the representation of the solution $\hat{Y}$ and boundedness of θ and r we deduce that $0 < k \leq \hat{Y}(t) \leq K$, $0 \leq t \leq T$. Hence, we conclude that there exists a unique solution (Y, Z) to the nonlinear BSDE (10.6) and $0 < k \leq Y(t) \leq K$, $0 \leq t \leq T$. We now prove the *BMO* property. By the Itô's formula we obtain

$$d\big(|Y(t)|^2\big) = 2Y(t)dY(t) + |Z(t)|^2dt,$$

and

$$\begin{aligned}&\int_t^T |Z(s)|^2ds \\ &\quad = 1 - |Y(t)|^2 \\ &\qquad + \int_t^T 2Y(s)\Big(2Y(s)r(s) - \frac{|Z(s)|^2}{Y(s)} - |\theta(s)|^2Y(s) - 2\theta(s)Z(s)\Big)ds \\ &\qquad - \int_t^T 2Y(s)Z(s)dW(s) \\ &\quad \leq K + K\int_t^T |Z(s)|ds - \int_t^T 2Y(s)Z(s)dW(s) \\ &\quad \leq K + K\frac{1}{\alpha}\int_t^T |Z(s)|^2ds + K\alpha - \int_t^T 2Y(s)Z(s)dW(s), \quad 0 \leq t \leq T,\end{aligned} \tag{10.9}$$

where we use boundedness of r, θ and Y, strict positivity of Y and inequality (3.10). Choosing α sufficiently large and taking the expectation, we derive

$$\mathbb{E}\Big[\int_t^T |Z(s)|^2 ds|\mathscr{F}_t\Big] \leq K, \quad 0 \leq t \leq T, \tag{10.10}$$

and the *BMO* property is proved. □

We will need higher moment estimates for the control process Z. From estimate (10.9) we deduce that for any $p \geq 2$ we have

$$\Big(\int_0^T |Z(s)|^2 ds\Big)^p \leq K + K\frac{1}{\alpha^p}\Big(\int_0^T |Z(s)|^2 ds\Big)^p + K\alpha^p$$
$$+ \Big|\int_0^T 2Y(s)Z(s)dW(s)\Big|^p.$$

We choose α sufficiently large, take the expected value, apply the Burkholder-Davis-Gundy inequality and we obtain the inequality

$$\mathbb{E}\Big[\Big(\int_0^T |Z(s)|^2 ds\Big)^p\Big] \leq K_p + K_p\mathbb{E}\Big[\Big(\int_0^T |Z(s)|^2 ds\Big)^{p/2}\Big]. \tag{10.11}$$

Starting with (10.10), by iteration we can derive the moment estimate

$$\mathbb{E}\Big[\Big(\int_0^T |Z(s)|^2 ds\Big)^p\Big] \leq K_p, \quad p \geq 1. \tag{10.12}$$

We use estimate (10.12) in the sequel.

We now investigate the second BSDE.

Proposition 10.2.2 *Assume that* (C1)–(C4) *from Chap. 7 hold.*

(a) *There exists a unique solution* $(\mathscr{Y}, \mathscr{Z}_1, \mathscr{Z}_2, \mathscr{U}) \in \mathbb{S}^2(\mathbb{R}) \times \mathbb{H}^2(\mathbb{R}) \times \mathbb{H}^2(\mathbb{R}) \times \mathbb{H}^2_N(\mathbb{R})$ *to the BSDE* (10.7). *The process* $\mathscr{Y}$ *has the representation*

$$\begin{aligned}\mathscr{Y}(t) &= \mathbb{E}^{\mathbb{Q}^*}\Big[e^{-\int_t^T r(s)ds}F \\ &\quad + \int_t^T e^{-\int_t^s r(u)du}\Big(H(s) + \int_{\mathbb{R}} G(s,z)Q(s,dz)\eta(s)\Big)ds|\mathscr{F}_t\Big] \\ &= \mathbb{E}^{\mathbb{Q}^*}\Big[\int_t^T e^{-\int_t^s r(u)du}dP(s)|\mathscr{F}_t\Big], \quad 0 \leq t \leq T,\end{aligned} \tag{10.13}$$

where the equivalent martingale measure $\mathbb{Q}^*$ *is given by*

$$\frac{d\mathbb{Q}^*}{d\mathbb{P}}\Big|\mathscr{F}_t = e^{-\int_0^t \theta(s)dW(s) - \frac{1}{2}\int_0^t |\theta(s)|^2 ds}, \quad 0 \leq t \leq T.$$

(b) *If*

$$\mathbb{E}\Big[|F|^p + \Big(\int_0^T |H(s)|^2 ds\Big)^{p/2} + \Big(\int_0^T \int_{\mathbb{R}} |G(s,z)\eta(s)|^2 Q(s,dz)ds\Big)^{p/2}\Big] < \infty,$$

for some $p > 2$, then

$$\mathbb{E}\Big[\sup_{0\le t\le T} |\mathscr{Y}(t)|^p\Big] < \infty. \tag{10.14}$$

Proof (a) The existence and the uniqueness follow from Theorem 3.1.1. Representation (10.13) can be derived by following the arguments from Proposition 3.3.1 and the fact that the compensator of N remains unchanged under $\mathbb{Q}^*$, see also the proof of Theorem 9.4.1.

(b) By the Burkholder-Davis-Gundy inequality, the dynamics of the BSDE (10.7), the Cauchy-Schwarz inequality and boundedness of r, θ we obtain

$$\begin{aligned}
&\mathbb{E}\Big[\Big(\int_0^T |\mathscr{Z}_1(s)|^2\Big)^{p/2}\Big] \\
&\le \mathbb{E}\Big[\Big(\int_0^T |\mathscr{Z}_1(s)|^2 ds + \int_0^T |\mathscr{Z}_2(s)|^2 ds + \int_0^T \int_{\mathbb{R}} |\mathscr{U}(s,z)|^2 N(ds,dz)\Big)^{p/2}\Big] \\
&\le K\mathbb{E}\Big[\sup_{0\le t\le T}\Big|\int_0^t \mathscr{Z}_1(s)dW(s) + \int_0^t \mathscr{Z}_2(s)dB(s) + \int_0^t\int_{\mathbb{R}} \mathscr{U}(s,z)\tilde{N}(ds,dz)\Big|^p\Big] \\
&= K\mathbb{E}\Big[\sup_{0\le t\le T}\Big|\mathscr{Y}(t) - \mathscr{Y}(0) + \int_0^t \Big(-\mathscr{Y}(s)r(s) + H(s) \\
&\quad + \int_{\mathbb{R}} G(s,z)Q(s,dz)\eta(s) - \mathscr{Z}_1(s)\theta(s)\Big)ds\Big|^p\Big] \\
&\le K\mathbb{E}\Big[(1+T^p)\sup_{0\le t\le T}|\mathscr{Y}(t)|^p + T^{p/2}\Big(\int_0^T |H(s)|^2 ds\Big)^{p/2} \\
&\quad + T^{p/2}\Big(\int_0^T\int_{\mathbb{R}} |G(s,z)\eta(s)|^2 Q(s,dz)ds\Big)^{p/2} + T^{p/2}\Big(\int_0^T |\mathscr{Z}_1(s)|^2 ds\Big)^{p/2}\Big].
\end{aligned}$$

Assume now that the time horizon T is sufficiently small. We get the inequality

$$\mathbb{E}\Big[\Big(\int_0^T |\mathscr{Z}_1(s)|^2\Big)^{p/2}\Big] \le \frac{KT^{p/2}}{1-KT^{p/2}} + \frac{K(1+T^p)}{1-KT^{p/2}}\mathbb{E}\Big[\sup_{0\le t\le T}|\mathscr{Y}(t)|^p\Big]. \tag{10.15}$$

From representation (10.7), the Cauchy-Schwarz inequality and boundedness of r and θ we deduce

$$
\begin{aligned}
|\mathscr{Y}(t)|^p &= \Big|\mathbb{E}\Big[F + \int_t^T \Big(-\mathscr{Y}(s)r(s) + H(s) \\
&\quad + \int_{\mathbb{R}} G(s,z)Q(s,dz)\eta(s) - \mathscr{Z}_1(s)\theta(s)\Big)ds|\mathscr{F}_t\Big]\Big|^p \\
&\leq \hat{K}\Big(\mathbb{E}\Big[|F|^2 + T^2 \sup_{0\leq s\leq T}|\mathscr{Y}(s)|^2 + T\int_0^T |H(s)|^2 ds \\
&\quad + T\int_0^T\int_{\mathbb{R}} |G(s,z)\eta(s)|^2 Q(s,dz)ds + T\int_0^T |\mathscr{Z}_1(s)|^2 ds|\mathscr{F}_t\Big]\Big)^{p/2}, \\
&\quad 0\leq t\leq T,
\end{aligned}
$$

which leads to the estimate

$$
\begin{aligned}
&\mathbb{E}\Big[\sup_{0\leq t\leq T}|\mathscr{Y}(t)|^p\Big] \\
&\leq \hat{K}\mathbb{E}\Big[|F|^p + T^{p/2}\Big(\int_0^T |H(s)|^2 ds\Big)^{p/2} \\
&\quad + T^{p/2}\Big(\int_0^T\int_{\mathbb{R}} |G(s,z)\eta(s)|^2 Q(s,dz)ds\Big)^{p/2} \\
&\quad + T^p \sup_{0\leq t\leq T}|\mathscr{Y}(t)|^p + T^{p/2}\Big(\int_0^T |\mathscr{Z}_1(s)|^2\Big)^{p/2}\Big] \\
&\leq \hat{K}\Big(1 + T^{p/2} + T^p\mathbb{E}\Big[\sup_{0\leq t\leq T}|\mathscr{Y}(t)|^p\Big] + T^{p/2}\mathbb{E}\Big[\Big(\int_0^T |\mathscr{Z}_1(s)|^2\Big)^{p/2}\Big]\Big) \\
&\leq \hat{K}\Big(1 + T^{p/2} + \frac{T^p}{1-KT^{p/2}} + \Big(T^p + T^{p/2}\frac{1+T^p}{1-KT^{p/2}}\Big)\mathbb{E}\Big[\sup_{0\leq t\leq T}|\mathscr{Y}(t)|^p\Big]\Big),
\end{aligned}
$$

where we use the Doob's martingale inequality, the Jensen's inequality and inequality (10.15). If T is sufficiently small, then we conclude that $\mathbb{E}[\sup_{0\leq t\leq T}|\mathscr{Y}(t)|^p] < K_{T,p}$. To prove estimate (10.14) for an arbitrary T, we divide the interval $[0,T]$ into sufficiently small subintervals $[T_i, T_{i+1}]$ and we consider the BSDEs

$$
\begin{aligned}
\mathscr{Y}(t) &= \mathscr{Y}(T_{i+1}) + \int_t^{T_{i+1}}\Big(-\mathscr{Y}(s-)r(s) + H(s) + \int_{\mathbb{R}} G(s,z)Q(s,dz)\eta(s) \\
&\quad - \mathscr{Z}_1(s)\theta(s)\Big)ds - \int_t^{T_{i+1}} \mathscr{Z}_1(s)dW(s) \\
&\quad - \int_t^{T_{i+1}} \mathscr{Z}_2(s)dB(s) - \int_t^{T_{i+1}}\int_{\mathbb{R}} \mathscr{U}(s,z)\tilde{N}(ds,dz), \quad T_i\leq t\leq T_{i+1}.
\end{aligned}
$$

We can establish the assertion $\mathbb{E}[\sup_{T_i \le t \le T_{i+1}} |\mathscr{Y}(t)|^p] < K_{i,p}$ on each subinterval. The global estimate for $\mathscr{Y}$ on $[0,T]$ follows by combining a finite number of local estimates. □

We now derive the solution the quadratic optimization problem (10.5).

Theorem 10.2.1 *Assume that* (C1)–(C4) *from Chap.* 7 *hold, and let*

$$\mathbb{E}\bigg[|F|^p + \Big(\int_0^T |H(s)|^2 ds\Big)^{p/2} + \Big(\int_0^T \int_{\mathbb{R}} |G(s,z)\eta(s)|^2 Q(s,dz)ds\Big)^{p/2}\bigg] < \infty,$$

for some $p > 2$. *We consider the quadratic hedging problem* (10.5). *The strategy of the form*

$$\begin{aligned}\pi^*(t) &= \frac{\mathscr{Z}_1(t)}{\sigma(t)}\\ &\quad - \Big(\frac{\mu(t)-r(r)}{\sigma^2(t)} + \frac{Z(t)}{Y(t)\sigma(t)}\Big)\big(X^*(t-)-\mathscr{Y}(t-)\big), \quad 0 \le t \le T, \quad (10.16)\end{aligned}$$

where (Y,Z) *and* $(\mathscr{Y},\mathscr{Z})$ *solve the BSDEs* (10.6) *and* (10.7), *and the process* X^* *is given by*

$$\begin{aligned}dX^*(t) &= \pi^*(t)\big(\mu(t)dt + \sigma(t)dW(t)\big) + \big(X^*(t-)-\pi^*(t)\big)r(t)dt,\\ &\quad - H(t)dt - \int_{\mathbb{R}} G(t,z)N(dt,dz), \quad X(0) = x > 0,\end{aligned}$$

are the optimal admissible hedging strategy $\pi^* \in \mathscr{A}$ *and the optimal hedging portfolio for the payment process* P.

Proof 1. The optimality. The proof is based on the method of completing the squares. Consider the BSDEs (10.6) and (10.7) written in the shorthand notation

$$\begin{aligned}dY(t) &= -f(t)dt + Z(t)dW(t), \quad Y(T) = 1,\\ d\mathscr{Y}(t) &= -f'(t)dt + \mathscr{Z}_1(t)dW(t) + \mathscr{Z}_2(t)dB(t)\\ &\quad + \int_{\mathbb{R}} \mathscr{U}(t,z)\tilde{N}(dt,dz), \quad \mathscr{Y}(T) = F,\end{aligned}$$

where the generators f and f' are appropriately defined. Key properties of the solutions to the BSDEs (10.6) and (10.7) are established in Propositions 10.2.1 and 10.2.2. We use these properties in the proof. We introduce the process

$$\hat{\mathscr{Y}}(t) = -2Y(t)\mathscr{Y}(t), \quad 0 \le t \le T.$$

The Itô's formula yields the dynamics

$$\begin{aligned}&d\hat{\mathscr{Y}}(t) = -\hat{f}(t)dt + \hat{\mathscr{Z}}_1(t)dW(t) + \hat{\mathscr{Z}}_2(t)dB(t) + \int_{\mathbb{R}} \hat{\mathscr{U}}(t,z)\tilde{N}(dt,dz),\\ &\hat{\mathscr{Y}}(T) = -2F,\end{aligned}$$

where

$$\begin{aligned}
\hat{f}(t) &= -2\big(Y(t)f'(t) + \mathscr{Y}(t)f(t) - \mathscr{Z}_1(t)Z(t)\big), \quad 0 \le t \le T,\\
\hat{\mathscr{Z}}_1(t) &= -2\big(Y(t)\mathscr{Z}_1(t) + \mathscr{Y}(t)Z(t)\big), \quad 0 \le t \le T,\\
\hat{\mathscr{Z}}_2(t) &= -2Y(t)\mathscr{Z}_2(t), \quad 0 \le t \le T,\\
\hat{\mathscr{U}}(t,z) &= -2Y(t)\mathscr{U}(t,z), \quad 0 \le t \le T,
\end{aligned} \tag{10.17}$$

and

$$\begin{aligned}
\hat{f}(t) = {} & \hat{\mathscr{Y}}(t)r(t) - 2Y(t)H(t) - 2Y(t)\int_{\mathbb{R}} G(t,z)Q(t,dz)\eta(t)\\
& - \theta(t)\frac{\hat{\mathscr{Y}}(t)(\mu(t) - r(t)) + \hat{\mathscr{Z}}_1(t)\sigma(t)}{\sigma(t)}\\
& - Z(t)\frac{\hat{\mathscr{Y}}(t)(\mu(t) - r(t)) + \hat{\mathscr{Z}}_1(t)\sigma(t)}{Y(t)\sigma(t)}, \quad 0 \le t \le T.
\end{aligned} \tag{10.18}$$

We choose an admissible strategy $\pi \in \mathscr{A}$ and we consider the investment portfolio process X^{π} given by (7.11). By the Itô's formula we derive the dynamics

$$\begin{aligned}
& d\big(Y(t)\big(X^{\pi}(t)\big)^2\big)\\
&= Y(t)\Big(2X^{\pi}(t-)\pi(t)\big(\mu(t)dt + \sigma(t)dW(t)\big)\\
&\quad + 2X^{\pi}(t-)\big(X^{\pi}(t-) - \pi(t)\big)r(t)dt - 2X^{\pi}(t-)H(t)dt\\
&\quad - \int_{\mathbb{R}} 2X^{\pi}(t-)G(t,z)N(dt,dz) + \big|\pi(t)\sigma(t)\big|^2 dt + \int_{\mathbb{R}} \big|G(t,z)\big|^2 N(dt,dz)\Big)\\
&\quad + \big|X^{\pi}(t-)\big|^2\big(-f(t)dt + Z(t)dW(t)\big) + 2Z(t)X^{\pi}(t)\pi(t)\sigma(t)dt,
\end{aligned}$$

and

$$\begin{aligned}
& d\big(\hat{\mathscr{Y}}(t)X^{\pi}(t)\big)\\
&= \hat{\mathscr{Y}}(t-)\Big(\pi(t)\big(\mu(t)dt + \sigma(t)dW(t)\big)\\
&\quad + \big(X^{\pi}(t-) - \pi(t)\big)r(t)dt - H(t)dt - \int_{\mathbb{R}} G(t,z)N(dt,dz)\Big)\\
&\quad + X^{\pi}(t-)\Big(-\hat{f}(t)dt + \hat{\mathscr{Z}}_1(t)dW(t) + \hat{\mathscr{Z}}_2(t)dB(t) + \int_{\mathbb{R}} \hat{\mathscr{U}}(t,z)\tilde{N}(dt,dz)\Big)\\
&\quad + \hat{\mathscr{Z}}_1(t)\pi(t)\sigma(t)dt - \int_{\mathbb{R}} \hat{\mathscr{U}}(t,z)G(t,z)N(dt,dz).
\end{aligned}$$

We can notice that the stochastic integrals with respect to W, B and $\tilde{N}$ are locally square integrable local martingales since X^{π} and $\hat{\mathscr{Y}}$ are càdlàg, Y is bounded, π, $Z\hat{\mathscr{Z}}_1$, $\hat{\mathscr{Z}}_2$ and $\hat{\mathscr{U}}$ are square integrable, see Theorem 2.3.1. Furthermore, the stochastic integrals with respect to N have locally integrable compensators since X^{π} and $\hat{\mathscr{Y}}$ are càdlàg, Y is bounded, G and $\hat{\mathscr{U}}$ are square integrable. Consequently, the compensated integrals are local martingales, see Theorem 2.3.2. Let $(\tau_n)_{n\geq 1}$ denote a localizing sequence for the local martingales such that $\tau_n \to T$, $n \to \infty$. We obtain the expectations

$$\begin{aligned}
&\mathbb{E}\big[Y(\tau_n)\big|X^{\pi}(\tau_n)\big|^2\big] \\
&= Y(0)x^2 + \mathbb{E}\bigg[\int_0^{\tau_n}\bigg\{Y(t)\bigg(2X^{\pi}(t-)\pi(t)\big(\mu(t)-r(t)\big) \\
&\quad + 2\big|X^{\pi}(t-)\big|^2 r(t) - 2X^{\pi}(t-)H(t) - \int_{\mathbb{R}} 2X^{\pi}(t-)G(t,z)Q(t,dz)\eta(t) \\
&\quad + \big|\pi(t)\sigma(t)\big|^2 + \int_{\mathbb{R}}\big|G(t,z)\big|^2 Q(t,dz)\eta(t)\bigg) \\
&\quad - \big|X^{\pi}(t-)\big|^2 f(t) + 2Z(t)X^{\pi}(t-)\pi(t)\sigma(t)\bigg\}dt\bigg],
\end{aligned} \tag{10.19}$$

and

$$\begin{aligned}
&\mathbb{E}\big[\hat{\mathscr{Y}}(\tau_n)X^{\pi}(\tau_n)\big] \\
&= \hat{\mathscr{Y}}(0)x + \mathbb{E}\bigg[\int_0^{\tau_n}\bigg\{\hat{\mathscr{Y}}(t-)\bigg(\pi(t)\big(\mu(t)-r(t)\big) + X^{\pi}(t-)r(t) \\
&\quad - H(t) - \int_{\mathbb{R}} G(t,z)Q(t,dz)\eta(t)\bigg) - X^{\pi}(t-)\hat{f}(t) + \hat{\mathscr{Z}}_1(t)\pi(t)\sigma(t) \\
&\quad - \int_{\mathbb{R}} G(t,z)\hat{\mathscr{U}}(t,z)Q(t,dz)\eta(t)\bigg\}dt\bigg].
\end{aligned} \tag{10.20}$$

From (10.19)–(10.20), after some tedious calculations, we deduce the formula

$$\begin{aligned}
&\mathbb{E}\big[Y(\tau_n)\big|X^{\pi}(\tau_n) - \mathscr{Y}(\tau_n)\big|^2\big] \\
&= \mathbb{E}\big[Y(\tau_n)\big|X^{\pi}(\tau_n)\big|^2 + \hat{\mathscr{Y}}(\tau_n)X^{\pi}(\tau_n) + Y(\tau_n)\big|\mathscr{Y}(\tau_n)\big|^2\big] \\
&= Y(0)x^2 + \hat{\mathscr{Y}}(0)x \\
&\quad + \mathbb{E}\bigg[\int_0^{\tau_n} Y(t)\big|\sigma(t)\big|^2\bigg\{\pi(t) + \bigg(\frac{\mu(t)-r(t)}{\sigma^2(t)} + \frac{Z(t)}{Y(t)\sigma(t)}\bigg)X^{\pi}(t-) \\
&\quad + \frac{\hat{\mathscr{Y}}(t)}{2Y(t)}\frac{\mu(t)-r(t)}{\sigma^2(t)} + \frac{\hat{\mathscr{Z}}_1(t)}{2Y(t)\sigma(t)}\bigg\}^2 dt
\end{aligned}$$

$$+\int_0^{\tau_n}\big|X^\pi(t)\big|^2\Big(-f(t)+2Y(t)r(t)-\frac{|Z(t)|^2}{Y(t)}-\big|\theta(t)\big|^2Y(t)-2\theta(t)Z(t)\Big)dt$$

$$+\int_0^{\tau_n}X^\pi(t)\Big(-\hat{f}(t)+\hat{\mathscr{Y}}(t)r(t)-2Y(t)H(t)-2Y(t)\int_{\mathbb{R}}G(t,z)Q(t,dz)\eta(t)$$

$$-\theta(t)\frac{\hat{\mathscr{Y}}(t)(\mu(t)-r(t))+\hat{\mathscr{Z}}_1(t)\sigma(t)}{\sigma(t)}$$

$$-Z(t)\frac{\hat{\mathscr{Y}}(t)(\mu(t)-r(t))+\hat{\mathscr{Z}}_1(t)\sigma(t)}{Y(t)\sigma(t)}\Big)dt\Big]$$

$$+\mathbb{E}\Big[Y(\tau_n)\big|\mathscr{Y}(\tau_n)\big|^2-\int_0^{\tau_n}\Big|\frac{\hat{\mathscr{Y}}(t)}{2Y(t)}\frac{\mu(t)-r(t)}{\sigma^2(t)}+\frac{\hat{\mathscr{Z}}_1(t)}{2Y(t)\sigma(t)}\Big|^2Y(t)\big|\sigma(t)\big|^2dt$$

$$+\int_0^{\tau_n}\Big(Y(t)\int_{\mathbb{R}}\big|G(t,z)\big|^2Q(t,dz)\eta(t)-\hat{\mathscr{Y}}(t-)H(t)$$

$$-\hat{\mathscr{Y}}(t-)\int_{\mathbb{R}}G(t,z)Q(t,dz)\eta(t)-\int_{\mathbb{R}}G(t,z)\hat{\mathscr{U}}(t,z)Q(t,dz)\eta(t)\Big)dt\Big],$$

and by the definition of the generators f and $\hat{f}$ from (10.6) and (10.18) we obtain

$$\begin{aligned}
&\mathbb{E}\big[Y(\tau_n)\big|X^\pi(\tau_n)-\mathscr{Y}(\tau_n)\big|^2\big]\\
&\quad=Y(0)x^2+\hat{\mathscr{Y}}(0)x\\
&\qquad+\mathbb{E}\Big[\int_0^{\tau_n}Y(t)\big|\sigma(t)\big|^2\Big\{\pi(t)+\Big(\frac{\mu(t)-r(t)}{\sigma^2(t)}+\frac{Z(t)}{Y(t)\sigma(t)}\Big)X^\pi(t-)\\
&\qquad+\frac{\hat{\mathscr{Y}}(t)}{2Y(t)}\frac{\mu(t)-r(t)}{\sigma^2(t)}+\frac{\hat{\mathscr{Z}}_1(t)}{2Y(t)\sigma(t)}\Big\}^2dt\Big]\\
&\qquad+\mathbb{E}\Big[Y(\tau_n)\big|\mathscr{Y}(\tau_n)\big|^2+\int_0^{\tau_n}\varphi(t)dt\Big],
\end{aligned}\tag{10.21}$$

where the process φ collects all terms independent of π. We let $n\to\infty$. We have to justify the interchange of the limit $n\to\infty$ and the expectation. First, we recall that Y is uniformly bounded, $\mathbb{E}[\sup_{t\in[0,T]}|\mathscr{Y}(t)|^2]<\infty$ since $\mathscr{Y}$ solves (10.7) and $\mathbb{E}[\sup_{t\in[0,T]}|X^\pi(t)|^2]<\infty$ by (7.12). From these properties we can conclude that $Y(\tau_n)|X^\pi(\tau_n)|^2$, $\hat{\mathscr{Y}}(\tau_n)X^\pi(\tau_n)$ and $Y(\tau_n)|\mathscr{Y}(\tau_n)|^2$ are bounded uniformly in n by an integrable random variable. Next, we deduce from the moment estimates (10.12) and (10.14) that the process $\hat{\mathscr{Z}}_1$ is square integrable. Finally, under our assumptions the process φ is integrable i.e. $\mathbb{E}[\int_0^T|\varphi(t)|dt]<\infty$. Hence, we apply the monotone convergence theorem to the first expectation on the right hand side of (10.21), the dominated convergence theorem to the second expectation on the right hand side of (10.21) and the dominated convergence theorem on the left hand side of (10.21). We end up with

$$\begin{aligned}
&\mathbb{E}\big[\big|X^{\pi}(T)-F\big|^2\big] \\
&\quad=\lim_{n\to\infty}\mathbb{E}\big[Y(\tau_n)\big|X^{\pi}(\tau_n)-\mathscr{Y}(\tau_n)\big|^2\big] \\
&\quad=Y(0)x^2+\hat{\mathscr{Y}}(0)x \\
&\qquad+\mathbb{E}\bigg[\int_0^T Y(t)\big|\sigma(t)\big|^2\bigg\{\pi(t)+\bigg(\frac{\mu(t)-r(t)}{\sigma^2(t)}+\frac{Z(t)}{Y(t)\sigma(t)}\bigg)X^{\pi}(t-) \\
&\qquad+\frac{\hat{\mathscr{Y}}(t)}{2Y(t)}\frac{\mu(t)-r(t)}{\sigma^2(t)}+\frac{\hat{\mathscr{Z}}_1(t)}{2Y(t)\sigma(t)}\bigg\}^2dt\bigg]+\mathbb{E}\bigg[F^2-\int_0^T\varphi(t)dt\bigg]. \qquad (10.22)
\end{aligned}$$

Since the last term in (10.22) does not depend on π, the optimal strategy can be immediately derived from (10.22) and (10.17).

2. The admissability. Let $\pi^*(t)=A(t)X^{\pi^*}(t-)+B(t)$. The dynamics of the investment portfolio (7.11) under the candidate strategy (10.16) is given by

$$\begin{aligned}
dX^{\pi^*}(t)&=X^{\pi^*}(t-)\big(r(t)+\big(\mu(t)-r(t)\big)A(t)\big)dt+X^{\pi^*}(t-)A(t)\sigma(t)dW(t) \\
&\quad+B(t)\big(\mu(t)-r(t)\big)dt+B(t)\sigma(t)dW(t) \\
&\quad-H(t)dt-\int_{\mathbb{R}}G(t,z)N(dt,dz). \qquad (10.23)
\end{aligned}$$

Since (10.23) is a linear forward SDE, there exists a unique càdlàg, adapted solution X^{π^*} to (10.23), see Theorem V.7 in Protter (2004). We can now conclude that π^* is a predictable process.

We are left with proving the square integrability of the strategy π^*. From (10.21) and integrability of the processes Y, $\mathscr{Y}$ and φ we can deduce the uniform estimate

$$\begin{aligned}
&\mathbb{E}\big[Y(\tau_n)\big|X^{\pi^*}(\tau_n)-\mathscr{Y}(\tau_n)\big|^2\big] \\
&\quad=Y(0)x^2+\hat{\mathscr{Y}}(0)x+\mathbb{E}\bigg[Y(\tau_n)\big|\mathscr{Y}(\tau_n)\big|^2+\int_0^{\tau_n}\varphi(t)dt\bigg]\le K,
\end{aligned}$$

and by Fatou's lemma we get

$$\begin{aligned}
K&\ge\lim_{n\to\infty}\mathbb{E}\big[Y(\tau_n\wedge t)\big|X^{\pi^*}(\tau_n\wedge t)-\mathscr{Y}(\tau_n\wedge t)\big|^2\big] \\
&\ge\mathbb{E}\big[Y(t)\big|X^{\pi^*}(t)-\mathscr{Y}(t)\big|^2\big],\quad 0\le t\le T.
\end{aligned}$$

Consequently, we can prove square integrability of the investment portfolio by noticing that

$$\begin{aligned}
\mathbb{E}\big[\big|X^{\pi^*}(t)\big|^2\big]&\le 2\mathbb{E}\big[\big|X^{\pi^*}(t)-\mathscr{Y}(t)\big|^2+\big|\mathscr{Y}(t)\big|^2\big] \\
&\le 2\mathbb{E}\bigg[\frac{1}{k}Y(t)\big|X^{\pi^*}(t)-\mathscr{Y}(t)\big|^2+\big|\mathscr{Y}(t)\big|^2\bigg]\le K,\quad 0\le t\le T,
\end{aligned}$$

where we use the lower bound $k \leq Y(t)$. Consider now the BSDE

$$\begin{aligned}\bar{Y}(t) &= X^{\pi^*}(T) + \int_t^T \big(-\bar{Y}(s-)r(s) - \bar{Z}(s)\theta(s)\big)ds \\ &\quad + \int_t^T H(s)ds + \int_t^T \int_{\mathbb{R}} G(s,z)N(ds,dz) \\ &\quad - \int_t^T \bar{Z}(s)dW(s) - \int_t^T \int_{\mathbb{R}} \bar{U}(s,z)\tilde{N}(ds,dz), \quad 0 \leq t \leq T. \quad (10.24)\end{aligned}$$

Since $\mathbb{E}[|X^{\pi^*}(T)|^2] < \infty$, there exists a unique solution $(\bar{Y}, \bar{Z}, \bar{U}) \in \mathbb{S}^2(\mathbb{R}) \times \mathbb{H}^2(\mathbb{R}) \times \mathbb{H}^2_N(\mathbb{R})$ to the BSDE (10.24). By (10.23) we get $\bar{Y}(t) = X^{\pi^*}(t)$, $\bar{Z}(t) = \pi^*(t)\sigma(t)$, $\bar{U}(t,z) = 0$. The admissibility of π^* has been proved. □

We remark that the arguments from the proof of Theorem 10.2.1 can be applied to solve a general linear quadratic control problem, see Lim (2004), Lim (2005) and Øksendal and Hu (2008).

Equation (10.22) gives the minimal hedging error under the optimal strategy π^*. We can easily find the optimal initial capital.

Proposition 10.2.3 *Under the assumptions of Propositions* 10.2.1, 10.2.2 *and Theorem* 10.2.1 *the initial value of the investment portfolio which minimizes the quadratic hedging error* (10.5) *takes the form*

$$x^* = \mathscr{Y}(0) = \mathbb{E}^{\mathbb{Q}^*}\left[\int_0^T e^{-\int_0^s r(u)du} dP(s)\right], \quad (10.25)$$

where the equivalent martingale measure $\mathbb{Q}^*$ *is given by*

$$\frac{d\mathbb{Q}^*}{d\mathbb{P}}\Big|\mathscr{F}_t = e^{-\int_0^t \theta(s)dW(s) - \frac{1}{2}\int_0^t |\theta(s)|^2 ds}, \quad 0 \leq t \leq T.$$

The price and the hedging strategy for the payment process P can be obtained by solving the nonlinear BSDE (10.6) and the linear BSDE (10.7). The price (10.25), the initial value of the hedging portfolio, is arbitrage-free. We notice that the risk premiums for the systematic insurance risk and the unsystematic insurance risk are equal to zero and the insurance risk is not priced under the measure $\mathbb{Q}^*$. In contrast to Sect. 10.1 where the pricing measure were assumed to be given, here the pricing measure $\mathbb{Q}^*$ is derived by solving the optimization problem (10.5). We remark that the pricing measure which arises from solving the quadratic pricing and hedging problem (10.5) is called a minimal variance measure, see Lim (2004) and Schweizer (2010). The optimal hedging strategy (10.16) consists of two terms: the first term is a delta hedging strategy and the second term is a correction factor taking into account discrepancies between the optimal hedging portfolio X^* and the market-consistent value $\mathscr{Y}$ of the insurance payment process (the reserve required for P). The second term adjusts the delta hedging strategy according to the shortfall or the surplus of the assets over the liabilities.

Example 10.2 We consider the classical Black-Scholes model with constant coefficients r, μ, σ and an insurer who issues a portfolio of n unit-linked endowment policies with a capital guarantee. We are again interested in pricing and hedging the claim $F = (n - J(T))(K - S(T))^+$ where J is the deaths counting process for the insurance portfolio. We consider a constant mortality intensity λ. In order to find the price and the optimal hedging strategy for the claim F, we have to solve the BSDEs

$$\begin{aligned} Y(t) = 1 + \int_t^T \left(2Y(s)r - \frac{|Z(s)|^2}{Y(s)} - \theta^2 Y(s) - 2\theta Z(s)\right)ds \\ - \int_t^T Z(s)dW(s), \quad 0 \le t \le T, \end{aligned} \tag{10.26}$$

$$\begin{aligned} \mathscr{Y}(t) = \big(n - J(T)\big)\big(K - S(T)\big)^+ + \int_t^T \big(-\mathscr{Y}(s-)r - \mathscr{Z}(s)\theta\big)ds \\ - \int_t^T \mathscr{Z}(s)dW(s) - \int_t^T \mathscr{U}(s)\tilde{N}(ds), \quad 0 \le t \le T. \end{aligned} \tag{10.27}$$

It is straightforward to notice that the solution to the BSDE (10.26) is of the form

$$\begin{aligned} Y(t) &= e^{(2r-\theta^2)(T-t)}, \quad 0 \le t \le T, \\ Z(t) &= 0, \quad 0 \le t \le T. \end{aligned}$$

Recalling the results from Example 9.5, we conclude that the solution to the BSDE (10.27) is given by the triple

$$\begin{aligned} \mathscr{Y}(t) &= \big(n - J(t)\big)e^{-\lambda(T-t)} \\ &\quad \cdot \Big(Ke^{-r(T-t)}\Phi\big(-d\big(t, S(t)\big) - \sigma\sqrt{T-t}\big) - S(t)\Phi\big(-d\big(t, S(t)\big)\big)\Big), \\ &\quad 0 \le t \le T, \\ \mathscr{Z}(t) &= -\big(n - J(t-)\big)e^{-\lambda(T-t)}\sigma S(t)\Phi\big(-d\big(t, S(t)\big)\big), \quad 0 \le t \le T, \\ \mathscr{U}(t) &= -e^{-\lambda(T-t)} \\ &\quad \cdot \Big(Ke^{-r(T-t)}\Phi\big(-d\big(t, S(t)\big) - \sigma\sqrt{T-t}\big) - S(t)\Phi\big(-d\big(t, S(t)\big)\big)\Big), \\ &\quad 0 \le t \le T. \end{aligned}$$

By Theorem 10.2.1 and Proposition 10.2.3 the optimal hedging strategy is given by the feedback formula

$$\begin{aligned} \pi^*(t) &= -\big(n - J(t-)\big)e^{-\lambda(T-t)}S(t)\Phi\big(-d\big(t, S(t)\big)\big) \\ &\quad - \frac{\mu - r}{\sigma^2}\big(X^*(t-) - \mathscr{Y}(t-)\big), \quad 0 \le t \le T, \end{aligned}$$

where the optimal hedging portfolio is determined by the process

$$X^*(t) = x^* e^{rt} + \int_0^t \pi^*(s)\big(\theta\sigma ds + \sigma dW(s)\big), \quad X^*(0) = x^*,$$

and the price of the claim, and the initial value of the optimal hedging portfolio, is equal to

$$x^* = \mathscr{Y}(0) = ne^{-\lambda T}\big(Ke^{-rT}\Phi\big(-d\big(0, S(0)\big) - \sigma\sqrt{T}\big) - S(0)\Phi\big(-d\big(0, S(0)\big)\big)\big).$$

The quadratic objective (10.5) can be generalized by keeping its mathematical tractability. We can solve the quadratic optimization problem

$$\min_{\pi} \mathbb{E}\left[\int_0^T \big(X^\pi(s) - \beta(s)\big)^2 ds + \big(X^\pi(T) - \xi\big)^2\right], \tag{10.28}$$

where a running cost is added. Under the objective (10.28) an investment strategy is chosen in such a way that the investment portfolio is as close as possible, in the mean-square sense, to the targets β and ξ. The targets β and ξ may represent solvency constraints or profit expectations, see Detemple and Rindisbacher (2008) and Delong (2010). We can also solve the Markowitz portfolio selection problem

$$\begin{aligned} &\min_{\pi} \mathrm{Var}\big[X^\pi(T)\big] \\ &\mathbb{E}\big[X^\pi(T)\big] = L, \end{aligned} \tag{10.29}$$

where the variance of the terminal wealth is minimized given the constraint on the expected return. We remark that the Markowitz portfolio selection is commonly used for investment decision making and asset-liability management, see Chap. 4 in Zenios and Ziemba (2006) and Sect. 23.2 in Adam (2007).

Notice that in order to apply the investment strategy (10.16) we have to estimate the drift μ. The advantage of the quadratic optimization under an equivalent martingale measure (10.1) over the quadratic optimization under the real-world measure (10.5) is that the drift of the stock does not enter the optimal strategy in the former case. It is known that the estimation of the drift is challenging. Consequently, the quadratic pricing and hedging under the real-world measure may be difficult to implement in practice, see Sect. 10.4.3 in Cont and Tankov (2004) for a discussion.

10.3 Quadratic Pricing and Hedging Under Local Risk-Minimization

We now investigate the objective of local risk-minimization which was developed in Schweizer (1991) and Schweizer (2008). Local risk-minimization is an important alternative to global quadratic hedging. We will see that the locally risk-minimizing strategy is derived by optimizing an objective formulated under the real-world measure but at the same time the optimal strategy does not depend on the drift of the stock.

Let $\hat{S}(t) = e^{-\int_0^t r(s)ds} S(t)$. The discounted stock price satisfies the dynamics

$$\hat{S}(t) = \hat{S}(0) + \int_0^t \hat{S}(u)\big(\mu(u) - r(u)\big)du + \int_0^t \hat{S}(u)\sigma(u)dW(u), \quad 0 \le t \le T.$$

We introduce a class of admissible investment strategies.

Definition 10.3.1 A strategy $\Pi := (\Pi(t), 0 \le t \le T)$ is called admissible, written $\Pi \in \mathscr{A}_s^{local}$, if it satisfies the conditions:

(i) $\Pi : \Omega \times [0, T] \to \mathbb{R}$ is a predictable process,
(ii) $\mathbb{E}[\int_0^T |\hat{S}(t)\Pi(t)\sigma(t)|^2 dt] < \infty$.

A strategy $\Gamma := (\Gamma(t), 0 \le t \le T)$ is called admissible, written $\Gamma \in \mathscr{A}_b^{local}$, if it satisfies the conditions:

(i) $\Gamma : \Omega \times [0, T] \to \mathbb{R}$ is an adapted process,
(ii) the process $\mathscr{X}^{\Pi,\Gamma}(t) = \Pi(t)\hat{S}(t) + \Gamma(t)$ is right-continuous and square integrable.

A strategy (Π, λ) is called admissible for the local risk-minimization problem, written $(\Pi, \lambda) \in \mathscr{A}^{local}$, if $\Pi \in \mathscr{A}_s^{local}$ and $\Gamma \in \mathscr{A}_b^{local}$

We remark that the strategy Π denotes the number of stock which are held in the investment portfolio, Γ denotes the position in the bank account and $\mathscr{X}$ is the discounted value of the investment portfolio. We point out that the investment portfolio $\mathscr{X}^{\Pi,\Gamma}$ is not self-financing under $(\Pi, \Gamma) \in \mathscr{A}^{local}$.

We define the cost process and the risk process of a hedging strategy.

Definition 10.3.2 Assume that (C1)–(C4) from Chap. 7 hold. The cost process of an admissible strategy $(\Pi, \Gamma) \in \mathscr{A}^{local}$ related to hedging the payment process P is given by

$$C^{\Pi,\Gamma}(t) = \int_0^t e^{-\int_0^s r(u)du} dP(s) + \mathscr{X}^{\Pi,\Gamma}(t) - \int_0^t \Pi(s)d\hat{S}(s), \quad 0 \le t \le T.$$

The risk process of an admissible strategy $(\Pi, \Gamma) \in \mathscr{A}^{local}$ related to hedging the payment process P is given by

$$R^{\Pi,\Gamma}(t) = \mathbb{E}\big[\big|C^{\Pi,\Gamma}(T) - C^{\Pi,\Gamma}(t)\big|^2 \big| \mathscr{F}_t\big], \quad 0 \le t \le T.$$

Since $(\Pi, \Gamma) \in \mathscr{A}^{local}$, the integral $\int_0^t \Pi(s)d\hat{S}(s)$ is well-defined and $C^{\Pi,\Gamma}$ is square integrable.

The cost process $C^{\Pi,\Gamma}$ describes accumulated discounted costs or profits (cash inflows or outflows) for the insurer who applies an investment strategy (Π, Γ), pays the benefits P and holds the investment portfolio $\mathscr{X}^{\Pi,\Gamma}$. If the payment process can be perfectly replicated by (Π, Γ), then

$$\mathscr{X}^{\Pi,\Gamma}(t) = \mathscr{X}(0) + \int_0^t \Pi(s)d\hat{S}(s) - \int_0^t e^{-\int_0^s r(u)du} dP(s), \quad 0 \le t \le T,$$

$$\mathscr{X}^{\Pi,\Gamma}(T) = 0,$$

and the cost process related to hedging P is equal to the initial premium $\mathscr{X}(0)$ - the cost of setting the replicating portfolio which matches the liability. If the payment process cannot be perfectly replicated, then the insurer has to inject capital or withdraw capital during the lifetime of the policy in order to match the assets with the liabilities. These inflows and outflows of capital from the investment portfolio are modelled by the cost process.

The idea of local risk-minimization is to find an admissible hedging strategy $(\Pi, \Gamma) \in \mathscr{A}^{local}$ which minimizes the risk process $R^{\Pi,\Gamma}$ for all $t \in [0, T]$. The precise definition of the hedging objective is very technical and it involves limit considerations and local perturbations of investment strategies, see Schweizer (1991) and Schweizer (2008). We can say that under the local risk-minimization we aim to find an asset portfolio which perfectly matches the liability with a minimal mean-square cost of matching. We point out that the hedging objective, the risk process, is evaluated under the real-world measure, as it should be.

We give the key result which characterizes a locally risk-minimizing strategy.

Theorem 10.3.1 *Assume that* (C1)–(C4) *from Chap. 7 hold. The payment process P admits an admissible locally risk-minimizing strategy $(\Pi^*, \Gamma^*) \in \mathscr{A}^{local}$ if and only if the discounted payment process has the representation*

$$\int_0^T e^{-\int_0^s r(u)du} dP(s) = \xi_0 + \int_0^T \zeta(s)d\hat{S}(s) + \mathscr{H}(T), \tag{10.30}$$

where $\zeta \in \mathscr{A}_s^{local}$, and $\mathscr{H}$ is a right-continuous, square integrable martingale which is strongly orthogonal to the martingale component of $\hat{S}$ and verifies $\mathscr{H}(0) = 0$. In this case, we define

$$\Pi^*(t) = \zeta(t), \quad 0 \le t \le T,$$

$$\mathscr{X}^*(t) = \xi_0 + \int_0^t \zeta(s)d\hat{S}(s) + \mathscr{H}(t) - \int_0^t e^{-\int_0^s r(u)du} dP(s), \quad 0 \le t \le T,$$

$$\Gamma^*(t) = \mathscr{X}^*(t) - \Pi^*(t)\hat{S}(t), \quad 0 \le t \le T.$$

Proof The result follows from Proposition 5.2 from Schweizer (2008). □

We remark that two square integrable martingales are strongly orthogonal if $\mathbb{E}[M^1(t)M^2(t)] = 0$.

Decomposition (10.30) is called the Föllmer-Schweizer decomposition. The locally risk-minimizing strategy (Π^*, Γ^*) is called mean-self-financing as the optimal cost process $C^*(t) = \xi_0 + \mathscr{H}(t)$ is a martingale, i.e. the average future cost is equal to zero. The strategy (Π^*, Γ^*) is also called 0-achieving as $\mathscr{X}^*(T) = 0$ and all claims are covered.

The next result shows that the Föllmer-Schweizer decomposition can be derived from a BSDE.

Proposition 10.3.1 *Assume that* (C1)–(C4) *from Chap.* 7 *hold.*

(a) *The Föllmer-Schweizer decomposition of the discounted payment process is given by*

$$\begin{aligned}
&\int_0^T e^{-\int_0^s r(u)du} dP(s) \\
&\quad = Y(0) + \int_0^T \frac{Z_1(s)}{\sigma(s)\hat{S}(s)} d\hat{S}(s) \\
&\qquad + \int_0^T Z_2(s) dB(s) + \int_0^T \int_{\mathbb{R}} U(s,z)\tilde{N}(ds,dz),
\end{aligned}$$

where $(Y, Z_1, Z_2, U) \in \mathbb{S}^2(\mathbb{R}) \times \mathbb{H}^2(\mathbb{R}) \times \mathbb{H}^2(\mathbb{R}) \times \mathbb{H}^2_N(\mathbb{R})$ *is the unique solution to the BSDE*

$$\begin{aligned}
Y(t) = &\int_0^T e^{-\int_0^s r(u)du} dP(s) - \int_t^T Z_1(s)\theta(s)ds \\
&- \int_t^T Z_1(s)dW(s) - \int_t^T Z_2(s)dB(s) \\
&- \int_t^T \int_{\mathbb{R}} U(s,z)\tilde{N}(ds,dz), \quad 0 \le t \le T.
\end{aligned} \tag{10.31}$$

(b) *The admissible locally risk-minimizing strategy* $\Pi^* \in \mathscr{A}^{local}$ *and the optimal discounted hedging portfolio for the payment process* P *are given by*

$$\begin{aligned}
\Pi^*(t) &= \frac{Z_1(t)}{\sigma(t)\hat{S}(t)}, \quad 0 \le t \le T, \\
\mathscr{X}^*(t) &= Y(t) - \int_0^t e^{-\int_0^s r(u)du} dP(s), \quad 0 \le t \le T.
\end{aligned}$$

Proof (a) Consider the BSDE

$$\begin{aligned}
Y(t) = &\int_0^T e^{-\int_0^s r(u)du} dP(s) + \int_t^T f(s)ds - \int_t^T Z_1(s)dW(s) \\
&- \int_t^T Z_2(s)dB(s) - \int_t^T \int_{\mathbb{R}} U(s,z)\tilde{N}(ds,dz), \quad 0 \le t \le T,
\end{aligned} \tag{10.32}$$

where the generator f will be specified in the sequel. Assume there exists a solution $(Y, Z_1, Z_2, U) \in \mathbb{S}^2(\mathbb{R}) \times \mathbb{H}^2(\mathbb{R}) \times \mathbb{H}^2(\mathbb{R}) \times \mathbb{H}^2_N(\mathbb{R})$ to (10.32). From the BSDE (10.32) we get the representation of the discounted payments

$$\begin{aligned}
&\int_0^T e^{-\int_0^s r(u)du} dP(s) \\
&\quad = Y(0) - \int_0^T f(s)ds + \int_0^T Z_1(s)dW(s) \\
&\qquad + \int_0^T Z_2(s)dB(s) + \int_0^T \int_{\mathbb{R}} U(s,z)\tilde{N}(ds,dz), \quad 0 \le t \le T, \qquad (10.33)
\end{aligned}$$

From property (5.5) we deduce that the stochastic integrals driven by B and $\tilde{N}$ are strongly orthogonal to the martingale component of $\hat{S}$, i.e. the stochastic integral driven by W. The discounted payment process (10.33) would now satisfy the Föllmer-Schweizer decomposition (10.30) if for some $\zeta \in \mathscr{A}_s^{local}$ we had

$$\int_0^T \zeta(s)d\hat{S}(s) = \int_0^T Z_1(s)dW(s) - \int_0^T f(s)ds. \qquad (10.34)$$

Since

$$\int_0^T \zeta(s)d\hat{S}(s) = \int_0^T \zeta(s)\hat{S}(s)\big(\mu(s) - r(s)\big)ds + \int_0^T \zeta(s)\hat{S}(s)\sigma(s)dW(s),$$

we should choose

$$\begin{aligned}
f^*(s) &= -Z_1(s)\theta(s), \quad 0 \le s \le T, \\
\zeta^*(s) &= \frac{Z_1(s)}{\sigma(s)\hat{S}(s)}, \quad 0 \le s \le T.
\end{aligned}$$

With this choice, there exists a unique solution $(Y, Z_1, Z_2, U) \in \mathbb{S}^2(\mathbb{R}) \times \mathbb{H}^2(\mathbb{R}) \times \mathbb{H}^2(\mathbb{R}) \times \mathbb{H}^2_N(\mathbb{R})$ to (10.32) and $\zeta^* \in \mathscr{A}_s^{local}$. From (10.33) and (10.34) we deduce the Föllmer-Schweizer decomposition.

(b) The formulas for Π^* and $\mathscr{X}^*$follow from item (a), (10.32)–(10.34) and Theorem 10.3.1. □

We state an important corollary.

Corollary 10.3.1 *Under the assumptions of Theorem* 10.3.1 *and Proposition* 10.3.1 *the optimal hedging portfolio process has the representation*

$$X^*(t) = \mathbb{E}^{\mathbb{Q}^*}\left[\int_t^T e^{-\int_t^s r(u)du} dP(s)|\mathscr{F}_t\right], \quad 0 \le t \le T, \qquad (10.35)$$

where

$$\frac{d\mathbb{Q}^*}{d\mathbb{P}}\Big|\mathscr{F}_t = e^{-\int_0^t \theta(s)dW(s) - \frac{1}{2}\int_0^t |\theta(s)|^2 ds}, \quad 0 \le t \le T,$$

and the optimal amount invested in the stock is given by

$$\pi^*(t) = \frac{Z_1(t)}{\sigma(t)} e^{\int_0^t r(s)ds}, \quad 0 \le t \le T. \tag{10.36}$$

Proof Since $X^*(t) = \mathscr{X}^*(t)e^{\int_0^t r(s)ds}$, the representation of X^* follows from Proposition 10.3.1 and the representation of the solution Y to the linear BSDE (10.31). Since $\pi^*(t) = \Pi^*(t)S(t)$, the formula for π^* follows immediately from Proposition 10.3.1. □

The hedging portfolio (10.35) and the optimal amount invested in the stock (10.36) can be obtained by solving the linear BSDE (10.31). The price of the insurance payment process, which is the initial value of the hedging portfolio, is arbitrage-free. The insurance risk is not priced under the measure $\mathbb{Q}^*$. The hedging strategy π^* is a delta-hedging strategy which is updated with the current information on the financial and the insurance risk. We notice that the optimal hedging portfolio and the optimal hedging strategy are characterized under the equivalent martingale measure $\mathbb{Q}^*$, hence the drift of the stock does not enter the solution. We point out that under the locally risk-minimizing strategy (Π^*, Γ^*) there is no mismatch between the assets and the liabilities. The hedging portfolio is forced to match the market-consistent value of the liabilities, see Corollary 10.3.1. This equivalence holds since the investment portfolio process $\mathscr{X}^*$ is not self-financing. By Proposition 10.3.1 we obtain

$$\begin{aligned} dX^*(t) &= r(t)e^{\int_0^t r(s)ds}\mathscr{X}^*(t)dt + e^{\int_0^t r(s)ds}d\mathscr{X}^*(t) \\ &= X^*(t)r(t)dt + e^{\int_0^t r(s)ds}\big(dY(t) - e^{-\int_0^t r(s)ds}dP(t)\big) \\ &= X^*(t)r(t)dt \\ &\quad + e^{\int_0^t r(s)ds}\big(\Pi^*(t)\hat{S}(t)\big(\mu(t) - r(t)\big)dt + \Pi^*(t)\hat{S}(t)\sigma(t)dW(t)\big) \\ &\quad - dP(t) + e^{\int_0^t r(s)ds}Z_2(t)dB(t) + \int_{\mathbb{R}} e^{\int_0^t r(s)ds}U(t,z)\tilde{N}(dt,dz). \end{aligned}$$

Using π^*, we derive the dynamics of the hedging portfolio

$$\begin{aligned} dX^*(t) &= \pi^*(t)\big(\mu(t)dt + \sigma(t)dW(t)\big) + \big(X^*(t) - \pi^*(t)\big)r(t)dt \\ &\quad - dP(t) + e^{\int_0^t r(s)ds}Z_2(t)dB(t) + \int_{\mathbb{R}} e^{\int_0^t r(s)ds}U(t,z)\tilde{N}(dt,dz). \end{aligned} \tag{10.37}$$

The stochastic integrals driven by B and $\tilde{N}$ are interpreted as cash inflows/outflows which guarantee that $X^*(t) = \mathbb{E}^{\mathbb{Q}^*}[\int_t^T e^{-\int_t^s r(u)du}dP(s)|\mathscr{F}_t]$ for all $0 \le t \le T$. As already noticed, the expected value of the cash inflows/outflows is zero

$$\mathbb{E}\left[\int_0^T e^{\int_0^t r(s)ds} Z_2(t)dB(t) + \int_0^T \int_{\mathbb{R}} e^{\int_0^t r(s)ds} U(t,z)\tilde{N}(dt,dz)\right] = 0,$$

and the hedging portfolio (10.37) is mean-self-financing. We remark that the locally risk-minimizing strategy (10.36) for the stock would coincide with the minimum mean-square error strategy for the stock derived in Theorem 10.1.1 if the measure $\mathbb{Q}^*$ from Corollary 10.3.1 were chosen in the global quadratic hedging problem. However, the investment strategy from Theorem 10.1.1 is obtained in the framework of self-financing portfolios and $X^*(t) \neq \mathbb{E}^{\mathbb{Q}^*}[\int_t^T e^{-\int_t^s r(u)du} dP(s)|\mathscr{F}_t]$ except at the inception of the contract.

10.4 Quadratic Pricing and Hedging Under an Instantaneous Mean-Variance Risk Measure

As discussed at the end of Sect. 10.2, the Markowitz mean-variance objective is often used by financial institutions. In this chapter we investigate a local version of the mean-variance objective.

In Sects. 10.1–10.2 where we were only interested in finding the price of the payment process P at the initial time $t = 0$. We are now interested in dynamic pricing of the payment process P over the period $[0, T]$. We assume that the price Y of the payment process P solves the BSDE

$$\begin{aligned} Y(t) = F + \int_t^T H(s)ds + \int_t^T \int_{\mathbb{R}} G(s,z)N(ds,dz) + \int_t^T f(s)ds \\ - \int_t^T Z_1(s)dW(s) - \int_t^T Z_2(s)dB(s) \\ - \int_t^T \int_{\mathbb{R}} U(s,z)\tilde{N}(ds,dz), \quad 0 \le t \le T, \end{aligned} \tag{10.38}$$

with a generator f. By Propositions 3.3.1 and 3.4.1 an arbitrage-free price of P, i.e. $Y(t) = \mathbb{E}^{\mathbb{Q}}[\int_t^T e^{-\int_t^s r(u)du} dP(s)|\mathscr{F}_t]$, $\mathbb{Q} \in \mathscr{Q}^m$, has to solve a BSDE of the form (10.38). Hence, it is reasonable to assume a priori that the price process satisfies (10.38). In the sequel we use a local mean-variance objective to derive the generator f of the BSDE (10.38), the price process and the hedging strategy.

We recall that the investment portfolio X^π under an admissible investment strategy $\pi \in \mathscr{A}$ is given by (7.11). We assume that there exists a solution $(Y, Z_1, Z_2, U) \in \mathbb{S}^2(\mathbb{R}) \times \mathbb{H}^2(\mathbb{R}) \times \mathbb{H}^2(\mathbb{R}) \times \mathbb{H}^2_N(\mathbb{R})$ to the BSDE (10.38). We introduce a surplus process $\mathscr{S} := (\mathscr{S}(t), 0 \le t \le T)$ which models the excess of the wealth of the insurer over the price of the payment process. We set

$$\mathscr{S}(t) = X^\pi(t) - Y(t), \quad 0 \le t \le T.$$

We can derive the dynamics of the surplus process. We get the equation

$$d\mathscr{S}(t) = X^{\pi}(t-)r(t)dt + \pi(t)\big(\mu(t) - r(t)\big)dt + f(t)dt$$
$$+ \big(\sigma(t)\pi(t) - Z_1(t)\big)dW(t) - Z_2(t)dB(t) - \int_{\mathbb{R}} U(t,z)\tilde{N}(dt,dz), \tag{10.39}$$

with the initial condition $\mathscr{S}(0) = 0$, since we should have $X(0) = Y(0)$. We remark that $\mathscr{S}$ is square integrable for any $\pi \in \mathscr{A}$. Since Y can be interpreted as the market-consistent reserve for the liabilities, the surplus process $\mathscr{S}$ models the excess of the assets over the liabilities—the net asset wealth. The surplus process $\mathscr{S}$ models the profit earned by the insurer. We recall that the net asset wealth is the key object investigated in Solvency II Directive, see European Commission QIS5 (2010). In Leitner (2007) the process $\mathscr{S}$ is called a tracking error.

We define the mean-variance Markowitz risk measure

$$\rho(\xi) = L\sqrt{\mathrm{Var}[\xi]} - \mathbb{E}[\xi], \tag{10.40}$$

where the parameter L is a risk aversion coefficient which sets the trade-off between variance minimization and expected return maximization. Following Leitner (2007), we apply the risk measure (10.40), with a time-varying risk aversion coefficient L, to the infinitesimal change in the surplus process $\mathscr{S}$. We investigate the instantaneous mean-variance risk measures

$$\rho\big(d\mathscr{S}(t)\big)/dt$$
$$= L(t)\sqrt{\mathbb{E}\big[d[\mathscr{S},\mathscr{S}](t)|\mathscr{F}_{t-}\big]/dt} - \mathbb{E}\big[d\mathscr{S}(t) - \mathscr{S}(t-)r(t)dt|\mathscr{F}_{t-}\big]/dt$$
$$= L(t)\sqrt{\big|\pi(t)\sigma(t) - Z_1(t)\big|^2 + \big|Z_2(t)\big|^2 + \int_{\mathbb{R}} \big|U(t,z)\big|^2 Q(t,z)\eta(t)}$$
$$- \big(Y(t-)r(t) + \pi(t)\big(\mu(t) - r(t)\big) + f(t)\big), \quad 0 \le t \le T. \tag{10.41}$$

The quadratic variation is now used for modelling the instantaneous variance. The moments in (10.41) are derived by Theorems 2.3.2–2.3.3. The goal is to find an admissible hedging strategy $\pi \in \mathscr{A}$ which minimizes the instantaneous risk measures $\rho(d\mathscr{S}(t))$ for all $t \in [0,T]$ and choose a generator f of the price dynamics Y which makes the instantaneous risk measures vanish $\rho(d\mathscr{S}(t)) = 0$ for all $t \in [0,T]$. This is a reasonable pricing and hedging objective. The insurer should be interesting in applying an investment strategy under which the expected excess return on the surplus over the risk-free return on the surplus is maximized. At the same time, the insurer should choose an investment strategy under which the return on the surplus (the tracking error) is not volatile. Hence, the mean-variance objective is used for choosing the hedging strategy. The insurance payment process is next priced by requiring that the instantaneous Sharpe ratio of the surplus process equals a prespecified target L, i.e. the insurance payment process is priced by requiring that

$$\begin{aligned} & \textit{Sharpe Ratio}\ \big(\mathscr{S}(t)\big) \\ & \quad = \frac{\mathbb{E}[d\mathscr{S}(t) - \mathscr{S}(t-)r(t)dt|\mathscr{F}_{t-}]/dt}{\sqrt{\mathbb{E}[d[\mathscr{S},\mathscr{S}](t)/dt|\mathscr{F}_{t-}]}} = L(t), \quad 0 \le t \le T. \end{aligned} \tag{10.42}$$

We can now interpret L as a process which controls the ratio of the expected earned surplus (the net asset wealth) to its deviation over time or as a profit, specified by a Sharpe ratio, which is demanded by the insurer who sells the insurance contract. We assume that $L(t) \ge \theta(t) + \varepsilon$, $\varepsilon > 0$. The former inequality is obvious since the investment in the payment process P carries an additional risk compared to the investment in the risky stock S and the insurer is interested in earning a risk premium strictly above θ, which is the risk premium earned by investing in the stock. For the instantaneous Sharpe ratio pricing we refer to Bayraktar and Young (2007), Young (2008), Bayraktar and Young (2008) and Bayraktar et al. (2009), for the connection with the cost of capital pricing we refer to Pelsser (2011).

Theorem 10.4.1 *Assume that* (C1)–(C4) *from Chap. 7 hold and let L be a predictable process such that $L(t) \ge \theta(t) + \varepsilon, \varepsilon > 0$, and $L(t) \le K$, $0 \le t \le T$. The admissible investment strategy $\pi^* \in \mathscr{A}$ which minimizes the risk measures* (10.41) *for all $t \in [0,T]$ and the generator f^* which makes the risk measures* (10.41) *vanish for all $t \in [0,T]$ take the form*

$$\begin{aligned} \pi^*(t) = & \frac{1}{\sigma(t)}\Bigg(Z_1(t) \\ & + \sqrt{\frac{|\theta(t)|^2}{|L(t)|^2 - |\theta(t)|^2}}\sqrt{|Z_2(t)|^2 + \int_{\mathbb{R}}|U(t,z)|^2 Q(t,dz)\eta(t)}\Bigg), \\ & 0 \le t \le T, \\ f^*(t) = & -Y(t-)r(t) - Z_1(t)\theta(t) \\ & + \sqrt{|L(t)|^2 - |\theta(t)|^2}\sqrt{|Z_2(t)|^2 + \int_{\mathbb{R}}|U(t,z)|^2 Q(t,dz)\eta(t)}, \quad 0 \le t \le T, \end{aligned} \tag{10.43}$$

where $(Y, Z_1, Z_2, U) \in \mathbb{S}^2(\mathbb{R}) \times \mathbb{H}^2(\mathbb{R}) \times \mathbb{H}^2(\mathbb{R}) \times \mathbb{H}^2_N(\mathbb{R})$ is the unique solution to the BSDE

$$\begin{aligned} Y(t) = F & + \int_t^T H(s)ds + \int_t^T\int_{\mathbb{R}} G(s,z)N(ds,dz) \\ & + \int_t^T\Big(-Y(s-)r(s) - Z_1(s)\theta(s) \end{aligned}$$

$$
+\sqrt{|L(s)|^2-|\theta(s)|^2}\sqrt{|Z_2(s)|^2+\int_{\mathbb{R}}|U_2(s,z)|^2Q(s,dz)\eta(s)}\Big)ds
$$

$$
-\int_t^T Z_1(s)dW(s)-\int_t^T Z_2(s)dB(s)
$$

$$
-\int_t^T\int_{\mathbb{R}} U(s,z)\tilde{N}(ds,dz), \quad 0\le t\le T. \tag{10.44}
$$

Proof The form of the generator and the strategy are deduced from properties of the function

$$
w(\pi)=L\sqrt{(\pi\sigma-z)^2+u^2+v^2}-ry-\pi(\mu-r)-f.
$$

It is straightforward to find a unique minimizer π^* of the function w and f^* such that $w(\pi^*)=0$. We now study the BSDE (10.44). By the Schwarz inequality

$$
\left|zz'+\int_{\mathbb{R}} u(z)u'(z)Q(t,dz)\eta\right|
$$

$$
\le\sqrt{|z|^2+\int_{\mathbb{R}}|u(z)|Q(t,dz)\eta}\sqrt{|z'|^2+\int_{\mathbb{R}}|u'(z)|^2Q(t,dz)\eta}, \tag{10.45}
$$

we can prove the following inequality

$$
\left|\sqrt{|z|^2+\int_{\mathbb{R}}|u(z)|^2Q(t,dz)\eta}-\sqrt{|z'|^2+\int_{\mathbb{R}}|u'(z)|^2Q(t,dz)\eta}\right|^2
$$

$$
\le|z-z'|^2+\int_{\mathbb{R}}|u(z)-u'(z)|^2Q(t,dz)\eta. \tag{10.46}
$$

Hence, the generator (10.44) is Lipschitz continuous in the sense of Theorem 3.1.1. From (3.22) and Theorem 3.1.1 we conclude that there exists a unique solution to the BSDE (10.44). It is clear that the investment strategy (10.43) is admissible. □

The price process of the payment process P solves the nonlinear BSDE (10.44). It is important to point out that the price process (10.44) may not be represented as the expected value of the future discounted claims under an equivalent martingale measure. We also remark that the price process (10.44) may not satisfy the property of monotonicity with respect to the claim in the sense that a more severe claim may have a lower price. Hence, the price derived in Theorem 10.4.1 may give rise to arbitrage opportunities. In the language of BSDEs the solution to the BSDE (10.44) does satisfy the comparison principle with respect to the terminal condition and the BSDE (10.44) does not have a measure solution. This problem, which arises for BSDEs with jumps, was pointed out in Sect. 3.2.

Example 10.3 We consider the classical Black-Scholes model with constant coefficients r, μ, σ and an insurer who issues an endowment policy. We are interested in pricing and hedging the claim $F = 1 - J(T)$ where J is the death counting process for the policy. We consider a constant mortality intensity λ and a constant Sharpe ratio L. In order to find the optimal price process and the optimal investment strategy for the claim F, we have to solve the BSDE

$$\begin{aligned} Y(t) &= 1 - J(T) \\ &\quad + \int_t^T \left(-Y(s-)r - Z(s)\theta + \sqrt{L^2 - \theta^2}\left|U(s)\right|\sqrt{(1 - J(s-))\lambda}\right)ds \\ &\quad - \int_t^T Z(s)dW(s) - \int_t^T U(s)\tilde{N}(ds), \quad 0 \le t \le T. \end{aligned} \tag{10.47}$$

Clearly, we can set $Z(t) = 0$, $0 \le t \le T$. Let us guess that $U(t) = -Y(t-)$, $Y(t) \ge 0$, $0 \le t \le T$. We end up with the equation

$$\begin{aligned} dY(t) &= Y(t-)rdt - \sqrt{L^2 - \theta^2}Y(t-)\sqrt{(1 - J(t-))\lambda}dt - Y(t-)\tilde{N}(dt), \\ Y(T) &= 1 - J(T). \end{aligned}$$

First, it is straightforward to derive the dynamics

$$\begin{aligned} &d\left(Y(t)e^{-\int_0^t (r - \sqrt{L^2-\theta^2}\sqrt{(1-J(s-))\lambda})ds}\right) \\ &\quad = -e^{-\int_0^t (r - \sqrt{L^2-\theta^2}\sqrt{(1-J(s-))\lambda})ds}Y(t-)\tilde{N}(dt), \\ &Y(T) = 1 - J(T). \end{aligned} \tag{10.48}$$

Integrating and taking the expected value, we can obtain the candidate solution

$$\begin{aligned} Y(t) &= \mathbb{E}\left[\left(1 - J(T)\right)e^{-\int_t^T (r - \sqrt{L^2-\theta^2}\sqrt{(1-J(s-))\lambda})ds}|\mathscr{F}_t\right] \\ &= \left(1 - J(t)\right)e^{-(r+\lambda-\sqrt{L^2-\theta^2}\sqrt{\lambda})(T-t)}, \quad 0 \le t \le T, \end{aligned} \tag{10.49}$$

where we use the fact that $J(T) = 0$ implies $J(t) = 0$, $0 \le t \le T$. Next, from (10.49) we derive the dynamics

$$dY(t) = -e^{-(r+\lambda-\sqrt{L^2-\theta^2}\sqrt{\lambda})(T-t)}dJ(t) + \left(r + \lambda - \sqrt{L^2 - \theta^2}\sqrt{\lambda}\right)Y(t-)dt,$$

and

$$\begin{aligned} &d\left(Y(t)e^{-\int_0^t (r - \sqrt{L^2-\theta^2}\sqrt{(1-J(s-))\lambda})ds}\right) \\ &\quad = e^{-\int_0^t (r - \sqrt{L^2-\theta^2}\sqrt{(1-J(s-))\lambda})ds}\left(-e^{-(r+\lambda-\sqrt{L^2-\theta^2}\sqrt{\lambda})(T-t)}dJ(t)\right. \\ &\qquad \left. + \left(r + \lambda - \sqrt{L^2 - \theta^2}\sqrt{\lambda}\right)Y(t-)dt\right) \end{aligned}$$

$$
\begin{aligned}
&- Y(t-)e^{-\int_0^t (r-\sqrt{L^2-\theta^2}\sqrt{(1-J(s-))\lambda})ds}\big(r - \sqrt{L^2-\theta^2}\sqrt{(1-J(t-))\lambda}\big)dt \\
&= e^{-\int_0^t (r-\sqrt{L^2-\theta^2}\sqrt{(1-J(s-))\lambda})ds}\big(-e^{-(r+\lambda-\sqrt{L^2-\theta^2}\sqrt{\lambda})(T-t)}\big(1-J(t-)\big)dJ(t) \\
&\quad + \lambda Y(t-)\big(1-J(t-)\big)\big)dt \\
&= -e^{-\int_0^t (r-\sqrt{L^2-\theta^2}\sqrt{(1-J(s-))\lambda})ds}Y(t-)\tilde{N}(dt),
\end{aligned}
$$

which agrees with (10.48). Our candidate solution (Y, Z, U), where Y is given by (10.49), $Z(t) = 0$, $U(t) = -Y(t-)$, $0 \le t \le T$, is square integrable. Hence, the unique solution to the BSDE (10.47) is found. By Theorem 10.4.1 the optimal price process is defined by

$$
Y(t) = \big(1 - J(t)\big)e^{-(r+\lambda-\sqrt{L^2-\theta^2}\sqrt{\lambda})(T-t)}, \quad 0 \le t \le T, \tag{10.50}
$$

the optimal investment strategy is given by

$$
\pi^*(t) = \big(1 - J(t-)\big)\frac{\theta\sqrt{\lambda}}{\sigma\sqrt{L^2-\theta^2}}e^{-(r+\lambda-\sqrt{L^2-\theta^2}\sqrt{\lambda})(T-t)}, \quad 0 \le t \le T.
$$

We have obtained the unique solution to the pricing and hedging problem under the instantaneous mean-variance risk measure. If $\lambda - \sqrt{L^2-\theta^2}\sqrt{\lambda} \le 0$, then the price of the endowment policy is larger than $e^{-r(T-t)}$ which is the price of the bond paying 1 at the maturity. The property of monotonicity of the pricing operator with respect to the claim is not satisfied. The insurer can buy the bond which hedges the payment from the endowment policy and earns a risk-free profit. We can conclude that the price (10.50) may give rise to arbitrage opportunities. To get the arbitrage-free price, we have to introduce the constraint $\lambda - \sqrt{L^2-\theta^2}\sqrt{\lambda} > 0$, or equivalently $L^2 < \lambda + \theta^2$.

We are now interested in pricing and hedging the claim $F = (1 - J(T))(K - S(T))^+$. In order to find the optimal price process and the optimal hedging strategy for the claim F, we have to solve the BSDE

$$
\begin{aligned}
Y(t) &= \big(1 - J(T)\big)\big(K - S(T)\big)^+ \\
&\quad + \int_t^T \big(-Y(s-)r - Z(s)\theta + \sqrt{L^2-\theta^2}\big|U(s)\big|\sqrt{\big(1-J(s-)\big)\lambda}\big)ds \\
&\quad - \int_t^T Z(s)dW(s) - \int_t^T U(s)\tilde{N}(ds), \quad 0 \le t \le T.
\end{aligned} \tag{10.51}
$$

We can conclude that the unique solution to the BSDE (10.51) is given by the triple

$$
\begin{aligned}
Y(t) &= \big(1 - J(t)\big)e^{-(\lambda-\sqrt{L^2-\theta^2}\sqrt{\lambda})(T-t)} \\
&\quad \cdot \big(Ke^{-r(T-t)}\Phi\big(-d\big(t, S(t)\big) - \sigma\sqrt{T-t}\big) - S(t)\Phi\big(-d\big(t, S(t)\big)\big)\big), \\
&\quad 0 \le t \le T,
\end{aligned}
$$

$$\begin{aligned}Z(t) &= -\big(1-J(t-)\big)e^{-(\lambda-\sqrt{L^2-\theta^2}\sqrt{\lambda})(T-t)}\sigma S(t)\Phi\big(-d\big(t,S(t)\big)\big), \quad 0\le t\le T,\\ U(t) &= -\big(1-J(t-)\big)e^{-(\lambda-\sqrt{L^2-\theta^2}\sqrt{\lambda})(T-t)}\\ &\quad\cdot\big(Ke^{-r(T-t)}\Phi\big(-d\big(t,S(t)\big)-\sigma\sqrt{T-t}\big)-S(t)\Phi\big(-d\big(t,S(t)\big)\big)\big),\\ &\quad 0\le t\le T,\end{aligned}$$

where we use (9.6) and (9.7). The optimal price process is defined by

$$\begin{aligned}Y(t) &= \big(1-J(t)\big)e^{-(\lambda-\sqrt{L^2-\theta^2}\sqrt{\lambda})(T-t)}\\ &\quad\cdot\big(Ke^{-r(T-t)}\Phi\big(-d\big(t,S(t)\big)-\sigma\sqrt{T-t}\big)\\ &\quad-S(t)\Phi\big(-d\big(t,S(t)\big)\big)\big), \quad 0\le t\le T,\end{aligned} \tag{10.52}$$

and the optimal hedging strategy is given by

$$\begin{aligned}\pi^*(t) &= -\big(1-J(t-)\big)e^{-(\lambda-\sqrt{L^2-\theta^2}\sqrt{\lambda})(T-t)}S(t)\Phi\big(-d\big(t,S(t)\big)\big)\\ &\quad+\big(1-J(t-)\big)e^{-(\lambda-\sqrt{L^2-\theta^2}\sqrt{\lambda})(T-t)}\frac{\theta\sqrt{\lambda}}{\sigma\sqrt{L^2-\theta^2}}\\ &\quad\cdot\big(Ke^{-r(T-t)}\Phi\big(-d\big(t,S(t)\big)-\sigma\sqrt{T-t}\big)-S(t)\Phi\big(-d\big(t,S(t)\big)\big)\big),\\ &\quad 0\le t\le T.\end{aligned}$$

The price (10.52) is arbitrage-free if $L^2<\lambda+\theta^2$.

We also give an example which shows that the price from Theorem 10.4.1 may be arbitrage-free without any additional assumptions on the parameters. It turns out that arbitrage-free prices arise for specific types of claims.

Example 10.4 Let the assumptions from Example 10.3 hold. We consider an insurer who issues a life insurance policy paying 1 at maturity of the contract provided that the policyholder dies within the duration of the contract. We are interested in pricing and hedging the claim $F=J(T)$ where J is the death counting process for the policy. In order to find the optimal price process and the optimal investment strategy for the claim F, we have to solve the BSDE

$$\begin{aligned}Y(t) &= J(T)\\ &\quad+\int_t^T\Big(-Y(s-)r-Z(s)\theta+\sqrt{L^2-\theta^2}\big|U(s)\big|\sqrt{\big(1-J(s-)\big)\lambda}\Big)ds\\ &\quad-\int_t^T Z(s)dW(s)-\int_t^T U(s)\tilde{N}(ds), \quad 0\le t\le T.\end{aligned} \tag{10.53}$$

Clearly, we can set $Z(t)=0$, $0\le t\le T$. Let us guess that $U(t)\ge 0$, $0\le t\le T$. We end up with the BSDE

$$\begin{aligned}Y(t) &= J(T)\\ &+\int_t^T\Big(-Y(s-)r+\sqrt{L^2-\theta^2}U(s)\sqrt{\big(1-J(s-)\big)\lambda}\Big)ds\\ &-\int_t^T U(s)\tilde{N}(ds),\quad 0\le t\le T.\end{aligned}\tag{10.54}$$

By Proposition 3.4.1 and a similar reasoning as in Example 10.3, the solution to the BSDE (10.54) is of the form

$$\begin{aligned}Y(t) &= e^{-r(T-t)}-\big(1-J(t)\big)e^{-(r+\lambda+\sqrt{L^2-\theta^2}\sqrt{\lambda})(T-t)},\quad 0\le t\le T,\\ U(t) &= e^{-(r+\lambda+\sqrt{L^2-\theta^2}\sqrt{\lambda})(T-t)},\quad 0\le t\le T,\end{aligned}\tag{10.55}$$

and we see that $U(t)\ge 0$, $0\le t\le T$. Our candidate solution (Y,Z,U), where (Y,Z) is given by (10.55), $Z(t)=0$, $0\le t\le T$, is square integrable. Hence, the unique solution to the BSDE (10.53) is found. By Theorem 10.4.1 the arbitrage-free optimal price process is defined by

$$Y(t)=e^{-r(T-t)}-\big(1-J(t)\big)e^{-(r+\lambda+\sqrt{L^2-\theta^2}\sqrt{\lambda})(T-t)},\quad 0\le t\le T,$$

and the optimal investment strategy is given by the formula

$$\pi^*(t)=\big(1-J(t-)\big)\frac{\theta\sqrt{\lambda}}{\sigma\sqrt{L^2-\theta^2}}e^{-(r+\lambda+\sqrt{L^2-\theta^2}\sqrt{\lambda})(T-t)},\quad 0\le t\le T.$$

By Theorem 10.4.1 the hedging strategy for the payment process P is characterized by the control processes of the nonlinear BSDE (10.44). The investment strategy (10.43) is studied in Chap. 12 where the results of this chapter are derived again under different pricing and hedging objectives. In Chap. 12 we also give conditions which guarantee that the price process (10.44) derived under the instantaneous mean-variance risk measure is arbitrage-free.

Instead of the mean-variance minimization, we can use the approach proposed by Bayraktar and Young (2007), Young (2008), Bayraktar and Young (2008), Bayraktar et al. (2009), and we can find a hedging strategy π^* which minimizes the instantaneous variation $\mathbb{E}[d[\mathscr{S},\mathscr{S}](t)|\mathscr{F}_{t-}]$ for all $t\in[0,T]$ together with a generator f^* which makes the instantaneous mean-variance risk measures vanish $\rho(d\mathscr{S}(t))=0$ for all $t\in[0,T]$. We obtain

$$\begin{aligned}\pi^*(t) &= \frac{1}{\sigma(t)}Z_1(t),\quad 0\le t\le T,\\ f^*(t) &= -Y(t-)r(t)-Z_1(t)\theta(t)\\ &\quad+L(t)\sqrt{\big|Z_2(t)\big|^2+\int_{\mathbb{R}}\big|U(t,z)\big|^2Q(t,dz)\eta(t)},\quad 0\le t\le T,\end{aligned}\tag{10.56}$$

where $(Y, Z_1, Z_2, U) \in \mathbb{S}^2(\mathbb{R}) \times \mathbb{H}^2(\mathbb{R}) \times \mathbb{H}^2(\mathbb{R}) \times \mathbb{H}^2_N(\mathbb{R})$ is the unique solution to the BSDE

$$\begin{aligned} Y(t) = F &+ \int_t^T H(s)ds + \int_t^T \int_{\mathbb{R}} G(s,z)N(ds,dz) \\ &+ \int_t^T \Bigg(-Y(s-)r(s) - Z_1(s)\theta(s) \\ &+ L(s)\sqrt{\left|Z_2(s)\right|^2 + \int_{\mathbb{R}} \left|U_2(s,z)\right|^2 Q(s,dz)\eta(s)}\Bigg)ds \\ &- \int_t^T Z_1(s)dW(s) - \int_t^T Z_2(s)dB(s) \\ &- \int_t^T \int_{\mathbb{R}} U(s,z)\tilde{N}(ds,dz), \quad 0 \le t \le T. \end{aligned} \tag{10.57}$$

Again, the price process (10.57) may lead to arbitrage opportunities.

Bibliographical Notes For a review on quadratic pricing and hedging we refer to Schweizer (2010). The proof of Theorem 10.1.1 is classical, see Sect. 10.4 in Cont and Tankov (2004) and Ankirchner and Imkeller (2008). Riccati backward stochastic differential equation is studied in Lim (2004), Kohlmann and Tang (2002), Kohlmann and Tang (2003). Higher moment estimates for a solution to a BSDE driven by a Brownian motion are investigated in El Karoui et al. (1997b). The proof of the optimality in Theorem 10.2.1 is taken from Delong (2010), see also Lim (2004) and Øksendal and Hu (2008) for the method of completing the squares. In the proof of the admissability in Theorem 10.2.1 we closely follow the arguments from Lim (2004). For Markowitz mean-variance optimization problems in finance and insurance we refer to Lim (2004), Øksendal and Hu (2008), Xie (2009), Xie et al. (2008), Kharroubi et al. (2012). It is worth mentioning that quadratic hedging problems in a financial model with assets driven by quasi-left continuous semimartingales with bounded jumps and in a financial model with assets driven by general semimartingales were solved in Kohlmann et al. (2010) and Jeanblanc et al. (2012) where the authors applied semimartingale BSDEs. The proof of Proposition 10.3.1 is inspired by Ankirchner and Heine (2012). For locally risk-minimizing strategies we refer to Ankirchner and Heine (2012), Dahl and Møller (2006), Dahl et al. (2008), Møller (2001) and Vandaele and Vanmaele (2008). Section 10.4 is taken from Delong (2012a).

Chapter 11
Utility Maximization and Indifference Pricing and Hedging

Abstract We investigate portfolio optimization and pricing and hedging of the insurance payment process under the exponential utility function. First, we find the investment strategy under which the expected exponential utility of the insurer's terminal wealth is maximized. We characterize the optimal value function of the optimization problem and the optimal investment strategy by a nonlinear BSDE. Next, we solve the exponential indifference pricing and hedging problem. We show that the indifference price and the indifference hedging strategy solve a nonlinear BSDE.

In this chapter we use the theory of decision making under uncertainty by von Neumann and Morgenstern which plays a fundamental role in economics. In the context of dynamic asset allocation, Merton (1969) was the first who found the optimal dynamic investment strategy for an agent maximizing the expected utility from the terminal wealth. Since then, utility maximization objectives have gained great popularity in portfolio optimization. In the case when an agent can dynamically trade in a financial market, Hodges and Neuberger (1989) introduced utility indifference arguments for pricing and hedging of financial claims, and Young and Zariphopoulou (2002) applied these arguments to price and hedge insurance claims. The utility indifference arguments have proved to be very useful for pricing and hedging of claims in incomplete markets.

In this chapter we solve the utility maximization problem and the indifference pricing and hedging problem for an insurer who dynamically trades in the financial market and faces the payment process. We should point out that the utility-based approach to pricing and hedging has a strong economic justification, see Carmona (2008). We use the exponential utility. From the point of view of decision making the exponential utility has many advantages, see Chaps. 1 and 5 in Denuit et al. (2001) and Sect. 10.3.3. in Cont and Tankov (2004).

11.1 Exponential Utility Maximization

The insurer faces the payment process (7.3). The insurer's investment portfolio X^{π} under an investment strategy π is given by (7.11). We assume that $r = 0$ or, in

Ł. Delong, *Backward Stochastic Differential Equations with Jumps and Their Actuarial and Financial Applications*, EAA Series, DOI 10.1007/978-1-4471-5331-3_11,

other words, we consider discounted quantities. We define a class of admissible investment strategies.

Definition 11.1.1 A strategy π is called admissible for the exponential utility maximization problem, written $\mathscr{A}^{exp}$, if $\pi \in \mathscr{A}$, see Definition 7.3.1, and

$$\text{the family } \{e^{-\alpha X^{\pi}(\tau)},\ \mathscr{F}\text{-stopping times } \tau\} \text{ is } \mathbb{P}\text{-uniformly integrable.}$$

The uniform integrability condition arises since we deal with exponential preferences.

We control the wealth of the insurer at time T. The goal is to find an admissible investment strategy $\pi \in \mathscr{A}^{exp}$ under which the expected exponential utility of the terminal wealth (the total return on the business) is maximized. We aim to solve the optimization problem

$$\sup_{\pi \in \mathscr{A}^{exp}} \mathbb{E}\big[-e^{-\alpha(X^{\pi}(T)-F)}\big], \tag{11.1}$$

where $\alpha > 0$ denotes the insurer's risk aversion coefficient. Notice that losses of the investment portfolio are heavier penalized than profits under the exponential utility, in contrast to the quadratic objective under which losses and profits are treated in a symmetric manner. Since the exponential utility can be related to the entropic risk measure, see Sect. 13.1, the goal is to find an investment strategy under which the entropic risk measure of the insurer's terminal wealth is minimized. We remark utility maximization and risk measure minimization can be used to derive ALM strategies, see Chap. 5 in Zenios and Ziemba (2006) and Sect. 23.2 in Adam (2007).

In this chapter we consider the filtration $\mathscr{F} = \mathscr{F}^W \vee \mathscr{F}^J$ generated by the Brownian motion W used for stock modelling and the step process J used for claims modelling. This restriction is introduced since the general case with two Brownian motions W and B would lead to a nonlinear BSDE for which existence of a solution has not been established yet. Despite the restriction, we can still consider general streams of liabilities. In particular, we can investigate traditional and equity-linked life and non-life claims under the unsystematic insurance risk, life insurance equity-linked claims under irrational lapses, see Example 7.7, weather derivatives, Example 7.9, and credit default claims, see Example 7.10.

Let us solve the optimization problem (11.1). We consider the BSDE

$$\begin{aligned} Y(t) = F &+ \int_t^T f(s)ds \\ &- \int_t^T Z(s)dW(s) - \int_t^T \int_{\mathbb{R}} U(s,z)\tilde{N}(ds,dz), \quad 0 \le t \le T, \end{aligned} \tag{11.2}$$

where the generator f will be specified in the sequel. We introduced the process $A^{\pi} := (A^{\pi}(t), 0 \le t \le T)$ defined by

$$A^{\pi}(t) = -e^{-\alpha(X^{\pi}(t)-Y(t))}, \quad 0 \le t \le T,\ \pi \in \mathscr{A}^{exp}.$$

We have

$$\mathbb{E}\big[-e^{-\alpha(X^{\pi}(T)-F)}\big]=\mathbb{E}\big[-e^{-\alpha(X^{\pi}(T)-Y(T))}\big]=\mathbb{E}\big[A^{\pi}(T)\big].$$

The process A^{π} plays a crucial role in establishing the optimality of a strategy. If A^{π} is a supermartingale for any $\pi\in\mathscr{A}^{exp}$, then we obtain the inequality

$$\mathbb{E}\big[A^{\pi}(T)\big]=\mathbb{E}\big[-e^{-\alpha(X^{\pi}(T)-Y(T))}\big]\le A(0), \tag{11.3}$$

and if A^{π^*} is a martingale for some $\pi^*\in\mathscr{A}^{exp}$, then we derive the equality

$$\mathbb{E}\big[A^{\pi^*}(T)\big]=\mathbb{E}\big[-e^{-\alpha(X^{\pi^*}(T)-Y(T))}\big]=A(0). \tag{11.4}$$

Combining (11.3) with (11.4), we get

$$\mathbb{E}\big[-e^{-\alpha(X^{\pi}(T)-Y(T))}\big]\le\mathbb{E}\big[-e^{-\alpha(X^{\pi^*}(T)-Y(T))}\big],$$

and we conclude that the strategy π^* is optimal and A^{π^*} is the optimal value function of the dynamic optimization problem (11.1). Therefore, we aim to find a generator f^* of the BSDE (11.2), independent of π, such that the process A^{π} is a supermartingale for all $\pi\in\mathscr{A}^{exp}$ and A^{π^*} is a martingale for some $\pi^*\in\mathscr{A}^{exp}$.

We show how to find (f^*,π^*). From (7.11) and (11.2) we obtain

$$\begin{aligned}
&-\alpha\big(X^{\pi}(t)-Y(t)\big)\\
&\quad=-\alpha\bigg(x+\int_0^t\pi(s)\mu(s)ds+\int_0^t\pi(s)\sigma(s)dW(s)\\
&\qquad-\int_0^t H(s)ds-\int_0^t\int_{\mathbb{R}}G(s,z)Q(s,dz)\eta(s)ds-\int_0^t\int_{\mathbb{R}}G(s,z)\tilde{N}(ds,dz)\\
&\qquad-Y(0)+\int_0^t f(s)ds-\int_0^t Z(s)dW(s)-\int_0^t\int_{\mathbb{R}}U(s,z)\tilde{N}(ds,dz)\bigg)\\
&\quad=-\alpha\big(x-Y(0)\big)-\alpha\int_0^t\pi(s)\mu(s)ds-\alpha\int_0^t\big(\pi(s)\sigma(s)-Z(s)\big)dW(s)\\
&\qquad+\alpha\int_0^t\bigg(H(s)+\int_{\mathbb{R}}G(s,z)Q(s,dz)\eta(s)\bigg)ds\\
&\qquad+\alpha\int_0^t\int_{\mathbb{R}}\big(G(s,z)+U(s,z)\big)\tilde{N}(ds,dz)-\alpha\int_0^t f(s)ds,\quad 0\le t\le T.
\end{aligned} \tag{11.5}$$

For $\pi\in\mathscr{A}^{exp}$ we define three processes:

$$M^{W,\pi}(t)=e^{-\alpha\int_0^t(\pi(s)\sigma(s)-Z(s))dW(s)-\frac{1}{2}\alpha^2\int_0^t(\pi(s)\sigma(s)-Z(s))^2ds},\quad 0\le t\le T,$$

$$M^N(t) = e^{\alpha \int_0^t \int_{\mathbb{R}} (G(s,z)+U(s,z))\tilde{N}(ds,dz)} \cdot e^{+\int_0^t \int_{\mathbb{R}} (\alpha(G(s,z)+U(s,z)) - e^{\alpha(G(s,z)+U(s,z))}+1)Q(s,dz)\eta(s)ds}, \quad 0 \le t \le T, \tag{11.6}$$

and

$$\begin{aligned} D^\pi(t) = & -\alpha\pi(t)\mu(t) + \frac{1}{2}\alpha^2\big(\pi(t)\sigma(t) - Z(t)\big)^2 \\ & -\alpha f(t) + \alpha\Big(H(t) + \int_{\mathbb{R}} G(t,z)Q(t,dz)\eta(t)\Big) \\ & -\int_{\mathbb{R}} \big(\alpha\big(G(t,z)+U(t,z)\big) - e^{\alpha(G(t,z)+U(t,z))} + 1\big)Q(t,dz)\eta(t), \\ & 0 \le t \le T. \end{aligned} \tag{11.7}$$

By (11.5)–(11.7) we derive the following relation

$$\begin{aligned} A^\pi(t) &= -e^{-\alpha(X^\pi(t) - Y(t))} \\ &= -M^{W,\pi}(t)M^N(t)e^{-\alpha(x-Y(0))+\int_0^t D^\pi(s)ds}, \quad 0 \le t \le T. \end{aligned} \tag{11.8}$$

We choose a strategy π^* which minimizes $D^\pi(s)$ for all $s \in [0,T]$, i.e. we solve

$$\min_\pi \Big\{ -\alpha\pi(s)\mu(s) + \frac{1}{2}\alpha^2\big(\pi(s)\sigma(s) - Z(s)\big)^2 \Big\}, \quad 0 \le s \le T.$$

We get the candidate strategy

$$\pi^*(s) = \frac{1}{\sigma(s)}\Big(Z(s) + \frac{\mu(s)}{\alpha\sigma(s)}\Big), \quad 0 \le s \le T. \tag{11.9}$$

Next, we choose a generator f^* in such a way that $D^{\pi^*}(s) = 0$ for all $s \in [0,T]$. We get

$$\begin{aligned} f^*(s) = & -\pi^*(s)\mu(s) + \frac{1}{2}\alpha\big(\pi^*(s)\sigma(s) - Z(s)\big)^2 \\ & + \Big(H(s) + \int_{\mathbb{R}} G(s,z)Q(s,dz)\eta(s)\Big) \\ & - \frac{1}{\alpha}\int_{\mathbb{R}} \big(\alpha\big(G(s,z)+U(s,z)\big) - e^{\alpha(G(s,z)+U(s,z))} + 1\big)Q(s,dz)\eta(s) \\ = & -\frac{\mu^2(s)}{2\alpha\sigma^2(s)} - \frac{\mu(s)}{\sigma(s)}Z(s) + H(s) \\ & + \int_{\mathbb{R}} \Big(\frac{1}{\alpha}\big(e^{\alpha(G(s,z)+U(s,z))} - 1\big) - U(s,z)\Big)Q(s,dz)\eta(s), \quad 0 \le s \le T. \end{aligned} \tag{11.10}$$

Notice that for any π we have $D^{\pi}(s) \geq 0$, $0 \leq s \leq T$. Moreover, f^* is independent of π. The above heuristic reasoning has now to be made more formal.

First, we recall Theorem 3.5 from Becherer (2006).

Proposition 11.1.1 *Consider the BSDE*

$$\begin{aligned} Y(t) = \xi &+ \int_t^T f\big(s, Y(s), Z(s), U(s,.)\big)ds \\ &- \int_t^T Z(s)dW(s) - \int_t^T \int_{\mathbb{R}} U(s,z)\tilde{N}(ds,dz), \quad 0 \leq t \leq T. \end{aligned} \tag{11.11}$$

We assume that

(i) *the filtration $\mathscr{F} = \mathscr{F}^W \vee \mathscr{F}^J$ where J is a step process,*
(ii) *the terminal value ξ is a bounded random variable,*
(iii) *the intensity process η of the step process J is bounded,*
(iv) *the generator $f : \Omega \times [0,T] \times \mathbb{R} \times \mathbb{R} \times L^2_Q(\mathbb{R}) \to \mathbb{R}$ is predictable and takes the form*

$$f(t,y,z,u) = \hat{f}(t,y,z,u) + \int_{\mathbb{R}} \tilde{f}\big(t, u(z)\big) Q(t,dz)\eta(t),$$

where $\hat{f} : \Omega \times [0,T] \times \mathbb{R} \times \mathbb{R} \times L^2_Q(\mathbb{R}) \to \mathbb{R}$ is Lipschitz continuous in the sense of Theorem 3.1.1 *and satisfies*

$$\begin{aligned} &\big|\hat{f}(\omega,t,y,z,u)\big| \leq K\big(1+|y|\big), \\ &\quad (y,z,u) \in \mathbb{R} \times \mathbb{R} \times L^2_Q(\mathbb{R}), \quad a.s., a.e.\ (\omega,t) \in \Omega \times [0,T], \end{aligned}$$

$\tilde{f} : \Omega \times [0,T] \times \mathbb{R} \to \mathbb{R}$ is locally Lipschitz continuous in u, uniformly in (ω,t), and satisfies

$$\begin{aligned} &\tilde{f}(\omega,t,u) \leq -u, \quad u \leq 0, \quad a.s., a.e.\ (\omega,t) \in \Omega \times [0,T], \\ &\tilde{f}(\omega,t,u) \geq -u, \quad u \geq 0, \quad a.s., a.e.\ (\omega,t) \in \Omega \times [0,T]. \end{aligned}$$

There exists a unique solution $(Y,Z,U) \in \mathbb{S}^\infty(\mathbb{R}) \times \mathbb{H}^2(\mathbb{R}) \times \mathbb{H}^2_N(\mathbb{R})$ to the BSDE (11.11). *Moreover, U is bounded ϑ-a.e. $(t,z) \in [0,T] \times \mathbb{R}$.*

We point out a useful property. From (11.11) we conclude

$$\int_{\mathbb{R}} \big|U(t,z)\big| N\big(\{t\}, dz\big) = \big|Y(t) - Y(t-)\big| \leq 2 \sup_{0 \leq t \leq T} \big|Y(t)\big|, \tag{11.12}$$

and, consequently, if Y is bounded, then U is bounded ϑ-a.e. $(t,z) \in [0,T] \times \mathbb{R}$.

We give the main result of this chapter.

Theorem 11.1.1 *Consider the filtration $\mathscr{F} = \mathscr{F}^W \vee \mathscr{F}^J$. Assume that* (C1)–(C4) *from Chap.* 7 *hold and let F, H, G, η be bounded and $r = 0$.*

(a) *There exists a unique solution $(Y, Z, U) \in \mathbb{S}^\infty(\mathbb{R}) \times \mathbb{H}^2(\mathbb{R}) \times \mathbb{H}^2_N(\mathbb{R})$ to the BSDE*

$$\begin{aligned} Y(t) = F + \int_t^T \Big(&-\frac{\mu^2(s)}{2\alpha\sigma^2(s)} - \frac{\mu(s)}{\sigma(s)} Z(s) + H(s) \\ &+ \int_{\mathbb{R}} \Big(\frac{1}{\alpha} \big(e^{\alpha(G(s,z)+U(s,z))} - 1\big) - U(s,z) \Big) Q(s,dz)\eta(s) \Big) ds \\ &- \int_t^T Z(s) dW(s) - \int_t^T \int_{\mathbb{R}} U(s,z)\tilde{N}(ds,dz), \quad 0 \le t \le T. \end{aligned} \tag{11.13}$$

(b) *The optimal admissible investment strategy $\pi^* \in \mathscr{A}^{\exp}$ for the utility maximization problem* (11.1) *is given by*

$$\pi^*(t) = \frac{1}{\sigma(t)} \Big(Z(t) + \frac{\mu(t)}{\alpha\sigma(t)} \Big), \quad 0 \le t \le T. \tag{11.14}$$

The optimal value function of the optimization problem (11.1) *at time $t = 0$ equals $-e^{-\alpha(x - Y(0))}$.*

Proof (a) Since $s \mapsto \frac{\mu(s)}{\sigma(s)}$ is bounded, we can define the equivalent probability measure

$$\frac{d\mathbb{Q}^0}{d\mathbb{P}}\Big|\mathscr{F}_t = e^{-\int_0^t \frac{\mu(s)}{\sigma(s)} dW(s) - \frac{1}{2}\int_0^t |\frac{\mu(s)}{\sigma(s)}|^2 ds}, \quad 0 \le t \le T.$$

We change the measure in (11.13) and we derive the BSDE

$$\begin{aligned} Y(t) = F + \int_t^T \Big(&-\frac{\mu^2(s)}{2\alpha\sigma^2(s)} + H(s) \\ &+ \int_{\mathbb{R}} \Big(\frac{1}{\alpha} \big(e^{\alpha(G(s,z)+U(s,z))} - 1\big) - U(s,z) \Big) Q(s,dz)\eta(s) \Big) ds \\ &- \int_t^T Z(s) dW^{\mathbb{Q}^0}(s) - \int_t^T \int_{\mathbb{R}} U(s,z)\tilde{N}^{\mathbb{Q}^0}(ds,dz), \quad 0 \le t \le T. \end{aligned} \tag{11.15}$$

Notice that the generator f of (11.15) can be divided into two parts:

$$f(s) = \hat{f}(s) + \int_{\mathbb{R}} \tilde{f}\big(U(s,z)\big) Q(s,dz)\eta(s), \quad 0 \le s \le T,$$

where

$$\hat{f}(s) = -\frac{\mu^2(s)}{\alpha\sigma^2(s)} + H(s) + \frac{1}{\alpha}\int_{\mathbb{R}} \big(e^{\alpha G(s,z)} - 1\big)Q(s,dz)\eta(s), \quad 0 \le s \le T,$$

$$\tilde{f}\big(U(s,z)\big) = \frac{1}{\alpha}\big(e^{\alpha(G(s,z)+U(s,z))} - e^{\alpha G(s,z)}\big) - U(s,z), \quad 0 \le s \le T,\ z \in \mathbb{R}.$$

Consequently, from Proposition 11.1.1 we can conclude that there exists a unique solution (Y, Z, U) to the BSDE (11.15) which verifies

$$\begin{aligned}
&\mathbb{E}^{\mathbb{Q}^0}\Big[\int_0^T |Z(s)|^2 ds\Big] < \infty,\\
&|Y(s)| \le K, \quad 0 \le s \le T,\\
&|U(s,z)| \le K, \quad \vartheta\text{-a.e. } (s,z) \in [0,T]\times\mathbb{R}.
\end{aligned} \tag{11.16}$$

Moreover, by (11.15), (5.5) and Theorem 2.3.3 we obtain

$$\begin{aligned}
&\mathbb{E}^{\mathbb{Q}^0}\Big[\Big|F + \int_t^T \Big(-\frac{\mu^2(s)}{2\alpha\sigma^2(s)} + H(s)\\
&\quad + \int_{\mathbb{R}}\Big(\frac{1}{\alpha}\big(e^{\alpha(G(s,z)+U(s,z))} - 1\big) - U(s,z)\Big)Q(s,dz)\eta(s)\Big)ds - Y(t)\Big|^2 |\mathscr{F}_t\Big]\\
&= \mathbb{E}^{\mathbb{Q}^0}\Big[\Big|\int_t^T Z(s)dW^{\mathbb{Q}^0}(s) + \int_t^T\int_{\mathbb{R}} U(s,z)\tilde{N}^{\mathbb{Q}^0}(ds,dz)\Big|^2 |\mathscr{F}_t\Big]\\
&= \mathbb{E}^{\mathbb{Q}^0}\Big[\int_t^T |Z(s)|^2 ds|\mathscr{F}_t\Big]\\
&\quad + \mathbb{E}^{\mathbb{Q}^0}\Big[\int_t^T\int_{\mathbb{R}} |U(s,z)|^2 Q(s,dz)\eta(s)ds|\mathscr{F}_t\Big], \quad 0 \le t \le T,
\end{aligned}$$

and from (11.16) and the boundedness assumptions for the coefficients we deduce the uniform bound

$$\mathbb{E}^{\mathbb{Q}^0}\Big[\int_\tau^T |Z(s)|^2 ds|\mathscr{F}_\tau\Big] \le K, \tag{11.17}$$

for any stopping time $\tau \in [0,T]$. Estimate (11.17) yields that the process $\int_0^t Z(s)dW^{\mathbb{Q}^0}(s)$ is a *BMO* $\mathbb{Q}^0$-martingale, see Definition 2.2.5. We consider the change of measure

$$\frac{d\mathbb{P}}{d\mathbb{Q}^0}\Big|\mathscr{F}_t = e^{\int_0^t \frac{\mu(s)}{\sigma(s)}dW^{\mathbb{Q}^0}(s) - \frac{1}{2}\int_0^t |\frac{\mu(s)}{\sigma(s)}|^2 ds}, \quad 0 \le t \le T,$$

which is defined by the stochastic exponential of the martingale $M(t) = \int_0^t \frac{\mu(s)}{\sigma(s)}dW^{\mathbb{Q}^0}(s)$. It is clear that the martingale M is a *BMO* $\mathbb{Q}^0$-martingale. Hence, by Theorem 3.6 in Kazamaki (1994), the *BMO* $\mathbb{Q}^0$-martingale $\int_0^t Z(s)dW(s)$ is also a *BMO* $\mathbb{P}$-martingale. Consequently, $\mathbb{E}^{\mathbb{P}}[\int_\tau^T |Z(s)|^2 ds|\mathscr{F}_\tau] \leq K$ for any stopping time $\tau \in [0,T]$. Finally, we can conclude that there exists a solution (Y,Z,U) to the BSDE (11.13) which verifies

$$\begin{aligned} &\mathbb{E}^{\mathbb{P}}\left[\int_0^T |Z(s)|^2 ds\right] < \infty, \\ &|Y(s)| \leq K, \quad 0 \leq s \leq T, \\ &|U(s,z)| \leq K, \quad \vartheta\text{-a.e. } (s,z) \in [0,T]\times\mathbb{R}. \end{aligned} \tag{11.18}$$

To show the uniqueness of a solution, assume there are two solutions (Y,Z,U), $(Y'Z',U') \in \mathbb{S}^\infty(\mathbb{R}) \times \mathbb{H}^2(\mathbb{R}) \times \mathbb{H}^2_N(\mathbb{R})$. From property (11.12) we conclude that U and U' are ϑ-a.e. bounded. Hence, the generator of the BSDE (11.13) is Lipschitz continuous in the sense of (A2) from Sect. 3.1 and the uniqueness of a solution follows from the a priori estimates (3.5) and (3.7).

(b) We prove the supermartingale property (11.3). We introduce the process

$$M^\pi(t) = M^{W,\pi}(t)M^N(t), \quad 0 \leq t \leq T,\ \pi \in \mathscr{A}^{exp},$$

and from (11.6) we obtain the dynamics

$$\begin{aligned} \frac{dM^\pi(t)}{M^\pi(t)} &= -\alpha\big(\pi(t)\sigma(t) - Z(t)\big)dW(t) \\ &\quad + \int_{\mathbb{R}} \big(e^{\alpha(G(t,z)+U(t,z))} - 1\big)\tilde{N}(dt,dz). \end{aligned} \tag{11.19}$$

For any $\pi \in \mathscr{A}^{exp}$ the process M^π is the stochastic exponential of a local martingale, hence it is a local martingale by (2.16). Moreover, since $e^{\alpha(G(t,z)+U(t,z))} - 1 > -1$, we conclude that the local martingale M^π is positive. We also notice that for any $\pi \in \mathscr{A}^{exp}$ the process D^π defined by (11.7) is a.s. integrable, i.e. $\int_0^T |D^\pi(s)|ds < \infty$ a.s. Recalling (11.8), we get

$$A^\pi(t) = -e^{-\alpha(X^\pi(t)-Y(t))} = -M^\pi(t)e^{-\alpha(x-Y(0))+\int_0^t D^\pi(s)ds}, \quad 0 \leq t \leq T. \tag{11.20}$$

Since M^π is a positive local martingale and $D^\pi(s) \geq 0$, we can derive

$$\begin{aligned} &\mathbb{E}\big[A^\pi(t\wedge\tau_n)|\mathscr{F}_s\big] \\ &\quad = \mathbb{E}\big[-M^\pi(t\wedge\tau_n)e^{-\alpha(x-Y(0))+\int_0^{t\wedge\tau_n} D^\pi(u)du}|\mathscr{F}_s\big] \\ &\quad \leq \mathbb{E}\big[-M^\pi(t\wedge\tau_n)|\mathscr{F}_s\big]e^{-\alpha(x-Y(0))+\int_0^{s\wedge\tau_n} D^\pi(u)du} \\ &\quad = -M^\pi(s\wedge\tau_n)e^{-\alpha(x-Y(0))+\int_0^{s\wedge\tau_n} D^\pi(u)du} = A^\pi(s\wedge\tau_n), \quad 0 \leq s \leq t \leq T, \end{aligned} \tag{11.21}$$

where $(\tau_n)_{n\geq 1}$ denotes a localizing sequence for the local martingale M^{π}. From the uniform integrability of the family $\{e^{-\alpha X^{\pi}(\tau)}, \mathscr{F}\text{-stopping times } \tau\}$ for $\pi \in \mathscr{A}^{exp}$ and boundedness of Y we conclude that the family $\{A^{\pi}(\tau), \mathscr{F}\text{-stopping times } \tau\}$ is uniformly integrable for $\pi \in \mathscr{A}^{exp}$. Taking the limit $n \to \infty$ in (11.21), we obtain the supermartingale property of A^{π} for any $\pi \in \mathscr{A}^{exp}$.

We prove the martingale property (11.4). We have $D^{\pi^*}(t) = 0$, $0 \leq t \leq T$. We next notice that $\pi^*(t)\sigma(t) - Z(t) = \frac{\mu(t)}{\alpha\sigma(t)}$ and the local martingale M^{π^*} is a square integrable martingale by Proposition 2.5.1. Hence, A^{π^*} is a martingale by (11.20). The supermartingale and martingale properties (11.3)–(11.4) yield the optimality of the candidate strategy π^*.

We are left with proving the admissability. The strategy π^* is predictable and square integrable. It is clear that there exists a unique adapted, càdlàg solution X^{π^*} to (7.11). By (11.20) we get

$$e^{-\alpha X^{\pi^*}(\tau)} = M^{\pi^*}(\tau)e^{-\alpha Y(\tau)-\alpha(x-Y(0))}.$$

Since M^{π^*} is uniformly integrable by Proposition 2.5.1 and Y is bounded, the family $\{e^{-\alpha X^{\pi^*}(\tau)}, \mathscr{F}\text{-stopping times } \tau\}$ is uniformly integrable. □

We remark that the boundedness assumptions are essential for establishing the existence of a unique solution to the nonlinear BSDE (11.13).

Theorem 11.1.1 shows that the optimal value function of the optimization problem (11.1) and the corresponding optimal investment strategy can be derived from the nonlinear BSDE (11.13). The optimal investment strategy π^* is independent of wealth. If we investigate the pure investment problem without the payment process and we set $\mu(t) = \mu$, $\sigma(t) = \sigma$, then the control process of the BSDE (11.13) satisfies $Z = 0$ and the optimal investment strategy is given by $\pi^*(t) = \frac{\mu}{\alpha\sigma^2}$, $0 \leq t \leq T$, which is the well-know optimal investment strategy under exponential preferences in the classical Black-Scholes model. A non-zero control process Z arises in the optimal investment strategy (11.14) if we allow for a random drift and volatility of the stock and equity-linked claims.

If we considered the insurance payment process (7.3) driven by three random noises (W, B, N), then the generator of the BSDE (11.13) would have quadratic and exponential terms for the control processes. In such a case we could not use Proposition 11.1.1. However, if we assumed that the step process J has only a finite number of jumps, which is the case for the deaths counting process for a life insurance portfolio, then we could use the recent results by Jiao et al. (2013) and Kharroubi and Lim (2012) to establish existence of a unique solution to a BSDE with quadratic and exponential terms in the generator.

11.2 Exponential Indifference Pricing and Hedging

We are interested in finding a price and a hedging strategy for the insurance payment process. We use the utility indifference arguments under which the expected utility of the insurer's wealth at a specified future date remains unchanged if the insurer decides to sell the contract today.

Let $\mathscr{V}^P$ and $\mathscr{V}^0$ denote the optimal value functions of the exponential utility maximization problems for the insurer who, respectively, faces the payment process P and is free from the liability. The optimal value function $\mathscr{V}^P$ follows from Theorem 11.1.1, and the optimal value function $\mathscr{V}^0$ for the pure investment problem also follows from Theorem 11.1.1 by setting $F = H = G = 0$. We can conclude that the optimal value functions $\mathscr{V}^P$ and $\mathscr{V}^0$ are characterized by the insurer's current capital and the solutions Y^P and Y^0 to the BSDEs

$$\begin{aligned} Y^P(t) &= F + \int_t^T \Big(-\frac{\mu^2(s)}{2\alpha\sigma^2(s)} - \frac{\mu(s)}{\sigma(s)} Z^P(s) + H(s) \\ &\quad + \int_{\mathbb{R}} \Big(\frac{1}{\alpha}\big(e^{\alpha(G(s,z)+U^P(s,z))} - 1\big) - U^P(s,z)\Big) Q(s,dz)\eta(s)\Big) ds \\ &\quad - \int_t^T Z^P(s)dW(s) - \int_t^T \int_{\mathbb{R}} U^P(s,z)\tilde{N}(ds,dz), \qquad (11.22) \\ &\quad 0 \le t \le T, \\ Y^0(t) &= \int_t^T \Big(\frac{-\mu^2(s)}{2\alpha\sigma^2(s)} - \frac{\mu(s)}{\sigma(s)} Z^0(s)\Big) ds - \int_t^T Z^0(s)dW(s), \quad 0 \le t \le T. \end{aligned}$$

The indifference price $\mathscr{Y}(t)$ makes the insurer indifferent at time t between selling the contract insuring the stream of claims P, collecting the premium $\mathscr{Y}(t)$, paying the future claims P and not selling the contract. The indifference price process $\mathscr{Y} := (\mathscr{Y}(t), 0 \le t \le T)$ is defined as a solution to the equation

$$\mathscr{V}^P\big(t, x + \mathscr{Y}(t)\big) = \mathscr{V}^0(t,x), \quad 0 \le t \le T, \qquad (11.23)$$

where x denotes the insurer's wealth at time t, and we equate the maximal expected utilities of the insurer's wealth at time T under two decisions. By Theorem 11.1.1 the indifference pricing principle takes the form

$$-e^{-\alpha(x+\mathscr{Y}(t)-Y^P(t))} = -e^{-\alpha(x-Y^0(t))}, \quad 0 \le t \le T,$$

and the price process satisfies

$$\mathscr{Y}(t) = Y^P(t) - Y^0(t), \quad 0 \le t \le T. \qquad (11.24)$$

First, we characterize the indifference price process $\mathscr{Y}$. The indifference hedging strategy will be next deduced from $\mathscr{Y}$.

Theorem 11.2.1 *Consider the filtration $\mathscr{F} = \mathscr{F}^W \vee \mathscr{F}^J$. Assume that* (C1)–(C4) *from Chap. 7 hold and let F, H, G, η be bounded and* $r = 0$.

(a) *There exists a unique solution* $(\mathscr{Y}, \mathscr{Z}, \mathscr{U}) \in \mathbb{S}^{\infty}(\mathbb{R}) \times \mathbb{H}^2(\mathbb{R}) \times \mathbb{H}^2_N(\mathbb{R})$ *to the BSDE*

$$\begin{aligned}\mathscr{Y}(t) = F + \int_t^T \bigg(& -\frac{\mu(s)}{\sigma(s)}\mathscr{Z}(s) + H(s) \\ & + \int_{\mathbb{R}} \Big(\frac{1}{\alpha}\big(e^{\alpha(G(s,z)+\mathscr{U}(s,z))} - 1\big) - \mathscr{U}(s,z)\Big) Q(s,dz)\eta(s)\bigg) ds \\ & - \int_t^T \mathscr{Z}(s)dW(s) - \int_t^T \int_{\mathbb{R}} \mathscr{U}(s,z)\tilde{N}(ds,dz), \quad 0 \le t \le T. \quad (11.25)\end{aligned}$$

(b) *The indifference price process* $\mathscr{Y} := (\mathscr{Y}(t), 0 \le t \le T)$ *of the payment process* P*, defined by* (11.23)*, solves the BSDE* (11.25)*. The indifference price process has the representation*

$$\mathscr{Y}(t) = \mathbb{E}^{\mathbb{Q}^*}\Big[\int_t^T dP(s)|\mathscr{F}_t\Big], \quad 0 \le t \le T, \quad (11.26)$$

where the equivalent martingale measure $\mathbb{Q}^*$ *is given by*

$$\frac{d\mathbb{Q}^*}{d\mathbb{P}}\Big|\mathscr{F}_t = M^*(t), \quad 0 \le t \le T,$$

$$\frac{dM^*(t)}{M^*(t-)} = -\frac{\mu(t)}{\sigma(t)}dW(t) + \int_{\mathbb{R}}\Big(\frac{e^{\alpha(G(t,z)+\mathscr{U}(t,z))} - 1}{\alpha(G(t,z)+\mathscr{U}(t,z))} - 1\Big)\tilde{N}(dt,dz).$$

Proof (a) The result can be proved by following the reasoning from Theorem 11.1.1.

(b) By Theorem 11.1.1 there exist unique solutions (Y^P, Z^P, U^P), $(Y^0, Z^0, U^0) \in \mathbb{S}^{\infty}(\mathbb{R}) \times \mathbb{H}^2(\mathbb{R}) \times \mathbb{H}^2_N(\mathbb{R})$ to the BSDEs (11.22). Substituting the dynamics of Y^P and Y^0 into (11.24) and introducing new control processes $\mathscr{Z} = Z^P - Z^0$ and $\mathscr{U} = U^P$, we can show that the indifference price process $\mathscr{Y}$ satisfies the BSDE (11.25). We prove the representation of $\mathscr{Y}$. Let $M^* = (M^*(t), 0 \le t \le T)$ be given by the dynamics

$$\frac{dM^*(t)}{M^*(t-)} = -\frac{\mu(t)}{\sigma(t)}dW(t) + \int_{\mathbb{R}}\Big(\frac{e^{\alpha(G(t,z)+\mathscr{U}(t,z))} - 1}{\alpha(G(t,z)+\mathscr{U}(t,z))} - 1\Big)\tilde{N}(dt,dz), \quad (11.27)$$

with $M^*(0) = 1$. Notice that the function $\varphi(y) = \frac{e^y - 1}{y} - 1$ is continuous, with $\varphi(0) = 0$, and satisfies $|\varphi(y)| \le e^{|y|} + 1$, $\varphi(y) > -1$, $y \in \mathbb{R}$. These properties of φ together with boundedness of $\mathscr{U}$, G and $\frac{\mu}{\sigma}$ yield that M^* is a positive, square integrable martingale, see Proposition 2.5.1. Hence, we conclude that $\mathbb{Q}^*$ is an equivalent probability measure. Moreover, the measure $\mathbb{Q}^*$ is an equivalent martingale measure since the process S is a $\mathbb{Q}^*$-martingale. We change the measure in (11.25) and the Girsanov's theorem yields the BSDE

$$\mathscr{Y}(t) = F + \int_t^T \Big(H(s) + \int_{\mathbb{R}} G(s,z)\big(1+\kappa^*(s,z)\big)Q(s,dz)\eta(s) \Big) ds$$
$$- \int_t^T \mathscr{Z}(s) dW^{\mathbb{Q}^*}(s) - \int_t^T \int_{\mathbb{R}} \mathscr{U}(s,z)\tilde{N}^{\mathbb{Q}^*}(ds,dz), \quad 0 \leq t \leq T, \tag{11.28}$$

where

$$\kappa^*(s,z) = \frac{e^{\alpha(G(s,z)+\mathscr{U}(s,z))} - 1}{\alpha(G(s,z)+\mathscr{U}(s,z))} - 1, \quad (s,z) \in [0,T] \times \mathbb{R}. \tag{11.29}$$

We deduce that the stochastic integrals in (11.28) are $\mathbb{Q}^*$-martingales, see (3.25)–(3.26). Taking the expected value, we obtain the representation

$$\begin{aligned}\mathscr{Y}(t) &= \mathbb{E}^{\mathbb{Q}^*}\Big[F + \int_t^T \Big(H(s) + \int_{\mathbb{R}} G(s,z)\big(1+\kappa^*(s,z)\big)Q(s,dz)\eta(s) \Big) ds | \mathscr{F}_t \Big] \\ &= \mathbb{E}^{\mathbb{Q}^*}\Big[F + \int_t^T \Big(H(s) + \int_{\mathbb{R}} G(s,z)N(ds,dz) \Big) ds | \mathscr{F}_t \Big], \\ &= \mathbb{E}^{\mathbb{Q}^*}\Big[\int_t^T dP(s) | \mathscr{F}_t \Big], \quad 0 \leq t \leq T, \end{aligned} \tag{11.30}$$

where we use the fact that $(1+\kappa^*(s,z))Q(s,dz)\eta(s)ds$ is the compensator of the random measure N under $\mathbb{Q}^*$. □

The indifference price process of the insurance payment process P solves the non-linear BSDE (11.25). Notice that the BSDE (11.25) has a Lipschitz generator as U is bounded. Representation (11.26) shows that the indifference price $\mathscr{Y}$ is arbitrage-free. The pricing measure $\mathbb{Q}^*$ results from solving the optimization problem (11.23). The unsystematic insurance risk is priced under $\mathbb{Q}^*$ and the insurance risk premium is given by (11.29). Notice that the pricing measure $\mathbb{Q}^*$ depends on the financial market, the insurance payment process and the risk aversion coefficient α.

We can derive the indifference hedging strategy.

Proposition 11.2.1 *Under the assumptions of Theorem* 11.2.1 *the indifference hedging strategy for the payment process P takes the form*

$$\Pi^*(t) = \frac{\mathscr{Z}(t)}{\sigma(t)}, \quad 0 \leq t \leq T. \tag{11.31}$$

Following Ankirchner et al. (2010) and Becherer (2006), we call the process $\Pi^*(t) = \frac{\mathscr{Z}(t)}{\sigma(t)} = \frac{Z^P(t)-Z^0(t)}{\sigma(t)}$ the indifference hedging strategy. The process Π^* represents the change in the optimal investment strategy π^*, for an insurer who aims to maximize the expected utility of the terminal wealth, resulting from selling the payment process P and hedging the claims. Since $\mathscr{Y}$ is an arbitrage-free price of the liability P and $\mathscr{Z}$ is the control process of the BSDE for $\mathscr{Y}$, we conclude that the investment strategy Π^* is a delta hedging strategy, see Sect. 9.2. From (11.27), (11.28) and (11.31) we deduce that the indifference price process satisfies the dynamics

$$\begin{aligned}\mathscr{Y}(t) &= \mathscr{Y}(0) + \int_0^t \Pi^*(s)\big(\mu(s) + \sigma(s)dW(s)\big) - P(t) \\ &\quad + \int_0^t \int_{\mathbb{R}} \big(\mathscr{U}(s,z) + G(s,z)\big)\tilde{N}^{\mathbb{Q}^*}(ds,dz) \\ &= \mathscr{Y}(0) + \int_0^t \Pi^*(s)\frac{dS(s)}{S(s)} - P(t) \\ &\quad + \int_0^t \int_{\mathbb{R}} \big(\mathscr{U}(s,z) + G(s,z)\big)\tilde{N}^{\mathbb{Q}^*}(ds,dz), \quad 0 \le t \le T.\end{aligned}$$

The strategy Π^* can be interpreted as the amount of wealth which should be invested in the stock S and the integral with respect to the random measure can be interpreted as the cash inflows/outflows, over the gains from the self-financing hedging strategy Π^*, which are needed to match the optimal investment portfolio X^{Π^*} with the market-consistent value of the liability $\mathscr{Y}$.

Example 11.1 We consider the classical Black-Scholes model with constant coefficients μ, σ and a non-life insurer who has a stop-loss contract on aggregate claims J. We are interested in pricing and hedging the claim $F = J(T) \wedge K = J(T) - (J(T) - K)^+$ where J is a compound Poisson process with intensity λ and jump size distribution q defined on a positive support. In order to find the indifference price process of the claim F, we have to solve the BSDE

$$\begin{aligned}\mathscr{Y}(t) &= J(T) \wedge K \\ &\quad + \int_t^T \left(-\frac{\mu}{\sigma}\mathscr{Z}(s) + \int_0^\infty \frac{1}{\alpha}\big((e^{\alpha\mathscr{U}(s,z)} - 1) - \mathscr{U}(s,z)\big)\lambda q(dz)\right)ds \\ &\quad - \int_t^T \mathscr{Z}(s)dW(s) - \int_t^T \int_0^\infty \mathscr{U}(s,z)\tilde{N}(ds,dz), \quad 0 \le t \le T. \end{aligned} \tag{11.32}$$

Clearly, we can set $\mathscr{Z}(t) = 0$, $0 \le t \le T$. By Proposition 11.2.1 the optimal hedging strategy for the claim F is $\Pi^*(t) = 0$, $0 \le t \le T$. This strategy agrees with our intuition since the claim is independent of the financial market and it cannot be hedged by investment into the stock. At the same time, from Theorem 11.1.1 we can conclude that the optimal investment strategy for an insurer who faces the claim F and maximizes the expected exponential utility from his terminal wealth is $\pi^*(t) = \frac{\mu}{\alpha\sigma^2}$, $0 \le t \le T$. By Proposition 3.4.3 the solution to the BSDE (11.32) is of the form

$$\mathscr{Y}(t) = \frac{1}{\alpha}\ln \mathbb{E}\big[e^{\alpha(J(T)\wedge K)}|\mathscr{F}_t^J\big], \quad 0 \le t \le T,$$

$$\mathscr{U}(t,z)=\frac{1}{\alpha}\ln\big(1+U(t,z)e^{-\alpha\mathscr{Y}(t-)}\big),\quad 0\le t\le T,\ z\in(0,\infty),$$

where U is derived from the predictable representation

$$e^{\alpha(J(T)\wedge K)}=\mathbb{E}\big[e^{\alpha(J(T)\wedge K)}\big]+\int_0^T\int_0^\infty U(t,z)\tilde{N}(dt,dz).$$

Let $\varphi:[0,T]\times[0,\infty)\to\mathbb{R}$ denote a measurable function such that

$$\varphi\big(t,J(t)\big)=\mathbb{E}\big[e^{\alpha(J(T)\wedge K)}|\mathscr{F}_t^J\big],\quad 0\le t\le T.$$

Applying the Malliavin calculus, Propositions 2.6.4 and 2.6.6 and Theorem 3.5.2, we get the formula

$$e^{\alpha(J(T)\wedge K)}=\mathbb{E}\big[e^{\alpha(J(T)\wedge K)}\big]+\int_0^T\int_0^\infty\big(\varphi\big(t,J(t-)+z\big)-\varphi\big(t,J(t-)\big)\big)\tilde{N}(dt,dz),$$

and the solution to the BSDE (11.32) is now completely characterized. By Theorem 11.2.1 we can also establish the representation

$$\mathscr{Y}(t)=\mathbb{E}^{\mathbb{Q}^*}\big[J(T)\wedge K|\mathscr{F}_t^J\big],\quad 0\le t\le T,$$

where the equivalent probability measure $\mathbb{Q}^*$ is defined by the Radon-Nikodym derivative with the kernel

$$\kappa^*(t,z)=\frac{e^{\alpha\mathscr{U}(t,z)}-1}{\alpha\mathscr{U}(t,z)}-1,\quad 0\le t\le T,\ z\in(0,\infty).$$

Bibliographical Notes This chapter is taken from Delong (2012b). The idea of solving the exponential utility optimization problem with the help of a BSDE was developed by Hu et al. (2005) for diffusion processes and extended by Becherer (2006) to jump processes. We closely follow their arguments in this chapter. We point out that the solution method from Hu et al. (2005) allows for introducing constraints on admissible investment strategies. Hu et al. (2005) also provided the optimal investment strategies for the power and logarithmic utilities. The utility maximization problem with a consumption process is studied by Cheridito and Hu (2011). We remark that Hu et al. (2005) and Cheridito and Hu (2011) consider diffusion models and their solutions rely on the existence result for quadratic BSDEs developed by Kobylanski (2000). Exponential utility maximization in models with jumps is much more difficult. We refer to Ankirchner et al. (2010), Bielecki et al. (2004), Morlais (2009), Lim and Quenez (2011), Jiao et al. (2013), Kharroubi and Lim (2012) where various financial problems are investigated and existence and uniqueness results for quadratic BSDEs with jumps are established. We also remark that the exponential utility maximization problem in a semimartingale financial model with a general continuous filtration was solved in Mania and Schweizer (2005) where the authors used BSDEs driven by local martingales. For exponential utility maximization and indifference pricing and hedging in insurance we refer to Ankirchner et al. (2010), Liang et al. (2011), Liu and Ma (2009), Ludkovski and Young (2008), Perera (2010), Wang (2007). The classical reference on utility maximization, portfolio optimization under risk preferences and indifference pricing and hedging is the book by Carmona (2008).

Chapter 12
Pricing and Hedging Under a Least Favorable Measure

Abstract We consider two optimization problems which take into account the uncertainty about the true probability (martingale) measure. First, we investigate pricing and hedging under model ambiguity. We find the hedging strategy which minimizes the expected terminal shortfall under a least favorable probability measure specifying the probability model for the risk factors and we set the price which offsets this worst shortfall. Next, we deal with no-good-deal pricing. We price the insurance payment process with a least favorable martingale measure under a Sharpe ratio constraint which excludes prices leading to extraordinarily high gains. Both pricing and hedging objectives lead to the same solution. We characterize the price and the hedging strategy by a nonlinear BSDE.

In this chapter we solve two optimization problems which take into account the uncertainty about the true probability (martingale) measure. We investigate pricing and hedging under model ambiguity and we deal with no-good-deal pricing. Both objectives have strong theoretical and practical justifications. In both cases the goal is to derive a price and a hedging strategy by optimizing the expectation of a payoff over a set of equivalent probability (martingale) measures. The least favorable measure is found and used for pricing and hedging. The connection between the objectives considered in this chapter and pricing and hedging under the instantaneous mean-variance risk measure considered in Sect. 10.4 is given.

12.1 Pricing and Hedging Under Model Ambiguity

In previous chapters we assumed that we know the true real-world probability measure (the true probability law) or we know the true parameters of the combined financial and insurance model. In real applications the true probabilities or the true values of parameters are uncertain and we face so-called model ambiguity.

We consider the financial model (7.1)–(7.2) and the insurance payment process (7.3). We assume that the process J is a point process and, consequently, the jump measure N of the point process J has the compensator $\vartheta(dt,\{1\}) = \eta(t)dt$. We allow for model ambiguity or Knightian uncertainty, see Chen and Epstein (2002).

Ł. Delong, *Backward Stochastic Differential Equations with Jumps and Their Actuarial and Financial Applications*, EAA Series, DOI 10.1007/978-1-4471-5331-3_12,

We introduce the set of equivalent probability measures

$$\begin{aligned}\mathscr{Q}^s &= \left\{\mathbb{Q}\sim\mathbb{P},\ \frac{d\mathbb{Q}}{d\mathbb{P}}\Big|\mathscr{F}_t = M^s(t),\ 0\leq t\leq T\right\},\\ \frac{dM^s(t)}{M^s(t-)} &= \psi(t)dW(t)+\phi(t)dB(t)+\kappa(t)\tilde{N}(dt),\quad M^s(0)=1,\end{aligned}\tag{12.1}$$

where ψ, ϕ and κ are predictable processes such that

$$\left|\psi(t)\right|^2+\left|\phi(t)\right|^2+\left|\kappa(t)\right|^2\eta(t)\leq\left|L(t)\right|^2\leq K,\quad \kappa(t)>-1,\quad 0\leq t\leq T,$$

and L is a predictable process. The purpose of the set $\mathscr{Q}^s$ is to represent different beliefs (different assumptions) about the parameters or the probability laws of the risk factors in our model. One way of determining the set $\mathscr{Q}^s$ for ambiguity modelling is to specify confidence sets around the estimates of the parameters and to take for $\mathscr{Q}^s$ the class of all measures that are consistent with these confidence sets. Then, the process L can be interpreted as an estimation error. Alternatively, the elements of $\mathscr{Q}^s$ can be interpreted as prior models which specify probabilities of future scenarios for the risk factors. Then, the process L can define the range of equivalent probabilities for every scenario.

Let us introduce the risk measure

$$\rho(\xi)=\sup_{\mathbb{Q}\in\mathscr{Q}^s}\mathbb{E}^{\mathbb{Q}}[-\xi].\tag{12.2}$$

The risk measure (12.2) measures the risk of a financial position ξ. We remark that $\xi>0$ is interpreted as a profit and $\xi<0$ as a loss. Under (12.2) we take the supremum of all expected shortfalls for all prior models and we are interested in the expected shortfall under the least favorable model (the least favorable assumptions).

We apply the conditional version of the risk measure (12.2) to the discounted surplus at time T (the net asset value at time T). We investigate the risk measures

$$\rho_t\left(e^{-\int_t^T r(s)ds}X^{\pi}(T)-e^{-\int_t^T r(s)ds}F\right),\quad 0\leq t\leq T,\tag{12.3}$$

where the investment portfolio X^{π} under an admissible investment strategy $\pi\in\mathscr{A}$ is given by (7.11). The goal is to find an admissible investment strategy π which minimizes the risk measures ρ_t for all $t\in[0,T]$ and a price Y which makes the risk measures vanish in the sense that

$$\rho_t\left(e^{-\int_t^T r(s)ds}X^{\pi}(T)-e^{-\int_t^T r(s)ds}F\right)=0,\quad 0\leq t\leq T,$$

under the condition that $X(t)=Y(t)$. The price and the hedging strategy are given by

$$\begin{aligned}Y(t)=\inf_{\pi\in\mathscr{A}}\Big\{\sup_{\mathbb{Q}\in\mathscr{Q}^s}\mathbb{E}^{\mathbb{Q}}\Big[-\Big(e^{-\int_t^T r(s)ds}X^{\pi}(T)-X(t)\\ -e^{-\int_t^T r(s)ds}F\Big)|\mathscr{F}_t\Big]\Big\},\quad 0\leq t\leq T.\end{aligned}$$

The hedging strategy leads to the lowest expected terminal shortfall of the assets below the liabilities under the least favorable probability measure from the set of prior models. The price offsets this expected terminal shortfall. The objective seems to be a sound pricing and hedging objective for insurers who are forced by regulators to carry stress-tests on model parameters and hold sufficient capital to withstand extreme scenarios. By applying the risk measure (12.3) the insurer protects the terminal net asset wealth under the scenario in which the worst model assumptions turn out to be the true assumptions. We remark that (12.3) is an example of robust utility optimization, see Schied (2005) and Schied (2006).

We now solve the optimization problem

$$\begin{aligned} Y(t) = \inf_{\pi}\Bigg\{ \sup_{\mathbb{Q}\in\mathscr{Q}^s} \mathbb{E}^{\mathbb{Q}}\Bigg[e^{-\int_t^T r(u)du} F + \int_t^T e^{-\int_t^s r(u)du} H(s)ds \\ + \int_t^T e^{-\int_t^s r(u)du} G(s)dJ(s) - \int_t^T e^{-\int_t^s r(u)du} \pi(s)\big(\big(\mu(s) - r(s)\big)ds \\ + \sigma(s)dW(s)\big)|\mathscr{F}_t\Bigg]\Bigg\}, \quad 0 \le t \le T. \end{aligned} \tag{12.4}$$

We deal with three BSDE:

$$\begin{aligned} Y^{\pi,\psi,\phi,\kappa}(t) = F + \int_t^T H(s)ds + \int_t^T G(s)dJ(s) \\ - \int_t^T \pi(s)\big(\mu(s) - r(s)\big)ds - \int_t^T \pi(s)\sigma(s)dW(s) \\ + \int_t^T \big(-Y^{\pi,\psi,\phi,\kappa}(s-)r(s) + Z_1^{\pi,\psi,\phi,\kappa}(s)\psi(s) + Z_2^{\pi,\psi,\phi,\kappa}(s)\phi(s) \\ + U^{\pi,\psi,\phi,\kappa}(s)\kappa(s)\eta(s)\big)ds \\ - \int_t^T Z_1^{\pi,\psi,\phi,\kappa}(s)dW(s) - \int_t^T Z_2^{\pi,\psi,\phi,\kappa}(s)dB(s) \\ - \int_t^T U^{\pi,\psi,\phi,\kappa}(s)\tilde{N}(ds), \quad 0 \le t \le T, \end{aligned} \tag{12.5}$$

where $\pi \in \mathscr{A}$ and $(\psi, \phi, \kappa) \in \mathscr{Q}^s$,

$$\begin{aligned} Y^{\pi,*}(t) = F + \int_t^T H(s)ds + \int_t^T G(s)dJ(s) \\ - \int_t^T \pi(s)\big(\mu(s) - r(s)\big)ds - \int_t^T \pi(s)\sigma(s)dW(s) + \int_t^T \Big(-Y^{\pi,*}(s-)r(s) \\ + L(s)\sqrt{\big|Z_1^{\pi,*}(s)\big|^2 + \big|Z_2^{\pi,*}(s)\big|^2 + \big|U^{\pi,*}(s)\big|^2\eta(s)}\Big)ds \end{aligned}$$

$$-\int_t^T Z_1^{\pi,*}(s)dW(s) - \int_t^T Z_2^{\pi,*}(s)dB(s)$$
$$-\int_t^T U^{\pi,*}(s)\tilde{N}(ds), \quad 0 \le t \le T, \tag{12.6}$$

where $\pi \in \mathscr{A}$, and

$$\begin{aligned}
Y^{*,*}(t) = F &+ \int_t^T H(s)ds + \int_t^T G(s)dJ(s) \\
&+ \int_t^T \Big(-Y^{*,*}(s-)r(s) - Z_1^{*,*}(s)\theta(s) \\
&+ \sqrt{|L(s)|^2 - |\theta(s)|^2}\sqrt{|Z_2^{*,*}(s)|^2 + |U^{*,*}(s)|^2\eta(s)}\Big)ds \\
&- \int_t^T Z_1^{*,*}(s)dW(s) - \int_t^T Z_2^{*,*}(s)dB(s) \\
&- \int_t^T U^{*,*}(s)\tilde{N}(ds), \quad 0 \le t \le T.
\end{aligned} \tag{12.7}$$

By Propositions 3.3.1 and 3.4.1 we can derive the representation

$$\begin{aligned}
Y^{\pi,\psi,\phi,\kappa}(t) = \mathbb{E}^{\mathbb{Q}^{\psi,\phi,\kappa}}\Big[-\Big(e^{-\int_t^T r(s)ds}X^\pi(T) - X(t) \\
- e^{-\int_t^T r(s)ds}F\Big)|\mathscr{F}_t\Big], \quad 0 \le t \le T,
\end{aligned}$$

where $\mathbb{Q}^{\psi,\phi,\kappa}$ is induced by $(\psi,\phi,\kappa) \in \mathscr{Q}^s$.

Theorem 12.1.1 *Let us assume that* (C1)–(C4) *from Chap.* 7 *hold and the jump measure N of the step process J has the compensator $\vartheta(dt,\{1\}) = \eta(t)dt$. We consider a predictable process L such that $L(t) \ge \theta(t) + \varepsilon$, $\varepsilon > 0$, and $L(t) \le K$, $0 \le t \le T$.*

(a) *There exist unique solutions* $(Y^{\pi,\psi,\phi,\kappa}, Z_1^{\pi,\psi,\phi,\kappa}, Z_2^{\pi,\psi,\phi,\kappa}, U^{\pi,\psi,\phi,\kappa})$, $(Y^{\pi,*}, Z_1^{\pi,*}, Z_2^{\pi,*}, U^{\pi,*}) \in \mathbb{S}^2(\mathbb{R}) \times \mathbb{H}^2(\mathbb{R}) \times \mathbb{H}^2(\mathbb{R}) \times \mathbb{H}^2_N(\mathbb{R})$ *to the BSDEs* (12.5) *and* (12.6) *with $\pi \in \mathscr{A}$ and $(\psi,\phi,\kappa) \in \mathscr{Q}^s$.*
(b) *For any $\pi \in \mathscr{A}$ and $(\psi,\phi,\kappa) \in \mathscr{Q}^s$ we have $Y^{\pi,\psi,\phi,\kappa}(t) \le Y^{\pi,*}(t)$, $0 \le t \le T$.*
(c) *For any $\pi \in \mathscr{A}$ such that*

$$\frac{L(t)U^{\pi,*}(t)}{\sqrt{|Z_1^{\pi,*}(t)|^2 + |Z_2^{\pi,*}(t)|^2 + |U^{\pi,*}(t)|^2\eta(t)}} \cdot \mathbf{1}\{U^{\pi,*}(t)\eta(t) \neq 0\} > -1, \quad 0 \le t \le T,$$

we have $\sup_{(\psi,\phi,\kappa)\in\mathscr{Q}^s} Y^{\pi,\psi,\phi,\kappa}(t) = Y^{\pi,*}(t)$, $0 \le t \le T$.

Proof (a) Choose $\pi \in \mathscr{A}$ and $(\psi,\phi,\kappa) \in \mathscr{Q}^s$. By (3.22), (10.46) and Theorem 3.1.1 there exist unique solutions to the BSDEs (12.5) and (12.6).

(b) Notice that the generator f of the BSDE (12.5) satisfies the property

$$f(t,y,z,u) - f(t,y,z,u') = \delta^{y,z,u,u'}(t)(u-u')\eta(t),$$

with $\delta^{y,z,u,u'}(t) = \kappa(t)$, for $(t,y,z,u), (t,y,z,u') \in [0,T] \times \mathbb{R} \times \mathbb{R} \times \mathbb{R}$. Recalling the arguments from the proof of Theorem 3.2.2 that led to (3.31), we obtain

$$\begin{aligned}
&Y^{\pi,*}(t) - Y^{\pi,\psi,\phi,\kappa}(t) \\
&\quad = \mathbb{E}^{\mathbb{Q}^{\psi,\phi,\kappa}}\Bigg[\int_t^T e^{-\int_t^s r(u)du}\Big(L(s)\sqrt{\big|Z_1^{\pi,*}(s)\big|^2 + \big|Z_2^{\pi,*}(s)\big|^2 + \big|U^{\pi,*}(s)\big|^2\eta(s)} \\
&\qquad - Z_1^{\pi,*}(s)\psi(s) - Z_2^{\pi,*}(s)\phi(s) - U^{\pi,*}(s)\kappa(s)\eta(s)\Big)ds \,|\mathscr{F}_t\Bigg], \quad 0 \le t \le T,
\end{aligned} \tag{12.8}$$

where $\mathbb{Q}^{\psi,\phi,\kappa}$ is induced by $(\psi,\phi,\kappa) \in \mathscr{Q}^s$. It is straightforward to check that the triple

$$\begin{aligned}
x^* &= \frac{\delta u}{\sqrt{u^2 + w^2 + v^2\eta}}\mathbf{1}\{u \neq 0\} \\
y^* &= \frac{\delta w}{\sqrt{u^2 + w^2 + v^2\eta}}\mathbf{1}\{w \neq 0\}, \\
z^* &= \frac{\delta v}{\sqrt{u^2 + w^2 + v^2\eta}}\mathbf{1}\{v\eta \neq 0\},
\end{aligned} \tag{12.9}$$

is the solution to the optimization problem

$$\begin{aligned}
&ux + wy + vz\eta \to \max \\
&x^2 + y^2 + z^2\eta \le \delta^2,
\end{aligned} \tag{12.10}$$

and the global maximum of (12.10) is equal to $\delta\sqrt{u^2 + w^2 + v^2\eta}$. Hence, from (12.8)–(12.10) we conclude that $Y^{\pi,*}(t) - Y^{\pi,\psi,\phi,\kappa}(t) \ge 0$ for all $t \in [0,T]$ and any $\pi \in \mathscr{A}$, $(\psi,\phi,\kappa) \in \mathscr{Q}^s$.

(c) Recalling (12.9), we define

$$\begin{aligned}
\psi^*(t) &= \frac{L(t)Z_1^{\pi,*}(t)}{\sqrt{|Z_1^{\pi,*}(t)|^2 + |Z_2^{\pi,*}(t)|^2 + |U^{\pi,*}(t)|^2\eta(t)}}\mathbf{1}\big\{Z_1^{\pi,*}(t) \neq 0\big\} \\
\phi^*(t) &= \frac{L(t)Z_2^{\pi,*}(t)}{\sqrt{|Z_1^{\pi,*}(t)|^2 + |Z_2^{\pi,*}(t)|^2 + |U^{\pi,*}(t)|^2\eta(t)}}\mathbf{1}\big\{Z_2^{\pi,*}(t) \neq 0\big\}, \\
\kappa^*(t) &= \frac{L(t)U^{\pi,*}(t)}{\sqrt{|Z_1^{\pi,*}(t)|^2 + |Z_2^{\pi,*}(t)|^2 + |U^{\pi,*}(t)|^2\eta(t)}}\mathbf{1}\big\{U^{\pi,*}(t)\eta(t) \neq 0\big\}.
\end{aligned} \tag{12.11}$$

The assumptions on π guarantee that $(\psi^*, \phi^*, \kappa^*) \in \mathcal{Q}^s$. The solution $(Y^{\pi,\psi^*,\phi^*,\kappa^*}, Z_1^{\pi,\psi^*,\phi^*,\kappa^*}, Z_2^{\pi,\psi^*,\phi^*,\kappa^*}, U^{\pi,\psi^*,\phi^*,\kappa^*})$ coincides with $(Y^{\pi,*}, Z_1^{\pi,*}, Z_2^{\pi,*}, U^{\pi,*})$ by uniqueness of solution to (12.5). Hence, we conclude that $\sup_{(\psi,\phi,\kappa)\in\mathcal{Q}^s} Y^{\pi,\psi,\phi,\kappa}(t) \geq Y^{\pi,\psi^*,\phi^*,\kappa^*}(t) = Y^{\pi,*}(t)$ for all $t \in [0,T]$ and any $\pi \in \mathcal{A}$. Since $\sup_{(\psi,\phi,\kappa)\in\mathcal{Q}^s} Y^{\pi,\psi,\phi,\kappa}(t) \leq Y^{\pi,*}(t)$ by item (b), the assertion of item (c) can be immediately deduced. □

We remark that the assumption on π from item (c) guarantees that the inequality $\kappa(t) > -1$ is strict in the optimum. Without this assumption, the least favorable measure cannot be found in the set $\mathcal{Q}^s$.

Theorem 12.1.2 *Let us assume that* (C1)–(C4) *from Chap.* 7 *hold and the jump measure N of the step process J has the compensator $\vartheta(dt,\{1\}) = \eta(t)dt$. We consider a predictable process L such that $L(t) \geq \theta(t) + \varepsilon$, $\varepsilon > 0$, and $L(t) \leq K$, $0 \leq t \leq T$.*

(a) *There exist unique solutions $(Y^{\pi,*}, Z_1^{\pi,*}, Z_2^{\pi,*}, U^{\pi,*})$, $(Y^{*,*}, Z_1^{*,*}, Z_2^{*,*}, U^{*,*}) \in \mathbb{S}^2(\mathbb{R}) \times \mathbb{H}^2(\mathbb{R}) \times \mathbb{H}^2(\mathbb{R}) \times \mathbb{H}_N^2(\mathbb{R})$ to the BSDEs* (12.6) *and* (12.7) *with $\pi \in \mathcal{A}$.*
(b) *Define the class of admissible strategies $\mathcal{A}^s$ which consists of strategies $\pi := (\pi(t), 0 \leq t \leq T)$ such that $\pi \in \mathcal{A}$ and*

$$\frac{L(t)U^{\pi,*}(t)}{\sqrt{|Z_1^{*,*}(t) - \pi(t)\sigma(t)|^2 + |Z_2^{*,*}(t)|^2 + |U^{\pi,*}(t)|^2\eta(t)}} \mathbf{1}\{U^{\pi,*}(t)\eta(t) \neq 0\} > -1,$$

$$0 \leq t \leq T,$$

$$\frac{L(t)U^{*,*}(t)}{\sqrt{|Z_1^{*,*}(t) - \pi(t)\sigma(t)|^2 + |Z_2^{*,*}(t)|^2 + |U^{*,*}(t)|^2\eta(t)}} \mathbf{1}\{U^{*,*}(t)\eta(t) \neq 0\} > -1,$$

$$0 \leq t \leq T.$$

For any $\pi \in \mathcal{A}^s$ we have $Y^{\pi,*}(t) \geq Y^{*,*}(t)$, $0 \leq t \leq T$.
(c) *Let Y denote the optimal value function of the optimization problem* (12.4) *under the new set of admissible strategies $\mathcal{A}^s$. If $U^{*,*}(t) \geq 0$ or $|L(t)|^2 < \eta(t) + |\theta(t)|^2$ on the set $\{\eta(t) > 0\}$, $0 \leq t \leq T$, then $\inf_{\pi\in\mathcal{A}^s} Y^{\pi,*}(t) = Y^{*,*}(t) = Y(t)$, $0 \leq t \leq T$, and the optimal admissible hedging strategy $\pi^* \in \mathcal{A}^s$ takes the form*

$$\pi^*(t) = \frac{1}{\sigma(t)}\left(Z_1^{*,*}(t) + \sqrt{\frac{|\theta(t)|^2}{|L(t)|^2 - |\theta(t)|^2}}\sqrt{\left|Z_2^{*,*}(t)\right|^2 + \left|U^{*,*}(t)\right|^2\eta(t)}\right),$$
$$0 \leq t \leq T. \tag{12.12}$$

Proof (a) Choose $\pi \in \mathcal{A}$. By (3.22), (10.46) and Theorem 3.1.1 there exist unique solutions to the BSDEs (12.6) and (12.7).

(b) We introduce the process

$$\hat{Z}_1^{\pi,*}(t) = Z_1^{\pi,*}(t) + \pi(t)\sigma(t), \quad 0 \leq t \leq T, \tag{12.13}$$

and we get the BSDE

$$\begin{aligned}
Y^{\pi,*}(t) = F &+ \int_t^T \Big(-Y^{\pi,*}(s-)r(s) + H(s)ds + G(s)dJ(s) - \pi(s)\big(\mu(s) - r(s)\big) \\
&+ L(s)\sqrt{\big|\hat{Z}_1^{\pi,*}(s) - \pi(s)\sigma(s)\big|^2 + \big|Z_2^{\pi,*}(s)\big|^2 + \big|U^{\pi,*}(s)\big|^2\eta(s)}\Big)ds \\
&- \int_t^T \hat{Z}_1^{\pi,*}(s)dW(s) - \int_t^T Z_2^{\pi,*}(s)dB(s) - \int_t^T U^{\pi,*}(s)\tilde{N}(ds), \\
&0 \le t \le T. \qquad (12.14)
\end{aligned}$$

Notice that the generator f of the BSDE (12.14) satisfies the property

$$\begin{aligned}
&f\big(t, Y^{*,*}(t), Z_1^{*,*}(t), Z_2^{*,*}(t), U^{\pi,*}(t)\big) - f\big(t, Y^{*,*}(t), Z_1^{*,*}(t), Z_2^{*,*}(t), U^{*,*}(t)\big) \\
&\quad = \delta^{Y^{*,*}, Z_1^{*,*}, Z_2^{*,*}, U^{\pi,*}, U^{*,*}}(t)\big(U^{\pi,*}(t) - U^{*,*}(t)\big)\eta(t), \quad 0 \le t \le T,
\end{aligned}$$

where

$$\begin{aligned}
&\delta^{Y^{*,*}, Z_1^{*,*}, Z_2^{*,*}, U^{\pi,*}, U^{*,*}}(t) \\
&\quad = L(t)\Big(\sqrt{\big|Z_1^{*,*}(t) - \pi(t)\sigma(t)\big|^2 + \big|Z_2^{*,*}(t)\big|^2 + \big|U^{\pi,*}(t)\big|^2\eta(t)} \\
&\qquad - \sqrt{\big|Z_1^{*,*}(t) - \pi(t)\sigma(t)\big|^2 + \big|Z_2^{*,*}(t)\big|^2 + \big|U^{*,*}(t)\big|^2\eta(t)}\Big) \\
&\qquad /\big(\big(U^{\pi,*}(t) - U^{*,*}(t)\big)\eta(t)\big)\mathbf{1}\big\{\big(U^{\pi,*}(t) - U^{*,*}(t)\big)\eta(t) \neq 0\big\}, \quad 0 \le t \le T.
\end{aligned}$$

From the Lipschitz property (10.46) and boundedness of L we deduce that the mapping $t \mapsto |\delta^{Y^{*,*}, Z_1^{*,*}, Z_2^{*,*}, U^{\pi,*}, U^{*,*}}(t)|^2\eta(t)$ is uniformly bounded. We also have

$$\begin{aligned}
&\delta^{Y^{*,*}, Z_1^{*,*}, Z_2^{*,*}, U^{\pi,*}, U^{*,*}}(t) \\
&\quad = L(t)\big(U^{\pi,*}(t) + U^{*,*}(t)\big) \\
&\qquad /\Big(\sqrt{\big|Z_1^{*,*}(t) - \pi(t)\sigma(t)\big|^2 + \big|Z_2^{*,*}(t)\big|^2 + \big|U^{\pi,*}(t)\big|^2\eta(t)} \\
&\qquad + \sqrt{\big|Z_1^{*,*}(t) - \pi(t)\sigma(t)\big|^2 + \big|Z_2^{*,*}(t)\big|^2 + \big|U^{*,*}(t)\big|^2\eta(t)}\Big) \\
&\qquad \cdot \mathbf{1}\big\{\big(U^{\pi,*}(t) - U^{*,*}(t)\big)\eta(t) \neq 0\big\} \\
&\quad = \alpha^{\pi} \frac{L(t)U^{\pi,*}(t)}{\sqrt{|Z_1^{*,*}(t) - \pi(t)\sigma(t)|^2 + |Z_2^{*,*}(t)|^2 + |U^{\pi,*}(t)|^2\eta(t)}} \\
&\qquad \cdot \mathbf{1}\big\{\big(U^{\pi,*}(t) - U^{*,*}(t)\big)\eta(t) \neq 0\big\}
\end{aligned}$$

$$+\left(1-\alpha^{\pi}\right)\frac{L(t)U^{*,*}(t)}{\sqrt{|Z_1^{*,*}(t)-\pi(t)\sigma(t)|^2+|Z_2^{*,*}(t)|^2+|U^{*,*}(t)|^2\eta(t)}}$$
$$\cdot\mathbf{1}\left\{\left(U^{\pi,*}(t)-U^{*,*}(t)\right)\eta(t)\neq 0\right\},$$

where

$$\alpha^{\pi}=\sqrt{\left|Z_1^{*,*}(t)-\pi(t)\sigma(t)\right|^2+\left|Z_2^{*,*}(t)\right|^2+\left|U^{\pi,*}(t)\right|^2\eta(t)}$$
$$/\left(\sqrt{\left|Z_1^{*,*}(t)-\pi(t)\sigma(t)\right|^2+\left|Z_2^{*,*}(t)\right|^2+\left|U^{\pi,*}(t)\right|^2\eta(t)}\right.$$
$$\left.+\sqrt{\left|Z_1^{*,*}(t)-\pi(t)\sigma(t)\right|^2+\left|Z_2^{*,*}(t)\right|^2+\left|U^{*,*}(t)\right|^2\eta(t)}\right)$$
$$\cdot\mathbf{1}\left\{\left(U^{\pi,*}(t)-U^{*,*}(t)\right)\eta(t)\neq 0\right\},$$

and $\alpha^{\pi}\in[0,1]$. We can conclude that $\delta^{Y^{*,*},Z_1^{*,*},Z_2^{*,*},U^{\pi,*},U^{*,*}}(t)>-1$, $0\leq t\leq T$, for any admissible strategy $\pi\in\mathscr{A}^s$. Recalling the arguments from the proof of Theorem 3.2.2 that led to (3.31), we obtain

$$\begin{aligned}
&Y^{*,*}(t)-Y^{\pi,*}(t)\\
&=\mathbb{E}^{\mathbb{Q}}\Bigg[\int_t^T e^{-\int_t^s r(u)du}\Big(\sqrt{\left|L(s)\right|^2-\left|\theta(s)\right|^2}\sqrt{\left|Z_2^{*,*}(s)\right|^2+\left|U^{*,*}(s)\right|^2\eta(s)}\\
&\quad -Z_1^{*,*}(s)\theta(s)+\pi(s)\big(\mu(s)-r(s)\big)\\
&\quad -L(s)\sqrt{\left|Z_1^{*,*}(s)-\pi(s)\sigma(s)\right|^2+\left|Z_2^{*,*}(s)\right|^2+\left|U^{*,*}(s)\right|^2\eta(s)}\Big)ds\,|\mathscr{F}_t\Bigg],\\
&\quad 0\leq t\leq T,
\end{aligned}\tag{12.15}$$

under some measure $\mathbb{Q}$. We now introduce the function

$$\varphi(\pi)=\sqrt{|L|^2-|\theta|^2}\sqrt{\left|z_2^{*,*}\right|^2+\left|u^{*,*}\right|^2\eta}-z_1^{*,*}\theta$$
$$+\pi(\mu-r)-L\sqrt{\left|z_1^{*,*}-\pi\sigma\right|^2+\left|z_2^{*,*}\right|^2+\left|u^{*,*}\right|^2\eta}.$$

By classical differential calculus we can find the global maximum of φ and we can show that $\varphi(\pi)\leq 0$. Hence, from (12.15) we conclude that $Y^{*,*}(t)-Y^{\pi,*}(t)\leq 0$ for all $t\in[0,T]$ and any $\pi\in\mathscr{A}^s$.

(c) It is easy to show that $\varphi(\pi^*)=0$ for π^* defined in (12.12). We have to check the admissibility of the candidate strategy π^*. Predictability and the square integrability of π^* are obvious. It is also clear that there exists an adapted, càdlàg solution X^{π^*} to (7.11). Hence, $\pi^*\in\mathscr{A}$. By uniqueness of solution to (12.14) the solution $(Y^{\pi^*,*},\hat{Z}_1^{\pi^*,*},Z_2^{\pi^*,*},U^{\pi^*,*})$ coincides with $(Y^{*,*},Z_1^{*,*},Z_2^{*,*},U^{*,*})$. From (12.12) we can now derive

$$Z_1^{*,*}(t) - \pi^*(t)\sigma(t) = \sqrt{\frac{|\theta(t)|^2}{|L(t)|^2 - |\theta(t)|^2}} \sqrt{\left|Z_2^{*,*}(t)\right|^2 + \left|U^{*,*}(t)\right|^2 \eta(t)}, \quad 0 \le t \le T,$$

and, consequently, $\pi^* \in \mathscr{A}^s$ if

$$\frac{L(t)U^{\pi^*,*}(t)}{\sqrt{|Z_1^{*,*}(t) - \pi^*(t)\sigma(t)|^2 + |Z_2^{*,*}(t)|^2 + |U^{\pi^*,*}(t)|^2\eta(t)}} \mathbf{1}\{U^{\pi^*,*}(t)\eta(t) \neq 0\}$$

$$= \frac{L(t)U^{*,*}(t)}{\sqrt{|Z_1^{*,*}(t) - \pi^*(t)\sigma(t)|^2 + |Z_2^{*,*}(t)|^2 + |U^{*,*}(t)|^2\eta(t)}} \mathbf{1}\{U^{*,*}(t)\eta(t) \neq 0\}$$

$$= \frac{\sqrt{|L(t)|^2 - |\theta(t)|^2} U^{*,*}(t)}{\sqrt{|Z_2^{*,*}(t)|^2 + |U^{*,*}(t)|^2\eta(t)}} \mathbf{1}\{U^{*,*}(t)\eta(t) \neq 0\} > -1, \quad 0 \le t \le T,$$

which holds if $U^{*,*}(t) \ge 0$ or $|L(t)|^2 < \eta(t) + |\theta(t)|^2$ on the set $\{\eta(t) > 0\}$. We can conclude that $Y^{*,*}(t) = Y^{\pi^*,*}(t) \ge \inf_{\pi \in \mathscr{A}^s} Y^{\pi,*}(t)$ for all $t \in [0, T]$. Since $Y^{*,*}(t) \le \inf_{\pi \in \mathscr{A}^s} Y^{\pi,*}(t)$ by item (b), the assertion of item (c) can be deduced. □

The additional constraints on the set of admissible strategies allow us to apply the comparison principle in the proof of Theorem 12.1.2 and we succeed in obtaining the optimal solution. We point out that the additional constraints are essential for deriving the arbitrage-free price (12.4) which satisfies the comparison principle (the property of monotonicity with respect to the claim and the Sharpe ratio).

Notice that the assumptions: $\theta(t) + \varepsilon \le L(t)$ and $|L(t)|^2 < \eta(t) + |\theta(t)|^2$ on the set $\{\eta(t) > 0\}$ may hold only if $\eta(t) \ge \varepsilon > 0$ on the set $\{\eta(t) > 0\}$. We remark that it is reasonable to assume that there exists a positive lower bound on the claim intensity, e.g. despite improvements in mortality, it is reasonable to assume that there exists a natural limit in these improvements and the mortality intensity process should be bounded away from zero. If $\theta(t) + \varepsilon \le L(t) < \sqrt{\eta(t)}$, then $\mathscr{A}^s = \mathscr{A}$.

Theorem 12.1.2 shows that the price process and the optimal hedging strategy, which are derived under the ambiguity risk minimization (12.4), can be characterized with the solution to the nonlinear BSDE (12.7). The price process $Y^{*,*}$ is arbitrage-free. From (12.7) we get

$$\begin{aligned} Y^{*,*}(t) = F &+ \int_t^T H(s)ds + \int_t^T G(s)dJ(s) \\ &+ \int_t^T \Bigg(-Y^{*,*}(s-)r(s) - Z_1^{*,*}(s)\theta(s) \\ &+ \frac{Z_2^{*,*}(s)}{\sqrt{|Z_2^{*,*}(s)|^2 + |U^{*,*}(s)|^2\eta(s)}} \sqrt{\left|L(s)\right|^2 - \left|\theta(s)\right|^2} Z_2^{*,*}(s) \end{aligned}$$

$$+\frac{U^{*,*}(s)}{\sqrt{|Z_2^{*,*}(s)|^2+|U^{*,*}(s)|^2\eta(s)}}\sqrt{\left|L(s)\right|^2-\left|\theta(s)\right|^2U^{*,*}(s)\eta(s)}\Big)ds$$
$$-\int_t^T Z_1^{*,*}(s)dW(s)-\int_t^T Z_2^{*,*}(s)dB(s)$$
$$-\int_t^T U^{*,*}(s)\tilde{N}(ds),\quad 0\le t\le T,$$

and by the Girsanov's theorem we obtain the representation

$$Y^{*,*}(t)=\mathbb{E}^{\mathbb{Q}^*}\Big[\int_t^T e^{-\int_t^s r(u)du}dP(s)|\mathscr{F}_t\Big],\quad 0\le t\le T, \tag{12.16}$$

where the equivalent martingale measure $\mathbb{Q}^*$ is given by

$$\frac{d\mathbb{Q}^*}{d\mathbb{P}}\Big|\mathscr{F}_t=M^*(t),\quad 0\le t\le T,$$
$$\frac{dM^*(t)}{M^*(t-)}=-\theta(t)dW(t)+\frac{Z_2^{*,*}(t)}{\sqrt{|Z_2^{*,*}(t)|^2+|U^{*,*}(t)|^2\eta(t)}}\sqrt{\left|L(t)\right|^2-\left|\theta(t)\right|^2}dB(t)$$
$$+\frac{U^{*,*}(t)}{\sqrt{|Z_2^{*,*}(t)|^2+|U^{*,*}(t)|^2\eta(t)}}\sqrt{\left|L(t)\right|^2-\left|\theta(t)\right|^2}\tilde{N}(dt). \tag{12.17}$$

Since we assume that $U^{*,*}(t)\ge 0$ or $|L(t)|^2<\eta(t)+|\theta(t)|^2$, the process given by (12.17) is a positive martingale and it defines an equivalent martingale measure. We can also prove that $Y^{*,*}$ satisfies the comparison principle and the property of monotonicity with respect to the claim, see Delong (2012a). The systematic and unsystematic insurance risks are priced under $\mathbb{Q}^*$ and the insurance risk premiums can be deduced from (12.17). The pricing measure $\mathbb{Q}^*$ depends on the financial market, the insurance payment process and the control process L. Since $Y^{*,*}$ is an arbitrage-free price of the liability and $Z^{*,*}$ is the control process of the BSDE for Y, we conclude that the optimal hedging strategy (12.12) is a delta hedging strategy with a correction term.

Recall now the hedging strategy (10.43) and the price (10.44) derived under the instantaneous mean-variance risk measure. There are obvious similarities between the results from Sect. 10.4 and the results of this chapter. This should not mislead us. We point out that the price process (10.44), which is obtained under the assumptions of Theorem 10.4.1, may give rise to arbitrage opportunities, see Example 10.3. This is never the case for the price process (12.7), which is obtained under the assumptions of Theorems 12.1.1 and 12.1.2. The stronger assumptions of Theorems 12.1.1 and 12.1.2 exclude those cases of (10.44) which lead to arbitrage prices, see Examples 10.3 and 10.4.

We conclude that there is an equivalence between arbitrage-free pricing and hedging under the instantaneous mean-variance risk measure and pricing and hedging under the ambiguous risk measure, which, by construction, always leads to an arbitrage-free price. Such an equivalence is beneficial for applications and interpretations. Firstly, it is straightforward to justify that $L(t) \geq \theta(t) + \varepsilon$ as the process L can be related to the Sharpe ratio of the surplus process. Secondly, from (10.43) and (10.56) we deduce that the correction term in the hedging strategy (10.43) or (12.12) arises if the insurer applies the mean-variance risk measure instead of the variance risk measure.

Example of pricing and hedging under model ambiguity are given in Examples 10.3 and 10.4.

12.2 No-Good-Deal Pricing

In the arbitrage-free financial and insurance model the payment process P should be priced by the expectation $\mathbb{E}^{\mathbb{Q}}[\int_0^T e^{-rt} dP(t)]$ under an equivalent martingale measure $\mathbb{Q} \in \mathscr{Q}^m$, see Sect. 9.1. Since the set of equivalent martingale measures $\mathscr{Q}^m$ is not a singleton, the first idea is to consider $\sup_{\mathbb{Q}\in\mathscr{Q}^m} \mathbb{E}^{\mathbb{Q}}[\int_0^T e^{-rt} dP(t)]$. Such a superhedging price is too high for practical applications, gives rise to arbitrage opportunities and cannot be used for pricing, see Sect. 9.3. The second idea is to consider $\sup_{\mathbb{Q}\in\mathscr{Q}'} \mathbb{E}^{\mathbb{Q}}[\int_0^T e^{-rt} dP(t)]$ where $\mathscr{Q}' \subset \mathscr{Q}^m$. In order to apply such a price, the form of the subset $\mathscr{Q}'$ of the set of equivalent martingale measures $\mathscr{Q}^m$ should be specified and justified. In this chapter we present a financial motivation for considering a particular subset of $\mathscr{Q}^m$ and we find the arbitrage-free price under the least favorable martingale measure from this subset.

We investigate the insurance payment process (7.3). We assume that the process J is a point process and, consequently, the jump measure N of the point process J has the compensator $\vartheta(dt, \{1\}) = \eta(t)dt$. From Propositions 3.3.1 and 3.4.1 we can conclude that an arbitrage-free price Y of the payment process P must satisfy the BSDE

$$\begin{aligned} Y(t) = F &+ \int_t^T H(s)ds + \int_t^T G(s)dJ(s) \\ &+ \int_t^T \big(-Y(s-)r(s) - Z_1(s)\theta(s) + Z_2(s)\phi(s) + U(s)\kappa(s)\eta(s)\big)ds \\ &- \int_t^T Z_1(s)dW(s) - \int_t^T Z_2(s)dB(s) \\ &- \int_t^T U(s)\tilde{N}(ds), \quad 0 \leq t \leq T, \end{aligned} \tag{12.18}$$

where ϕ and κ denote the insurance risk premiums and $(\phi, \kappa) \in \mathscr{Q}^m$. We aim to constrain the possible values of $(\phi, \kappa) \in \mathscr{Q}^m$ in a financially sensible way. A reasonable constraint on the risk premiums can be derived by referring to the theory of no-good-deal pricing.

Sharpe ratios are often used to characterize investment opportunities. An investment opportunity with an extraordinarily high Sharpe ratio is called a good deal and the theory of finance argues that good deals cannot survive in competitive markets, see Cochrane and Saá-Requejo (2000) and Björk and Slinko (2006). Empirical studies also support the fact that Sharpe ratios of investment opportunities in competitive markets take restricted range of values, see Lo (2002). These arguments lay the foundation for no-good-deal pricing.

We assume that the stock and the insurance contract are traded in the market. We remark that market-consistent valuation of insurance liabilities assumes that insurance obligations can be transferred between parties, see V.2.–V.3. in European Commission QIS5 (2010). The insurer has two risky opportunities: he can invest in the risky stock or he can sell the risky insurance payment process, collect a premium and back the liability with the risk-free investment in the bank account. We define the instantaneous Sharpe ratios of the investment in the stock S and the investment in the insurance contract Y:

$$\begin{aligned}
\text{Sharpe Ratio}_t(S) &= \frac{\mathbb{E}[dS(t) - S(t)r(t)dt|\mathscr{F}_{t-}]/dt}{\sqrt{\mathbb{E}[d[S,S](t)|\mathscr{F}_{t-}]/dt}} = \theta(t), \quad 0 \leq t \leq T,\\
\text{Sharpe Ratio}_t(Y) &= \frac{\mathbb{E}[d\mathscr{S}(t) - \mathscr{S}(t-)r(t)dt|\mathscr{F}_{t-}]/dt}{\sqrt{\mathbb{E}[d[\mathscr{S},\mathscr{S}](t)|\mathscr{F}_{t-}]dt}}\\
&= \frac{\mathbb{E}[-dY(t) + Y(t-)r(t)dt|\mathscr{F}_{t-}]/dt}{\sqrt{\mathbb{E}[d[Y,Y](t)|\mathscr{F}_{t-}]dt}}\\
&= \frac{-Z_1(t)\theta(t) + Z_2(t)\phi(t) + U(t)\kappa(t)\eta(t)}{\sqrt{|Z_1(t)|^2 + |Z_2(t)|^2 + |U(t)|^2\eta(t)}}, \quad 0 \leq t \leq T,
\end{aligned}$$

where we use the Sharpe ratio of the surplus $\mathscr{S}$ which is earned by the insurer who takes the short position in the insurance contract and the long position in the bank account, see (10.41) and (10.42). By the Schwarz inequality we get

$$\begin{aligned}
&\left|-Z_1(t)\theta(t) + Z_2(t)\phi(t) + U(t)\kappa(t)\eta(t)\right|\\
&\leq \sqrt{\left|Z_1(t)\right|^2 + \left|Z_2(t)\right|^2 + \left|U(t)\right|^2\eta(t)}\sqrt{\left|\theta(t)\right|^2 + \left|\phi(t)\right|^2 + \left|\kappa(t)\right|^2\eta(t)},
\end{aligned} \tag{12.19}$$

which yields the inequality

$$\left|\text{Sharpe Ratio}_t(Y)\right| \leq \sqrt{\left|\theta(t)\right|^2 + \left|\phi(t)\right|^2 + \left|\kappa(t)\right|^2\eta(t)}, \quad 0 \leq t \leq T.$$

The goal is to find a least favorable martingale measure for pricing the insurance payment process under the constraint that the Sharpe ratio of the insurance contract is within the no-good-deal range given by the market. In order to guarantee that in the combined financial and insurance market the instantaneous Sharpe ratios are bounded by a process L, which itself should be bounded to exclude good deals, we have to introduce the constraint

$$\left|\theta(t)\right|^2+\left|\phi(t)\right|^2+\left|\kappa(t)\right|^2\eta(t)\le\left|L(t)\right|^2\le K,\quad 0\le t\le T. \tag{12.20}$$

The bound L for the Sharpe ratio of the insurance contract should be greater than θ since the risk premium θ can be earned by investing in the stock. The insurance contract carries an additional risk and the risk premium for the insurance contract has to be strictly above θ. Hence, we should consider $L(t)\ge\theta(t)+\varepsilon$, $\varepsilon>0$.

We now define the no-good-deal price of the insurance payment process P. We choose a process L which represents the bound on possible gains in the market measured by instantaneous Sharpe ratios. We introduce the set of equivalent martingale measures

$$\mathscr{Q}^{ngd}=\left\{\mathbb{Q}\sim\mathbb{P},\ \frac{d\mathbb{Q}}{d\mathbb{P}}\Big|\mathscr{F}_t=M^{ngd}(t),\ 0\le t\le T\right\},$$

$$\frac{dM^{ngd}(t)}{M^{ngd}(t-)}=-\theta(t)dW(t)+\phi(t)dB(t)+\kappa(t)\tilde{N}(dt),\quad M^{ngd}(0)=1, \tag{12.21}$$

where ϕ and κ are predictable processes such that

$$\left|\phi(t)\right|^2+\left|\kappa(t)\right|^2\eta(t)dt\le\left|L(t)\right|^2-\left|\theta(t)\right|^2,\quad\kappa(t)>-1,\quad 0\le t\le T.$$

The no-good-deal price is defined by

$$Y(t)=\sup_{\mathbb{Q}\in\mathscr{Q}^{ngd}}\mathbb{E}^{\mathbb{Q}}\Big[e^{-\int_t^T r(s)ds}F+\int_t^T e^{-\int_t^s r(u)du}G(s)dJ(s)$$
$$+\int_t^T e^{-\int_t^s r(u)du}H(s)ds|\mathscr{F}_t\Big],\quad 0\le t\le T. \tag{12.22}$$

We price the insurance payment process with a least favorable martingale measure under the Sharpe ratio constraint which excludes too good prices leading to extraordinarily high gains. The constraint (12.20) has a financial justification and leads to the well-defined optimization problem (12.22).

Following the steps from the proofs of Theorems 12.1.1 and 12.1.2, we can derive the no-good-deal price.

Theorem 12.2.1 *Let us assume that* (C1)–(C4) *from Chap.* 7 *hold and the jump measure N of the step process J has the compensator $\vartheta(dt,\{1\})=\eta(t)dt$. We consider a predictable process L such that $L(t)\ge\theta(t)+\varepsilon$, $\varepsilon>0$, and $L(t)\le K$, $0\le t\le T$. Let us investigate the BSDE*

$$Y(t)=F+\int_t^T H(s)ds+\int_t^T G(s)N(ds)+\int_t^T\Big(-Y(s-)r(s)-Z_1(s)\theta(s)$$
$$+\sqrt{\left|L(s)\right|^2-\left|\theta(s)\right|^2}\sqrt{\left|Z_2(s)\right|^2+\left|U(s)\right|^2\eta(s)}\Big)ds-\int_t^T Z_1(s)dW(s)$$

$$-\int_t^T Z_2(s)dB(s) - \int_t^T U(s)\tilde{N}(ds), \quad 0 \le t \le T. \tag{12.23}$$

If $U(t) \ge 0$ or $|L(t)|^2 < \eta(t) + |\theta(t)|^2$ on the set $\{\eta(t) > 0\}$, $0 \le t \le T$, then the no-good-deal price process (12.22) *is the unique solution to the BSDE* (12.23). *The no-good-deal price process has the representation*

$$Y(t) = \mathbb{E}^{\mathbb{Q}^*}\left[\int_t^T e^{-\int_t^s r(u)du} dP(s)|\mathscr{F}_t\right], \quad 0 \le t \le T,$$

where the least favorable martingale measure $\mathbb{Q}^*$ is given by

$$\frac{d\mathbb{Q}^*}{d\mathbb{P}}\Big|\mathscr{F}_t = M^*(t), \quad 0 \le t \le T,$$

$$\frac{dM^*(t)}{M^*(t-)} = -\theta(t)dW(t) + \frac{Z_2(t)}{\sqrt{|Z_2(t)|^2 + |U(t)|^2\eta(t)}}\sqrt{|L(t)|^2 - |\theta(t)|^2}dB(t)$$
$$+ \frac{U(t)}{\sqrt{|Z_2(t)|^2 + |U(t)|^2\eta(t)}}\sqrt{|L(t)|^2 - |\theta(t)|^2}\tilde{N}(dt).$$

The no-good-deal price (12.23) coincides with the price (12.7) derived in Theorem 12.1.2 under the ambiguity risk minimization.

Bibliographical Notes This chapter presents the results from Delong (2012a). Model ambiguity and Knightian uncertainty were introduced by Chen and Epstein (2002) to financial modelling. Risk measures based on least favorable probability measures and robust optimization problems are studied in Leitner (2007), Becherer (2009), Schied (2005), Schied (2006) and McNeil et al. (2005). Becherer (2009) found the hedging strategy which minimizes the expected terminal shortfall under a least favorable probability measure and solved the no-good-deal pricing problem in a diffusion model by applying BSDEs. We use the arguments from Becherer (2009) and we solve the pricing and hedging problems in the model with jumps. Cochrane and Saá-Requejo (2000) introduced the theory of no-good-deal pricing into the financial literature, and Björk and Slinko (2006) were the first to investigate a Lévy-driven asset model. Robust pricing and no-good-deal pricing in insurance is also investigated by Pelsser (2011). Utility maximization and optimal investment and consumption problems under model ambiguity in a jump-diffusion financial market consisting of assets modelled with Itô-Lévy processes are studied by Øksendal and Sulem (2011), Øksendal and Sulem (2012) and Laeven and Stadje (2012). We remark that the min-max problem (12.4) is an example of a stochastic differential game in which two players choose strategies, the first player wishes to minimize a pay-off and the second player wishes to maximize the same pay-off. Stochastic differential games and related BSDEs are considered in Hamadène and Lepeltier (1995), El Karoui and Hamadène (2003), Øksendal and Sulem (2011), Øksendal and Sulem (2012).

Chapter 13
Dynamic Risk Measures

Abstract We investigate dynamic risk measures which describe the riskiness of financial positions taken by investors. We deal with dynamic risk measures which are modelled by g-expectations. We study properties of dynamic risk measures and we show that properties of dynamic risk measures are determined by the generator of the BSDE defining the g-expectation and the risk measure. We discuss methods for choosing the generator of a g-expectation. We also solve a problem of optimal risk sharing between two parties and we find the optimal derivative for the risk transfer. Properties of the prices (risk measures) derived in previous chapters are investigated.

In actuarial mathematics we can find many principles for pricing insurance risks. Properties of these actuarial premium principles have been well studied, see Chap. 5 in Denuit et al. (2001). Since the publication by Artzner et al. (1999), the notion of a risk measures have gained importance in financial and insurance mathematics. A risk measure provides a general description of the riskiness of a financial position taken by an investor.

Quantifying the risk of financial positions is the key task of risk management. Financial practice indicates that an operator used for quantifying the risk should be nonlinear. It is straightforward to notice that only a nonlinear risk measure can reflect a diversification effect. Static risk measures, which quantify the risk of a financial position over a specified period of time, provide useful information for risk management. However, we should be more interested in dynamic risk measures which can quantify the riskiness of a financial position continuously during a specified period of time. Since the information about the future liability arrives continuously in time, the investor should continuously update the capital requirement for the liability. In this chapter we investigate dynamic risk measures modelled by g-expectations. We study properties of dynamic risk measures and discuss methods for choosing the generator of a g-expectation. We also solve a problem of optimal risk sharing between two parties.

13.1 Dynamic Risk Measures by g-Expectations

Let ξ represent a financial position taken by an investor. By a financial position we mean the terminal value of an investment portfolio or the terminal net asset

Ł. Delong, *Backward Stochastic Differential Equations with Jumps and Their Actuarial and Financial Applications*, EAA Series, DOI 10.1007/978-1-4471-5331-3_13,

value of an insurance company (the terminal surplus). A positive value $\xi > 0$ is interpreted as a profit and a negative value $\xi < 0$ as a loss. The goal is to summarize the information about the riskiness of the financial position ξ in a single number. A risk measures $\rho(\xi)$ is used for that purpose. We give a definition of a dynamic risk measure.

Definition 13.1.1 A family $(\rho_t)_{0\leq t\leq T}$ of mappings $\rho_t : \mathbb{L}^2(\Omega, \mathscr{F}_T, \mathbb{P}; \mathbb{R}) \to \mathbb{L}^2(\Omega, \mathscr{F}_t, \mathbb{P}; \mathbb{R})$ such that $\rho_T(\xi) = -\xi$ is called a dynamic risk measure.

The measure $\rho_t(\xi)$ quantifies at time t the risk of the position ξ which is going to be liquidated (or reported at a balance sheet) at time T. It is clear that the risk of the terminal pay-off ξ should be consistently quantified during the time period $[0, T]$. The risk ξ is called acceptable at time t if $\rho_t(\xi) \leq 0$.

A reasonable risk measure should satisfy properties which agree with financial practice and views of investors. We introduce the key properties (sometimes called the axioms of risk measures) which should be satisfied by a risk measure:

- **Convexity**

$$\rho_t\big(c\xi_1 + (1-c)\xi_2\big) \leq c\rho_t(\xi_1) + (1-c)\rho_t(\xi_2),$$
$$0 \leq t \leq T,\ c \in (0,1),\ \xi_1, \xi_2 \in \mathbb{L}^2(\Omega, \mathscr{F}_T, \mathbb{P}; \mathbb{R}).$$

- **Monotonicity**

$$\xi_1 \geq \xi_2 \quad \Rightarrow \quad \rho_t(\xi_1) \leq \rho_t(\xi_2), \quad 0 \leq t \leq T,\ \xi_1, \xi_2 \in \mathbb{L}^2(\Omega, \mathscr{F}_T, \mathbb{P}; \mathbb{R}).$$

- **Cash invariance**

$$\rho_t(c) = -c, \quad 0 \leq t \leq T,\ c \in \mathbb{R}.$$

- **Translation invariance**

$$\rho_t(\xi_1 + \xi_2) = \rho_t(\xi_1) - \xi_2,$$
$$0 \leq t \leq T,\ \xi_1 \in \mathbb{L}^2(\Omega, \mathscr{F}_T, \mathbb{P}; \mathbb{R}),\ \xi_2 \in \mathbb{L}^2(\Omega, \mathscr{F}_t, \mathbb{P}; \mathbb{R}).$$

- **Sub-linearity: sub-additivity and positively homogeneity**

$$\rho_t(\xi_1 + \xi_2) \leq \rho_t(\xi_1) + \rho_t(\xi_2), \qquad \rho_t(c\xi) = c\rho_t(\xi),$$
$$0 \leq t \leq T,\ c > 0,\ \xi_1, \xi_2, \xi \in \mathbb{L}^2(\Omega, \mathscr{F}_T, \mathbb{P}; \mathbb{R}).$$

- **Time-consistency**

$$\rho_s(\xi) = \rho_s\big(-\rho_t(\xi)\big), \quad 0 \leq s \leq t \leq T,\ \xi_1, \xi_2 \in \mathbb{L}^2(\Omega, \mathscr{F}_T, \mathbb{P}; \mathbb{R}).$$

All these properties have their financial interpretations, see Goovaerts et al. (2003) for the actuarial point of view. The convexity implies that diversification through "weighted" sums of positions (portfolios) reduces the risk. Let λ denote a fraction of the wealth invested in ξ_1. Given the total wealth available, it is better to pool two portfolios and diversify the wealth between ξ_1 and ξ_2. The monotonicity reflects the common rule that a position is less risky if it yields a higher pay-off in all scenarios. The cash invariance means that the riskiness of a constant pay-off is equal to the opposite of that pay-off. In order to protect the loss c, the insurer has to keep the reserve in the amount of c. The translation invariance implies that the riskiness of a position is only affected by the uncertainty of this position and additive components that are determined by the current information are treated as constant pay-offs. The translation invariance allows for the interpretation of $\rho_t(\xi)$ as the amount of money which makes the position ξ acceptable. Indeed, we have

$$\rho_t\big(\xi + \rho_t(\xi)\big) = \rho_t(\xi) - \rho_t(\xi) = 0.$$

Consequently, a risk measure is also called a capital requirement. The sub-linearity again encourages diversification: the riskiness of two combined portfolios is smaller than the riskiness of two separate portfolios. The positive homogeneity arises since the riskiness of a liquid position should be proportional to the volume of the risk taken. The property of positive homogeneity assumes that there is no liquidity risk in the market. The time-consistency implies that in order to quantify the risk of a position at time s, we can first quantify the risk of the position at an intermediate point of time $t > s$ and next quantify the risk from time t to time s.

Notice that the cash invariance excludes the time value of money. The time-value of money is a celebrated property in finance. The minimal requirement which allows for the time-value of money is a cash sub-additivity, see El Karoui and Ravanelli (2009). We introduce the property

- **Cash sub-additivity**

  the mapping $c \mapsto \rho_t(\xi + c) + c$ is non-decreasing on $\mathbb{R}$, $0 \le t \le T$, $\xi \in \mathbb{L}^2(\mathbb{R})$.

The cash sub-additivity yields $\rho_t(\xi + c) \ge \rho_t(\xi) - c$ for $c > 0$, which means that if c units are added as a profit at time T, then the capital reserve for the new position is reduced by less than c units. We also have $\rho_t(\xi - c) \le \rho_t(\xi) + c$ for $c > 0$, which means that if c units are added as a liability at time T, then the capital reserve for the new position is increased by less than c units.

We define two important families of risk measures, see Artzner et al. (1999) and Föllmer and Schied (2002).

Definition 13.1.2 A dynamic risk measure $\rho := (\rho_t)_{0 \le t \le T}$ is called

(i) coherent: if it is monotone, translation invariant and sub-linear,
(ii) convex: if it is convex and $\rho_t(0) = 0$, $0 \le t \le T$.

Since g-expectations are useful for modelling filtration-consistent nonlinear expectations, we can use g-expectations to model dynamic risk measures. It is clear that the generator of the g-expectation determines properties of the dynamic risk measure.

Proposition 13.1.1 *Let $\mathscr{E}_g$ be a g-expectation and set $\rho_t(\xi) = \mathscr{E}_g[-\xi|\mathscr{F}_t]$, $0 \leq t \leq T$. Then ρ is a monotone, time-consistent dynamic risk measure. In addition,*

(a) *if g is sublinear in (z, u) and independent of y, then ρ is a coherent dynamic risk measure.*
(b) *if g is convex in (y, z, u), then ρ is a convex dynamic risk measure.*

Proof The results follow from Theorem 6.2.1 and Proposition 6.2.3. □

Proposition 13.1.2 *Let $\mathscr{E}_g$ be a g-expectation and set $\rho_t(\xi) = \mathscr{E}_g[-\xi|\mathscr{F}_t]$, $0 \leq t \leq T$. If $y \mapsto g(\omega, t, y, z, u)$ is non-increasing a.s. for all $(t, z, u) \in [0, T] \times \mathbb{R} \times L^2_Q$, then ρ is a cash sub-additive dynamic risk measure.*

Proof Consider the BSDE

$$Y^\xi(t) = -\xi + \int_t^T g\big(s, Y^\xi(s), Z^\xi(s), U^\xi(s)\big)ds - \int_t^T Z^\xi(s)dW(s) - \int_t^T \int_{\mathbb{R}} U^\xi(s, z)\tilde{N}(ds, dz), \quad 0 \leq t \leq T, \quad (13.1)$$

which describes the evolution of the risk measure $\rho_t(\xi) = Y^\xi(t)$. Notice that $Y^{\xi,c}(t) = \rho_t(\xi + c) + c = Y^{\xi+c} + c$ satisfies the BSDE

$$Y^{\xi,c}(t) = -\xi + \int_t^T g\big(s, Y^{\xi,c}(s) - c, Z^{\xi,c}(s), U^{\xi,c}(s)\big)ds - \int_t^T Z^{\xi,c}(s)dW(s) - \int_t^T \int_{\mathbb{R}} U^{\xi,c}(s, z)\tilde{N}(ds, dz), \quad 0 \leq t \leq T. \quad (13.2)$$

The comparison principle yields

$$Y^{\xi,c_1}(t) \geq Y^{\xi,c_2}(t), \quad 0 \leq t \leq T, \ c_1 \geq c_2.$$ □

Notice that the generator of a BSDE defining a cash sub-additive risk measure depends on the current level of the risk measure, which is never the case for a translation invariant (and coherent) risk measure, see Proposition 13.1.1 and Theorems 3.1 and 3.2 in Jiang (2008).

Example 13.1 Let us give examples of well-known dynamic risk measures which can be derived from g-expectations. We consider the Brownian filtration $\mathscr{F}^W$. In finance we often use the following risk measures:

- The linear expectation $\rho_t(\xi) = \mathbb{E}^{\mathbb{Q}}[-e^{-\int_t^T \alpha(s)ds}\xi|\mathscr{F}_t^W]$ can derived from the linear BSDE

$$Y(t) = -\xi + \int_t^T \big(-\alpha(s)Y(s) + \beta(s)Z(s)\big)ds - \int_t^T Z(s)dW(s), \quad 0 \leq t \leq T, \quad (13.3)$$

where $\frac{d\mathbb{Q}}{d\mathbb{P}} = e^{\int_0^T \beta(s)dW(s) - \frac{1}{2}\int_0^T |\beta(s)|^2 ds}$, and β is an $\mathscr{F}^W$-predictable, bounded process,

- The ambiguous risk measure $\rho_t(\xi) = \sup_{\mathbb{Q}\in\mathscr{Q}} \mathbb{E}^{\mathbb{Q}}[-\xi|\mathscr{F}_t^W]$ where $\mathscr{Q} = \{\mathbb{Q} \sim \mathbb{P} : \frac{d\mathbb{Q}}{d\mathbb{P}} = e^{\int_0^T \phi(s)dW(s) - \frac{1}{2}\int_0^T |\phi(s)|^2 ds}$, ϕ is $\mathscr{F}^W$-predictable, $|\phi(t)| \le \beta(t) \le K\}$ can be derived from the nonlinear BSDE

$$Y(t) = -\xi + \int_t^T \beta(s)\big|Z(s)\big|ds - \int_t^T Z(s)dW(s), \quad 0 \le t \le T, \qquad (13.4)$$

- The entropic risk measure $\rho_t(\xi) = \frac{1}{\beta}\ln \mathbb{E}[e^{-\beta\xi}|\mathscr{F}_t^W]$ can be derived from the nonlinear BSDE

$$Y(t) = -\xi + \frac{1}{2}\int_t^T \beta\big|Z(s)\big|^2 ds - \int_t^T Z(s)dW(s), \quad 0 \le t \le T. \qquad (13.5)$$

In Chaps. 9–12 we derived prices of the insurance payment process under various objectives. We can view these prices as examples of g-expectations. We remark that prices are closely related to risk measures since a price is just the opposite of a risk measure. We can now study properties of the g-expectations which were used in previous chapters to price the insurance payment process. We point out that the properties of monotonicity and time-consistency for prices should be now defined in accordance with the pricing axioms, see Chap. 6 in Barrieu and El Karoui (2005).

Proposition 13.1.3

(a) *The prices from Theorems* 9.2.1 *and* 9.4.1 *derived under perfect replication are linear, monotone, time-consistent and cash sub-additive.*
(b) *The price for a bounded claim from Theorem* 9.3.3 *derived under superhedging is nonlinear, convex, monotone, sub-linear, time-consistent and cash sub-additive.*
(c) *The prices from Theorems* 10.1.1, 10.2.3 *and Corollary* 10.3.1 *derived under quadratic loss minimization are linear, monotone, time-consistent and cash sub-additive.*
(d) *The price from Theorem* 11.2.1 *derived under exponential indifference pricing is nonlinear, convex, monotone, cash invariant, translation invariant and time-consistent.*
(e) *The price from Theorem* 12.2.1 *derived under no-good-deal pricing and the price from Theorem* 12.1.2 *derived under ambiguity risk minimization is nonlinear, convex, monotone, sub-linear, time-consistent and cash sub-additive.*

Proof The assertions follow from the arbitrage-free representations of the prices, the comparison principles and Propositions 13.1.1 and 13.1.2. The time-consistency of the superhedging price follows from Theorem 6.2 and Proposition 9.1 in Delbaen (2006). □

We point out that BSDEs and g-expectations are also useful for modelling recursive stochastic differential utilities, see Duffie and Epstein (1997), Lazrak and

Quenez (2003), Skiadas (2003). A recursive utility extends the classical notion of a utility by disentangling the risk aversion and the intertemporal substitutability of preferences.

13.2 Generators of Dynamic Risk Measures

In order to define a dynamic risk measure by a BSDE, we need to specify the generator. The choice of the generator of a g-expectation is a crucial point in modelling dynamic risk measures. We present two approaches which may be helpful in deriving the generator. Under the first approach we should specify a pricing and hedging objective. Then, the solution of the optimization problem determines the generator and the g-expectation. We followed this approach in Chaps. 9–12. Under the second approach we directly specify the generator of a BSDE. To follow the second approach, we have to be able to interpret the generator of a g-expectation.

Let us focus on the Brownian filtration and the BSDE

$$Y(t) = -\xi + \int_t^T g\big(s, Y(s), Z(s)\big)ds - \int_t^T Z(s)dW(s), \quad 0 \le t \le T. \tag{13.6}$$

Heuristically, the BSDE (13.6) yields the relation

$$\mathbb{E}\big[dY(t)|\mathscr{F}_t\big] = -g\big(t, Y(t), Z(t)\big)dt, \quad 0 \le t \le T. \tag{13.7}$$

From (13.7) we deduce that the coefficient g describes investors' beliefs about the expected infinitesimal change of the risk measure. Consequently, we can interpret g as an instantaneous risk measure or a local preference-based pricing rule. We also have the following relation

$$\mathbb{E}\big[d[Y,Y](t)|\mathscr{F}_t\big] = \big|Z(t)\big|^2 dt, \quad 0 \le t \le T, \tag{13.8}$$

and we can interpret Z as an intensity of variability of the risk measure. Moreover, by the comparison principle we can also establish the following property

$$g \le g' \quad \Rightarrow \quad \mathscr{E}_g[-\xi|\mathscr{F}_t] \le \mathscr{E}_{g'}[-\xi|\mathscr{F}_t], \quad 0 \le t \le T, \ \xi \in \mathbb{L}^2(\mathbb{R}),$$

from which we conclude that the greater the generator g is, the more conservative the corresponding risk measure $\mathscr{E}_g$ is.

In order to define a dynamic risk measure, we can proceed as follows. First, we choose an instantaneous risk measure g which should be justified in a financially sensible way by referring to (13.7) and (13.8). Given the local valuation rule g, we solve the BSDE (13.6) with the generator g to obtain a global valuation rule Y. Such an approach to constructing dynamic risk measures and recursive utilities was developed by Barrieu and El Karoui (2005), Rosazza Gianin (2006), El Karoui and Ravanelli (2009) and Lazrak and Quenez (2003).

Example 13.2 For the expectation $\mathbb{E}[-e^{-r(T-t)}\xi|\mathscr{F}_t]$ the instantaneous risk measure takes the form

$$\mathbb{E}\big[dY(t)|\mathscr{F}_t\big] = rY(t)dt, \quad 0 \le t \le T. \tag{13.9}$$

It is known that investors have to deal with interest rate ambiguity. It is easy to modify the generator (13.9) so that it allows for interest rate ambiguity. We can introduce the instantaneous risk measure

$$\mathbb{E}\big[dY(t)|\mathscr{F}_t\big] = \sup_{\alpha(t)\le r(t)\le\beta(t)} \big\{r(t)Y(t)\big\}dt, \quad 0 \le t \le T. \tag{13.10}$$

Under (13.10) a dynamic risk measure is locally driven by the worst case scenario for the interest rate. Following the proof of Proposition 3.3.2, we conclude that the local valuation rule (13.10) yields the global valuation rule

$$\rho_t(\xi) = \operatorname{ess} \sup_{\alpha(t)\le r(t)\le\beta(t)} \mathbb{E}\big[-e^{-\int_t^T r(s)ds}\xi|\mathscr{F}_t\big], \quad 0 \le t \le T.$$

In spite of the interpretations which can be deduced from (13.7) and (13.8), it is still difficult to come up with interpretable generators. Since the generator of a g-expectation is interpreted as an infinitesimal risk measure, we can try to relate the generator of a dynamic risk measure to static risk measures over short period of time. If we succeeded in obtaining the generator as a limit of static risk measures, then the generator could be justified by the interpretations of static risk measures and the limiting procedure. Such a construction would be helpful since static risk measures are well-studied and have clear interpretations. We now present the construction by Stadje (2010).

We consider a partition $0 = t_0 < t_1 < \cdots < t_n = T$ and we denote $\Delta t_{i+1} = t_{i+1} - t_i$. Let $(B_j^n)_{j=1,\dots,n}$ be a sequence of independent Bernoulli random variables such that $\mathbb{P}(B_j^n = 1) = \mathbb{P}(B_j^n = -1) = 0.5$. We introduce the random walk

$$R^n(0) = 0, \qquad R^n(t_i) = \sum_{j=1}^{i} \sqrt{\Delta t_j} B_j^n, \quad i = 1, \dots, n, \tag{13.11}$$

and the natural filtration $\mathscr{F}_{t_i}^n = \sigma(R(t_1), \dots, R(t_i))$. Let $\rho_{t_i,t_{i+1}} : \mathbb{L}^2(\Omega, \mathscr{F}_{t_{i+1}}^n, \mathbb{P}; \mathbb{R}) \mapsto \mathbb{L}^2(\Omega, \mathscr{F}_{t_i}^n, \mathbb{P}; \mathbb{R})$, for $i = 0, 1, \dots, n-1$, be one-period static risk measures which are assumed to be monotone and translation invariant. The risk measure $\rho_{t_i,t_{i+1}}$ quantifies the risk over the period $[t_i, t_{i+1}]$. We define the scaled and tilted static risk measures

$$\hat{\rho}_{t_i,t_{i+1}}(\xi) = -(1 - \sqrt{\Delta t_{i+1}})\mathbb{E}\big[\xi|\mathscr{F}_{t_i}^n\big] + \Delta t_{i+1}\rho_{t_i,t_{i+1}}\left(\frac{\xi}{\sqrt{\Delta t_{i+1}}}\right),$$
$$\xi \in \mathbb{L}^2\big(\Omega, \mathscr{F}_{t_{i+1}}^n, \mathbb{P}\big), \quad i = 0, 1, \dots, n-1. \tag{13.12}$$

Using the one-period risk measures $\hat{\rho}_{t_i,t_{i+1}}$, we define the multi-period risk measure

$$\rho_T^n(\xi^n) = -\xi^n$$
$$\rho_{t_i}^n(\xi^n) = \hat{\rho}_{t_i,t_{i+1}}\big(-\rho_{t_{i+1}}^n(\xi^n)\big), \quad i = 0, \dots, n-1,$$

for $\xi^n \in \mathbb{L}^2(\Omega, \mathscr{F}_T^n, \mathbb{P})$. The risk measure $\rho_{t_i}^n(\xi^n)$ quantifies the risk of an $\mathscr{F}_T^n$-measurable position ξ^n at time t_i. From the predictable representation property for one-dimensional Bernoulli random walk we deduce that for an $\mathscr{F}_{t_i+1}^n$-measurable $\rho_{t_{i+1}}^n$ there exist $\mathscr{F}_{t_i}^n$-measurable random variables $\alpha^n(t_i)$ and $\psi^n(t_i)$ such that

$$\rho_{t_{i+1}}^n(\xi^n) = \alpha^n(t_i) + \psi^n(t_i)\Delta R^n(t_{i+1}).$$

By (13.12), the predictable representation property and the translation invariance of $\rho_{t_i,t_{i+1}}$ we derive

$$\begin{aligned}
\Delta\rho_{t_i}^n(\xi^n) &= \rho_{t_{i+1}}^n(\xi^n) - \rho_{t_i}^n(\xi^n) = \rho_{t_{i+1}}^n(\xi^n) - \hat{\rho}_{t_i,t_{i+1}}\big(-\rho_{t_{i+1}}^n(\xi^n)\big) \\
&= \alpha^n(t_i) + \psi^n(t_i)\Delta R^n(t_{i+1}) - \hat{\rho}_{t_i,t_{i+1}}\big(-\alpha^n(t_i) - \psi^n(t_i)\Delta R^n(t_{i+1})\big) \\
&= \alpha^n(t_i) + \psi^n(t_i)\Delta R^n(t_{i+1}) \\
&\quad - \left((1 - \sqrt{\Delta t_{i+1}})\alpha^n(t_i) + \Delta t_{i+1}\left(\frac{\alpha^n(t_i)}{\sqrt{\Delta t_{i+1}}} + \rho_{t_i,t_{i+1}}\big(-\psi^n(t_i)B_{i+1}^n\big)\right)\right) \\
&= -g^n\big(t_i, \psi^n(t_i)\big)\Delta t_{i+1} + \psi^n(t_i)\Delta R^n(t_{i+1}), \quad i = 0, 1, \dots, n-1
\end{aligned} \tag{13.13}$$

where we introduce the discrete-time generator

$$g^n(t_i, z) = \rho_{t_i,t_{i+1}}\big(-zB_{i+1}^n\big), \quad i = 0, \dots, n-1, \ z \in \mathbb{R}. \tag{13.14}$$

From (13.13) we can conclude that the risk measure ρ_t^n satisfies the backward stochastic difference equation

$$\begin{aligned}
\rho_{t_i}^n(\xi^n) &= -\xi^n + \sum_{j=i}^{n-1} g^n\big(t_j, \psi^n(t_j)\big)\Delta t_{j+1} \\
&\quad - \sum_{j=i}^{n-1} \psi^n(t_j)\Delta R^n(t_{j+1}), \quad i = 0, 1, \dots, n-1.
\end{aligned} \tag{13.15}$$

We set $\psi^n(s) = \psi^n(t_i)$, $\rho_s^n = \rho_{t_i}^n$, $g^n(s, \psi^n(s)) = g^n(t_i, \psi^n(t_i))$, $t_i \le s < t_{i+1}$.

We have glued the one-period static risk measures $\rho_{t_i,t_{i+1}}$ and we have obtained the multi-period, discrete-time risk measure $\rho_{t_i}^n$. Now, we are interested in convergence to a continuous-time risk measure. We state Theorem 6.4 from Stadje (2010).

Theorem 13.2.1 *Choose a partition $0 = t_0 < t_1 < \cdots < t_n = T$. Let $\mathscr{F}^W$ be the filtration generated by a Brownian motion and let $\mathscr{F}^n$ be the filtration generated by* *the random walk* (13.11). *Let* $\rho_{t_i,t_{i+1}} : \mathbb{L}^2(\Omega, \mathscr{F}^n_{t_{i+1}}, \mathbb{P}; \mathbb{R}) \mapsto \mathbb{L}^2(\Omega, \mathscr{F}^n_{t_i}, \mathbb{P}; \mathbb{R})$ *be a family of monotone and translation invariant operators satisfying* $\rho_{t_i,t_{i+1}}(0) = 0$. *We consider the backward stochastic difference equation* (13.15) *with the generator* g^n *given by* (13.14). *We assume that*

(i) $\xi^n \in \mathscr{F}^n_T$, $\xi \in \mathscr{F}^W_T$ *and* $\xi^n, \xi \in \mathbb{L}^2(\mathbb{R})$,
(ii) $\lim_{n\to\infty} \mathbb{E}[|\xi^n - \xi|^2] = 0$,
(iii) *there exists a function* $g : \Omega \times [0, T] \times \mathbb{R} \to \mathbb{R}$ *such that*

$$\lim_{n\to\infty} \mathbb{E}\Big[\sup_{0\le t\le T} \big|g^n(t, z) - g(t, z)\big|^2\Big] = 0, \quad z \in \mathbb{R},$$

$$\big|g(t, z) - g\big(t, z'\big)\big| \le K\big|z - z'\big|, \quad (t, z), \big(t, z'\big) \in [0, T] \times \mathbb{R},$$

$$\mathbb{E}\Big[\int_0^T \big|g(t, 0)\big|^2 dt\Big] < \infty.$$

If $\lim_{n\to\infty} \sup_{j=1,\ldots,n} \Delta t_j = 0$, *then*

$$\lim_{n\to\infty} \mathbb{E}\Big[\sup_{0\le t\le T} \big|\rho^n_t(\xi^n) - Y(t)\big|^2\Big] = 0, \qquad \lim_{n\to\infty} \mathbb{E}\Big[\int_0^T \big|\psi^n(t-) - Z(t)\big|^2 dt\Big] = 0,$$

where $(Y, Z) \in \mathbb{S}^2(\mathbb{R}) \times \mathbb{H}^2(\mathbb{R})$ *is the unique solution to the BSDE*

$$Y(t) = -\xi + \int_t^T g\big(s, Z(s)\big)ds - \int_t^T Z(s)dW(s), \quad 0 \le t \le T.$$

As far as the approximation of ξ with ξ^n is concerned, we can rely on two possibilities: we can choose $\xi^n = \mathbb{E}[\xi|\mathscr{F}^n_T]$ or in the case when $\xi = \varphi(W(T))$ we can set $\xi^n = \varphi(R^n(T))$.

Theorem 13.2.1 shows that it is possible to construct a generator g of a dynamic risk measure by means of a discrete-time analog g^n and one-period static risk measures $\rho_{t_i,t_{i+1}}$ whose behaviors are better-understood. It turns out that the generator can be formally interpreted as an infinitesimal version of a static risk measure. We point out that tilting and scaling (13.12) is necessary. Without the tilting and scaling procedure the operator $\rho_{t_i,t_{i+1}}(-\rho_{t_{i+1}}(\xi^n))$ would blow up when more and more time instances are taken into account, see Proposition 5.1 in Stadje (2010). Stadje (2010) derives the generators of dynamic risk measures which correspond to the semi-deviation pricing principle, the Value-at-Risk pricing principle, the Average-Value-at-Risk pricing principle and the Gini pricing principle.

13.3 Optimal Risk Sharing

We investigate a problem of optimal risk sharing between two parties who value their future risky positions with dynamic risk measures. Since Borch (1962) and

Arrow (1963), optimal risk sharing problems have become classical topics in economics, insurance and finance.

Consider two agents A and B. Agent A takes a risky position ξ. In order to reduce the risk exposure, agent A issues a financial derivative F and tries to sell F to agent B for a price π. The derivative F transfers the risk from agent A to agent B and after the transaction the risky position ξ is shared between agents A and B. Agent B is willing to accept the risk if the risk transfer is favorable to him. In the insurance context, agent A plays the role of an insured who is interested in having an insurance protection against claims and agent B plays the role of an insurer. Agents A and B may also represent an insurer and a reinsurer, and F may describe a reinsurance treaty. We aim to find an optimal derivative F (an optimal insurance contract) and a price π of this derivative. In other words, the goal is to establish an optimal risk transfer (F, π) between agents A and B.

Let us start with a static model. We assume that agents A and B value their positions with translation invariant and convex risk measures ρ^A and ρ^B. Agent A aims to reduce his risk exposure, hence he chooses F and π which minimize the risk measure of his position after the transaction

$$\min_{F,\pi} \rho^A(\xi - F + \pi). \tag{13.16}$$

Agent B is interested in the transaction provided that the risk of his position does not increase after the transaction, hence he requires

$$\rho^B(F - \pi) \leq \rho^B(0). \tag{13.17}$$

By the translation invariance property of the risk measures ρ^A, ρ^B we can conclude that the optimal price is given by

$$\pi^* = \rho^B(0) - \rho^B(F),$$

and the optimal contract solves the optimization problem

$$\inf_{F \in \mathbb{L}^2(\mathbb{R})} \{\rho^A(\xi - F) + \rho^B(F)\}, \quad \xi \in \mathbb{L}^2(\mathbb{R}).$$

We now move to a dynamic model. We consider the natural Brownian filtration $\mathscr{F}^W$. We are interested in solving the following optimization problem

$$\rho_t^A \Box \rho_t^B(\xi) = \operatorname{ess} \inf_{F \in \mathbb{L}^2(\mathbb{R})} \{\rho_t^A(\xi - F) + \rho_t^B(F)\}, \quad 0 \leq t \leq T,\ \xi \in \mathbb{L}^2(\mathbb{R}), \tag{13.18}$$

where ρ^A and ρ^B are translation invariant and convex dynamic risk measures defined by g-expectations. We can assume that the generators g^A and g^B of the BSDEs defining the risk measures depend only on (t, z), are convex in z and satisfy $g^A(t, 0) = g^B(t, 0) = 0$, see Definition 6.2.3 and Proposition 6.2.3 as well as Theorems 3.1 and 3.2 in Jiang (2008).

We will use the operation of infimal-convolution for convex functions, see Barrieu and El Karoui (2005) and Barrieu and El Karoui (2004).

Definition 13.3.1 Let α and β be two closed convex functions. The infimal-convolution of α and β is defined by

$$\alpha\Box\beta(z) = \inf_x\{\alpha(z-x) + \beta(x)\}, \quad z \in \mathbb{R}.$$

Let α be a closed convex function. The recession function associated with α is defined by

$$\alpha_{0^+}(z) = \lim_{c\to 0^+} c\alpha\left(\frac{z}{c}\right), \quad z \in \mathbb{R}.$$

We recall the key result on the inf-convolution, see Proposition 8.1 in Barrieu and El Karoui (2005).

Lemma 13.3.1 *Let α and β be two closed convex functions. If*

$$\alpha_{0^+}(z) + \beta_{0^+}(-z) > 0, \quad z \neq 0,$$

then the infimal-convolution $\alpha\Box\beta$ is exact and the infimum is attained for some x^*, i.e.

$$\alpha\Box\beta(z) = \alpha(z - x^*) + \beta(x^*), \quad z \in \mathbb{R}.$$

We now solve the optimal risk sharing problem (13.18). We consider three BSDEs

$$\rho_t^A(\xi - F) = -\xi + F + \int_t^T g^A(s, Z^A(s))ds - \int_t^T Z^A(s)dW(s), \quad 0 \le t \le T, \tag{13.19}$$

$$\rho_t^B(F) = -F + \int_t^T g^B(s, Z^B(s))ds - \int_t^T Z^B(s)dW(s), \quad 0 \le t \le T, \tag{13.20}$$

$$\rho_t^{A,B}(\xi) = -\xi + \int_t^T g^A\Box g^B(s, Z^{A,B}(s))ds - \int_t^T Z^{A,B}(s)dW(s), \quad 0 \le t \le T. \tag{13.21}$$

Theorem 13.3.1 *Consider the optimization problem* (13.18) *and the BSDEs* (13.19)–(13.21). *We assume*

(i) ξ *and* F *are* $\mathscr{F}^W$*-measurable and* $\xi, F \in \mathbb{L}^2(\mathbb{R})$,
(ii) *the generators* $g^A : [0,T] \times \mathbb{R} \to \mathbb{R}$ *and* $g^B : [0,T] \times \mathbb{R} \to \mathbb{R}$ *are Lipschitz continuous and convex in* z, *uniformly in* t, *they satisfy* $g^A(t,0) = g^B(t,0) = 0$,

$0 \le t \le T$, *and*

$$g^A_{0+}(t,z) + g^B_{0+}(t,-z) > 0, \quad z \neq 0,\ 0 \le t \le T. \tag{13.22}$$

The following results hold:

(a) *There exist unique solutions* $(\rho^A, Z^A) \in \mathbb{S}^2(\mathbb{R}) \times \mathbb{H}^2(\mathbb{R})$, $(\rho^B, Z^B) \in \mathbb{S}^2(\mathbb{R}) \times \mathbb{H}^2(\mathbb{R})$, $(\rho^{A,B}, Z^{A,B}) \in \mathbb{S}^2(\mathbb{R}) \times \mathbb{H}^2(\mathbb{R})$ *to the BSDEs* (13.19)–(13.21).

(b) *We have*

$$\rho_t^{A,B}(\xi) \le \rho_t^A \Box \rho_t^B(\xi) \le \rho_t^A(\xi - F) + \rho_t^B(F), \quad 0 \le t \le T.$$

(c) *If there exists a progressively measurable and square integrable process* Z^* *such that*

$$Z^*(t) = \arg\min_x \bigl\{ g^A\bigl(t, Z^{A,B}(t) - x\bigr) + g^B(t,x) \bigr\}, \quad 0 \le t \le T,$$

then there exists an optimal solution $F^* \in \mathbb{L}^2(\mathbb{R})$ *of the optimization problem* (13.18) *given by*

$$F^* = \int_0^T g^B\bigl(s, Z^*(s)\bigr)ds - \int_0^T Z^*(s)dW(s).$$

Moreover, we have

$$\rho_t^{A,B}(\xi) = \rho_t^A \Box \rho_t^B(\xi) = \rho_t^A\bigl(\xi - F^*\bigr) + \rho_t^B\bigl(F^*\bigr), \quad 0 \le t \le T.$$

Proof (a) By Theorem 3.1.1 there exist unique solutions (ρ^A, Z^A) and (ρ^B, Z^B) to the BSDEs (13.19)–(13.20). From Sect. 9.3.4 in Barrieu and El Karoui (2005) we recall that the mapping $z \mapsto g^A \Box g^B(t,z)$ is Lipschitz. Hence, by Theorem 3.1.1 there also exists a unique solution $(\rho^{A,B}, Z^{A,B})$ to the BSDE (13.21).

(b) We set $\rho_t = \rho_t^A(\xi - F) + \rho_t^B(F)$ and $Z(t) = Z^A(t) + Z^B(t)$ and we derive the BSDE

$$\rho_t = -\xi + \int_t^T \Bigl(g^A\bigl(s, Z(s) - Z^B(s)\bigr) + g^B\bigl(s, Z^B(s)\bigr)\Bigr)ds - \int_t^T Z(s)dW(s), \quad 0 \le t \le T. \tag{13.23}$$

The inequality

$$g^A \Box g^B(t,z) \le g^A\bigl(t, z - Z^B(t)\bigr) + g^B\bigl(t, Z^B(t)\bigr), \quad z \in \mathbb{R},\ 0 \le t \le T,$$

and the comparison principle yield

$$\rho_t^{A,B}(\xi) \le \rho_t = \rho_t^A(\xi - F) + \rho_t^B(F), \quad 0 \le t \le T.$$

We can conclude that $\rho_t^{A,B}(\xi) \le \rho_t^A \Box \rho_t^B(\xi), 0 \le t \le T$.

(c) From Lemma 13.3.1 we deduce that there exists a measurable process Z^*. For the square integrability of Z^* and the BMO property of $\int Z^*(t)dW(t)$ we refer to Theorem 8.4 in Barrieu and El Karoui (2005). We introduce the process

$$V(t) = \int_0^t g^B\big(s, Z^*(s)\big)ds - \int_0^t Z^*(s)dW(s), \quad 0 \le t \le T.$$

We conclude that $V \in \mathbb{S}^2(\mathbb{R})$ and the pair $(-V, Z^*)$ solves

$$-V(t) = -F^* + \int_t^T g^B\big(s, Z^*(s)\big)ds - \int_t^T Z^*(s)dW(s), \quad 0 \le t \le T,$$

where

$$F^* = \int_0^T g^B\big(s, Z^*(s)\big)ds - \int_0^T Z^*(s)dW(s).$$

Hence, the pair $(-V, Z^*) \in \mathbb{S}^2(\mathbb{R}) \times \mathbb{H}^2(\mathbb{R})$ is the unique solution to the BSDE (13.20) with the terminal condition F^*. We have $\rho_t^B(F^*) = -V(t)$, $Z^B(t) = Z^*(t)$, $0 \le t \le T$. The BSDE (13.23) yields the dynamics

$$\begin{aligned}\rho_t = -\xi &+ \int_t^T \big(g^A\big(s, Z(s) - Z^*(s)\big) + g^B\big(s, Z^*(s)\big)\big)ds \\ &- \int_t^T Z(s)dW(s), \quad 0 \le t \le T.\end{aligned} \tag{13.24}$$

Since there exists a unique solution to (13.24), we must have

$$\begin{aligned}Z(t) &= Z^{A,B}(t), \quad 0 \le t \le T, \\ \rho_t &= \rho_t^A\big(\xi - F^*\big) + \rho_t^B\big(F^*\big) = \rho_t^{A,B}(\xi), \quad 0 \le t \le T,\end{aligned}$$

where we use the solution to the BSDE (13.21) and the definitions of Z^* and $g^A \square g^B$. We now conclude that $\rho_t^{A,B}(\xi) = \rho_t^A(\xi - F^*) + \rho_t^B(F^*) \ge \rho_t^A \square \rho_t^B(\xi)$, $0 \le t \le T$. Combining the last statement with the assertion of item (b), we get $\rho_t^{A,B}(\xi) = \rho_t^A \square \rho_t^B(\xi)$, $0 \le t \le T$. The optimality and the admissability of F^* can now be established. □

Condition (13.22) has an economic interpretation in terms of a conservative seller price and a conservative buyer price, see Sect. 4.2 in Barrieu and El Karoui (2005). By the translation invariance of ρ^A and ρ^B the optimal derivative F^* is determined uniquely up to a constant.

Going through the proof of Theorem 13.3.1, we can notice that we need an existence result for the BSDEs (13.19)–(13.21) and a comparison principle to derive the optimal structure of the derivative. We remark that existence and comparison results also hold for BSDEs with generators having a quadratic growth in z, see

Theorem 5.3 in Barrieu and El Karoui (2005). In particular, the assertions of Theorem 13.3.1 remain true if we consider the entropic risk measure (13.5), see Theorem 8.2 in Barrieu and El Karoui (2005) and Barrieu and El Karoui (2004).

For a special family of risk measures we can derive an explicit form of the optimal derivative.

Proposition 13.3.1 *Let the assertions of Theorem* 13.3.1 *hold. Assume that the generators g^A and g^B belong to the family of functions g^c that satisfy the so-called tolerance property*

$$g^c(t,z) = cg\left(t, \frac{1}{c}z\right), \quad 0 \le t \le T,\ z \in \mathbb{R},\ c > 0. \tag{13.25}$$

The random variable

$$F^* = \frac{c_B}{c_A + c_B}\xi,$$

solves the optimization problem (13.18).

Proof We first recall Proposition 3.5 from Barrieu and El Karoui (2005) which says that a function g satisfying (13.25) also satisfies the following property

$$g^{c_A} \Box g^{c_B}(t,z) = g^{c_A + c_B}(t,z), \quad 0 \le t \le T,\ z \in \mathbb{R}. \tag{13.26}$$

We guess that $Z^*(t) = \frac{c_B}{c_A + c_B} Z^{A,B}(t)$, where Z^* is defined in Theorem 13.3.1 and $Z^{A,B}$ solves the BSDE (13.21). We verify our guess. From (13.25) and (13.26) we obtain

$$\begin{aligned}
&g^{c_A}\big(t, Z^{A,B}(t) - Z^*(t)\big) + g^{c_B}\big(t, Z^*(t)\big) \\
&= g^{c_A}\left(t, \frac{c_A}{c_A + c_B} Z^{A,B}(t)\right) + g^{c_B}\left(t, \frac{c_B}{c_A + c_B} Z^{A,B}(t)\right) \\
&= c_A g\left(t, \frac{1}{c_A + c_B} Z^{A,B}(t)\right) + c_B g\left(t, \frac{1}{c_A + c_B} Z^{A,B}(t)\right) \\
&= (c_A + c_B) g\left(t, \frac{1}{c_A + c_B} Z^{A,B}(t)\right) \\
&= g^{c_A + c_B}\big(t, Z^{A,B}(t)\big) = g^{c_A} \Box g^{c_B}\big(t, Z^{A,B}(t)\big), \quad 0 \le t \le T,
\end{aligned}$$

and we conclude that our guess is indeed optimal. We can now derive F^*. Theorem 13.3.1, properties (13.25)–(13.26) and the third BSDE (13.21) yield

$$\begin{aligned}
F^* &= \int_0^T g^{c_B}\left(t, \frac{c_B}{c_A + c_B} Z^{A,B}(t)\right) dt - \int_0^T \frac{c_B}{c_A + c_B} Z^{A,B}(t) dW(t) \\
&= \frac{c_B}{c_A + c_B}\left(\int_0^T (c_A + c_B) g\left(t, \frac{1}{c_A + c_B} Z^{A,B}(t)\right) dt - \int_0^T Z^{A,B}(t) dW(t)\right)
\end{aligned}$$

$$
\begin{aligned}
&= \frac{c_B}{c_A + c_B}\left(\int_0^T g^{c_A}\Box g^{c_B}\big(Z^{A,B}(t)\big)dt - \int_0^T Z^{A,B}(t)dW(t)\right) \\
&= \frac{c_B}{c_A + c_B}\big(\xi + \rho_0^{c_A,c_B}(\xi)\big).
\end{aligned}
$$

The result is proved as F^* is determined uniquely up to a constant. □

Proposition 13.3.1 specifies dynamic risk measures which lead to proportional risk sharing rules. In particular, the quadratic generator of the entropic risk measure (13.5) in the natural Brownian filtration satisfies the tolerance property (13.25).

Bibliographical Notes For an introduction to dynamic risk measures modelled by g-expectations driven by Brownian motions we refer to Barrieu and El Karoui (2005) and Rosazza Gianin (2006). For g-expectations driven by Brownian motions and Poisson random measures we refer to Quenez and Sulem (2012), and for g-expectations in general probability spaces we refer to Cohen (2011). Peng (2006) proposes a test to verify whether the real market pricing mechanism is generated by a g-expectation. Example 13.2 is taken from El Karoui and Ravanelli (2009). The proof of Theorem 13.3.1 is taken from Barrieu and El Karoui (2005). The assertion of Proposition 13.3.1 can be found in Barrieu and El Karoui (2005). Optimal insurance and reinsurance contracts under the expected utility theory are discussed in Bühlmann (1970) and Denuit et al. (2001). Applications of FBSDEs to risk-sharing problems, optimal contracts problems and principal-agent problems are investigated in Cvitanić and Zhang (2012).

Part III
Other Classes of Backward Stochastic Differential Equations

Chapter 14
Other Classes of BSDEs

Abstract We investigate three classes of backward stochastic differential equations which can be useful for applications. First, we introduce a time-delayed BSDE in which the terminal condition and the generator depend on the past values of the solution. Next, we consider a reflected BSDE in which the solution is constrained to stay above a barrier. Finally, we deal with a constrained BSDE in which all components of the solution are forced to satisfy a constraint.

In this last chapter we investigate three classes of BSDEs. We deal with time-delayed BSDEs, reflected BSDEs and constrained BSDEs. We discuss key theoretical properties of these equations and we point out their actuarial and financial applications. BSDEs considered in this chapter extend the range of applications of BSDEs which we investigated in Part II.

14.1 Time-Delayed Backward Stochastic Differential Equations

A time-delayed backward stochastic differential equation is an equation of the form

$$\begin{aligned} Y(t) = \xi(Y_T, Z_T, U_T) + \int_t^T f(s, Y_s, Z_s, U_s)ds \\ - \int_t^T Z(s)dW(s) - \int_t^T \int_{\mathbb{R}} U(s,z)\tilde{N}(ds,dz), \quad 0 \le t \le T, \end{aligned} \tag{14.1}$$

where $Y_s := (Y(s+v))_{-T\le v\le 0}$, $Z_s := (Z(s+v))_{-T\le v\le 0}$ and $U_s := (U(s+v,.))_{-T\le v\le 0}$, $0 \le s \le T$. We set $Z(t) = U(t,z) = 0$ and $Y(t) = Y(0)$ for $t < 0, z \in \mathbb{R}$. Given a terminal condition ξ and a generator f, we are interested in finding a triple $(Y, Z, U) \in \mathbb{S}^2(\mathbb{R}) \times \mathbb{H}^2(\mathbb{R}) \times \mathbb{H}^2_N(\mathbb{R})$ which satisfies (14.1). The novel feature of the BSDE (14.1) is that the generator f and the terminal condition ξ depend on the past values of the solution.

Time-delayed BSDEs (14.1) are related to fully coupled FBSDEs considered in Sect. 4.3. A fully coupled BSDE only allows for a delay generated by a forward SDE, whereas under a time-delayed BSDE we can investigate more general types

Ł. Delong, *Backward Stochastic Differential Equations with Jumps and Their Actuarial and Financial Applications*, EAA Series, DOI 10.1007/978-1-4471-5331-3_14,

of delays. Under a time-delayed BSDE we can consider fixed time delays $Y_t = Y(t-r)$, $Z_t = Z(t-r)$, $U_t = \int_{\mathbb{R}} U(t-r,z)\delta(t-r,z)Q(t-r,dz)\eta(t-r)$, delays of the integral form $Y_t = \int_0^t Y(s)ds$, $Z_t = \int_0^t Z(s)ds$, $U_t = \int_0^t \int_{\mathbb{R}} U(s,z)\delta(s,z) \times Q(s,dz)\eta(s)ds$, and the running supremum $Y_t = \sup_{0\le s\le t} Y(s)$.

We can prove the following result on existence and uniqueness of a solution to the time-delayed BSDE (14.1).

Theorem 14.1.1 *Assume that*

(i) *the generator* $f: \Omega \times [0,T] \times \mathbb{S}^2(\mathbb{R}) \times \mathbb{H}^2(\mathbb{R}) \times \mathbb{H}^2_N(\mathbb{R})$ *and the terminal value* $\xi : \Omega \times [0,T] \times \mathbb{S}^2(\mathbb{R}) \times \mathbb{H}^2(\mathbb{R}) \times \mathbb{H}^2_N(\mathbb{R})$ *are adapted, measurable and they satisfy*

$$\mathbb{E}\big[\big|f(t,Y_t,Z_t,U_t) - f\big(t,Y'_t,Z'_t,U'_t\big)\big|^2\big]$$
$$\le K_1\mathbb{E}\bigg[\sup_{0\le u\le t}\big|Y(u)-Y'(u)\big|^2 + \int_{-T}^0 \big|Z(t+u)-Z'(t+u)\big|^2\alpha(du)$$
$$+ \int_{-T}^0\int_{\mathbb{R}} \big|U(t+u,z)-U'(t+u,z)\big|^2 Q(t+u,dz)\eta(t+u)\alpha(du)\bigg],$$
$$\mathbb{E}\big[\big|\xi(Y_T,Z_T,U_T) - \xi\big(Y'_T,Z'_T,U'_T\big)\big|^2\big]$$
$$\le K_2\mathbb{E}\bigg[\sup_{0\le u\le T}\big|Y(u)-Y'(u)\big|^2 + \int_0^T \big|Z(u)-Z'(u)\big|^2 du$$
$$+ \int_0^T\int_{\mathbb{R}} \big|U(u,z)-U'(u,z)\big|^2 Q(u,dz)\eta(u)du\bigg],$$

with a probability measure α *defined on* $\mathscr{B}([-T,0])$, *for a.e.* $t \in [0,T]$ *and for all* (Y,Z,U),$(Y',Z',U') \in \mathbb{S}^2(\mathbb{R}) \times \mathbb{H}^2(\mathbb{R}) \times \mathbb{H}^2_N(\mathbb{R})$,

(ii) $\mathbb{E}[\int_0^T |f(t,0,0,0)|^2dt] < \infty$ *and* $\mathbb{E}[|\xi(0,0,0)|^2] < \infty$,

(iii) $f(\omega,t,y,0,0) = 0$ *for* $(\omega,y) \in \Omega \times \mathbb{R}$ *and* $t < 0$.

For sufficiently small time horizon T *and sufficiently small Lipschitz constant* K_2 *or for sufficiently small Lipschitz constants* K_1 *and* K_2 *the time-delayed BSDE* (14.1) *has a unique solution* $(Y,Z,U) \in \mathbb{S}^2(\mathbb{R}) \times \mathbb{H}^2(\mathbb{R}) \times \mathbb{H}^2_N(\mathbb{R})$.

Proof The result can be established by applying the arguments of the fixed point procedure from the proof of Theorem 3.1.1, for details we refer to Theorem 2.1 from Delong and Imkeller (2010b). □

We point out that we cannot expect that a time-delayed BSDE has a unique solution for arbitrary K_1, K_2 and T under the Lipschitz assumptions of Theorem 14.1.1, see Delong and Imkeller (2010a). This feature was already pointed out at the end of Sect. 4.3 in the context of fully coupled BSDEs.

Time-delayed BSDEs (and coupled FBSDEs) may arise in insurance and finance in an attempt to find an investment strategy and an investment portfolio which replicate a liability depending on the investment strategy and the investment portfolio. In some applications, an investment portfolio may serve simultaneously as the underlying security on which the liability is contingent and as a replicating portfolio for that liability. We give two examples.

Example 14.1 Let us consider a Black-Scholes financial model consisting of a risky stock and a risk-free bank account. We investigate the perfect replication problem for a large investor. We recall that an investor is called large if his investment decisions affect market prices of securities. Let us assume that in our model the investor's strategy π and the investment portfolio X^π affect the drift μ of the stock and the interest rate r. Such a feedback may arise because of the size of transactions made by the large investor or because other agents in the market believe that the large investor has superior information. Consequently, we deal with the dynamics

$$\begin{aligned}
dX^\pi(t) &= \pi(t)\frac{dS(t)}{S(t)} + \big(X^\pi(t) - \pi(t)\big)\frac{dS_0(t)}{S_0(t)}dt, \quad X(0) = x > 0,\\
\frac{dS_0(t)}{S_0(t)} &= r\big(t, \pi(t), X^\pi(t)\big)dt, \quad S_0(0) = 1,\\
\frac{dS(t)}{S(t)} &= \mu\big(t, \pi(t), X^\pi(t)\big)dt + \sigma(t)dW(t), \quad S(0) = s > 0,
\end{aligned}$$

where r, μ, σ are $\mathscr{F}^W$-predictable processes. The goal is to find an admissible replicating strategy $\pi \in \mathscr{A}$ for a claim $\hat{F}(S(T))$. It is straightforward to notice that the problem of finding a replicating strategy is equivalent to the problem of solving the coupled FBSDE

$$\begin{aligned}
X(t) &= \hat{F}\big(S(T)\big) + \int_t^T \Big(-\pi(s)\big(\mu\big(s, \pi(s), X(s)\big) - r\big(s, \pi(s), X(s)\big)\big)\\
&\quad - X(s)r\big(s, \pi(s), X(s)\big)\Big)ds - \int_t^T \pi(s)\sigma(s)dW(s), \quad 0 \le t \le T, \qquad (14.2)\\
S(t) &= s + \int_0^t S(s)\mu\big(s, \pi(s), X(s)\big)ds + \int_0^t S(s)\sigma(s)dW(s), \quad 0 \le t \le T.
\end{aligned}$$

Under further assumptions the coupled FBSDE (14.2) fits into the setting of Sect. 4.3. Notice that the coupled FBSDE (14.2) is a time-delayed BSDE (14.1) since the terminal condition of the backward component of (14.2) depends on (π_T, X_T). We remark that the solution to the replication problem for a large investor in a more general financial model was found by Cvitanić and Ma (1996) where the authors proved existence of a solution to a fully coupled FBSDE.

Example 14.2 The key feature of a participating contract is that it provides a guaranteed rate of return together with a bonus which is linked, by a so-called profit-sharing rule, to the performance of an asset portfolio held and managed by the insurer. Such a construction implies that the insurer's asset allocation π and the asset portfolio

X^π affect the final pay-off from the policy. According to Solvency II Directive, see V.2.2 in European Commission QIS5 (2010), the insurance reserve must include an estimate of the value of the liability arising under the contract including all possible guarantees, profits and bonuses. When valuating the liability under a participating contract, the future asset allocation strategy for the asset portfolio should be taken into account, see TP.2.92 in European Commission QIS5 (2010). On the other hand, the insurer's asset portfolio must match the reserve (the value of the liability) and the assets held by the insurer must finance the liability which depends on the past and future performance of the asset portfolio and the allocation strategy. The assets and the liabilities interact with each other. An appropriate investment strategy should be identified to match the assets with the liabilities and fulfill the obligation. For applications of time-delayed BSDEs in this field we refer to Delong (2012c).

Let us consider an example of the investment problem from Example 14.2. We assume that a financial institution issues a wealth management product which offers a ratchet option as a capital guarantee. Under the ratchet option any intermediate investment gain earned by the financial institution is locked in as the liability and guaranteed to be paid back at maturity, i.e. under the ratchet option the highest value of the investment portfolio, which is managed by the financial institution, is paid. Ratchet options of this type are offered as wealth-dependent guarantees (capital protections) by investment funds, variable annuities and pension plans.

We consider the financial market (7.1)–(7.2). Let $\pi \in \mathscr{A}$ be an admissible investment strategy, see Definition 7.3.1, and let the investment portfolio $X^\pi := (X^\pi(t), 0 \le t \le T)$ be given by the dynamics

$$dX^\pi(t) = \pi(t)\big(\mu(t)dt + \sigma(t)dW(t)\big) + \big(X^\pi(t) - \pi(t)\big)r(t)dt.$$

The ratchet option takes the form

$$\xi = \sup_{s\in[0,T]} \big\{X^\pi(s)\big\}.$$

The goal is to find an admissible investment strategy under which the pay-off from the ratchet option can be delivered by managing the assets in the investment portfolio. It is easy to notice that this investment problem can be solved by finding a solution to the time-delayed BSDE

$$\begin{aligned} X(t) = {} & \sup_{s\in[0,T]} \big\{X(s)\big\} + \int_t^T \big(-X(s)r(s) - \pi(s)\sigma(s)\theta(s)\big)ds \\ & - \int_t^T \pi(s)\sigma(s)dW(s), \quad 0 \le t \le T. \end{aligned} \tag{14.3}$$

Let us investigate the time-delayed BSDE (14.3). We introduce the bond price process

$$D(t) = \mathbb{E}^{\mathbb{Q}^*}\big[e^{-\int_t^T r(s)ds} | \mathscr{F}_t^W\big], \quad 0 \le t \le T, \tag{14.4}$$

where

$$\frac{d\mathbb{Q}^*}{d\mathbb{P}}\Big|_{\mathscr{F}_t^W} = e^{-\int_0^t \theta(s)dW(s)-\frac{1}{2}\int_0^t |\theta(s)|^2 ds}, \quad 0 \le t \le T,$$

is the unique equivalent martingale measure for the financial model considered. We can deduce from the predictable representation property and positivity of the bond price D that there exists an $\mathscr{F}^W$-predictable process $\sigma^D := (\sigma^D(t), 0 \le t \le T)$ such that

$$\frac{dD(t)}{D(t)} = r(t)dt + \sigma^D(t)\theta(t)dt + \sigma^D(t)dW(t). \tag{14.5}$$

First, we prove a preliminary result, which is interesting in its own right.

Proposition 14.1.1 *Assume that* (C1)–(C2) *from Chap. 7 hold and let the interest rate r be a strictly positive process. Let the bond price* (14.4) *satisfy the dynamics* (14.5) *with a bounded volatility σ^D. Choose an $\mathscr{F}^W$-predictable process $\varphi := (\varphi(t), 0 \le t \le T)$ such that*

$$\mathbb{E}\Big[\int_0^T |\varphi(t)\sigma(t)|^2 dt\Big] < \infty,$$

and consider a process $\hat{S} := (\hat{S}(t), 0 \le t \le T)$ given by the dynamics

$$d\hat{S}(t) = \varphi(t)\frac{dS(t)}{S(t)} + \big(\hat{S}(t) - \varphi(t)\big)r(t)dt, \quad \hat{S}(0) = \hat{s} > 0.$$

There exists a unique square integrable solution $\mathscr{X} \in \mathbb{S}^2(\mathbb{R})$ to the forward SDE

$$\begin{aligned} d\mathscr{X}(t) = {} & \sup_{0\le s\le t}\{\mathscr{X}(s)\}D(t)\frac{dD(t)}{D(t)} \\ & + \Big(\mathscr{X}(t) - \sup_{0\le s\le t}\{\mathscr{X}(s)\}D(t)\Big)\mathbf{1}\{\hat{S}(t) > 0\}\frac{d\hat{S}(t)}{\hat{S}(t)}, \quad \mathscr{X}(0) = x > 0, \end{aligned} \tag{14.6}$$

which satisfies $\mathscr{X}(t) \ge \sup_{0\le s\le t}\{\mathscr{X}(s)\}D(t)$, $0 \le t \le T$.

Proof We introduce the discounted processes $V(t) = \frac{\mathscr{X}(t)}{D(t)}$ and $R(t) = \frac{\hat{S}(t)}{D(t)}$. By the Itô's formula we derive

$$dV(t) = \Big(V(t) - \sup_{0\le s\le t}\{V(s)D(s)\}\Big)\mathbf{1}\{R(t) > 0\}\frac{dR(t)}{R(t)}, \tag{14.7}$$

and

$$\begin{aligned} dR(t) = {} & \Big(-R(t)\theta(t)\sigma^D(t) + R(t)\big|\sigma^D(t)\big|^2 + \frac{\varphi(t)}{D(t)}\theta(t)\sigma(t) - \frac{\varphi(t)}{D(t)}\sigma(t)\sigma^D(t)\Big)dt \\ & + \Big(\frac{\varphi(t)}{D(t)}\sigma(t) - R(t)\sigma^D(t)\Big)dW(t). \end{aligned} \tag{14.8}$$

Let us denote $A(t)=\sup_{0\le s\le t}\{\frac{V(s)D(s)}{D(0)}\}$. We notice that $dA(t)\neq 0$ if and only if $A(t)=\frac{V(t)D(t)}{D(0)}$. We rewrite (14.7) in the form

$$dV(t)=\big(V(t)-A(t)D(0)\big)\mathbf{1}\big\{R(t)>0\big\}\frac{dR(t)}{R(t)}. \tag{14.9}$$

Let us consider the sequence of stopping times $\tau_n=\tau_n^D\wedge\tau_n^R$ where $\tau_n^D=\inf\{t: D(t)=1-\frac{1}{n}\}$ and $\tau_n^R=\inf\{t:R(t)=\frac{1}{n}\}$. We first solve (14.9) on $[0,\tau_n]$. By the Itô's formula we get

$$\begin{aligned}
d\left(\frac{V(t)}{A(t)}\right)&=\left(\frac{V(t)}{A(t)}-D(0)\right)\frac{dR(t)}{R(t)}-V(t)\frac{dA(t)}{A^2(t)}\\
&=\left(\frac{V(t)}{A(t)}-D(0)\right)\frac{dR(t)}{R(t)}-\frac{D(0)}{D(t)}\frac{dA(t)}{A(t)},
\end{aligned}$$

and

$$\begin{aligned}
d\left(\log\left(\frac{V(t)}{A(t)}-D(0)\right)\right)&=\frac{dR(t)}{R(t)}-\frac{1}{2}\frac{d[R,R](t)}{R^2(t)}-\frac{1}{\frac{V(t)}{A(t)}-D(0)}\frac{D(0)}{D(t)}\frac{dA(t)}{A(t)}\\
&=\frac{dR(t)}{R(t)}-\frac{1}{2}\frac{d[R,R](t)}{R^2(t)}-\frac{1}{1-D(t)}\frac{dA(t)}{A(t)}.
\end{aligned}\tag{14.10}$$

In order to establish (14.10), we first use the localizing sequence $\tau_m=\inf\{t:\frac{V(t)}{A(t)}-D(0)=\frac{1}{m}\}$ and we next let $m\to\infty$. Consequently, $V(t)>A(t)D(0)$, $0\le t\le\tau_n$. From (14.10) we deduce

$$\begin{aligned}
&\log\left(\frac{V(t)}{A(t)}-D(0)\right)-\log\left(\frac{D(0)}{D(t)}-D(0)\right)\\
&=\log\big(1-D(0)\big)-\log\left(\frac{D(0)}{D(t)}-D(0)\right)+\log R(t)-\log R(0)\\
&\quad-\int_0^t\frac{1}{1-D(s)}\frac{dA(s)}{A(s)},\quad 0\le t\le\tau_n.
\end{aligned}$$

Referring to the Skorohod equation, see Lemma 6.14 in Karatzas and Shreve (1988), we can obtain unique processes $(L,\hat{L})$ such that

$$\begin{aligned}
L(t)&=\log\big(1-D(0)\big)-\log\left(\frac{D(0)}{D(t)}-D(0)\right)\\
&\quad+\log R(t)-\log R(0),\quad 0\le t\le\tau_n,\\
\hat{L}(t)&=\int_0^t\frac{1}{1-D(s)}\frac{dA(s)}{A(s)}=\sup_{0\le s\le t}L(t),\quad 0\le t\le\tau_n,\\
L(t)-\hat{L}(t)&=\log\left(\frac{V(t)}{A(t)}-D(0)\right)-\log\left(\frac{D(0)}{D(t)}-D(0)\right),\quad 0\le t\le\tau_n.
\end{aligned}\tag{14.11}$$

We clearly have $L(0) = \hat{L}(0) = 0$ and $\hat{L}(t) \geq 0$, $0 \leq t \leq \tau_n$. From (14.11) we can conclude that there exists a unique solution to (14.9) on $[0, \tau_n]$ which is of the form

$$\begin{aligned} V(t) &= A(t)\left[D(0) + \big(1 - D(0)\big)\frac{R(t)}{R(0)}e^{-\hat{L}(t)}\right], \quad 0 \leq t \leq \tau_n, \\ A(t) &= V(0)e^{\int_0^t (1-D(s))d\hat{L}(s)}, \quad 0 \leq t \leq \tau_n. \end{aligned} \tag{14.12}$$

Let $\tau_\infty = \lim_{n\to\infty} \tau_n^D \wedge \tau_n^R$. Taking the limits, we can consider progressively measurable processes $(L, \hat{L})$ and (A, V) on $[0, \tau_\infty]$ defined by (14.11) and (14.12). It is straightforward to observe that $A(t) \leq V(0)e^{\hat{L}(t)}$, $0 \leq t \leq \tau_\infty$. Hence, by (14.12) we obtain the estimate

$$0 \leq V(t) - A(t)D(0) \leq \big(1 - D(0)\big)V(0)\frac{R(t)}{R(0)}, \quad 0 \leq t \leq \tau_\infty. \tag{14.13}$$

We now investigate the process

$$\hat{V}(t) = V(0) + \int_0^t \big(V(s) - A(s)D(0)\big)\mathbf{1}\big\{R(s) > 0\big\}\frac{dR(s)}{R(s)}, \quad 0 \leq t \leq \tau_\infty,$$

which coincides on $[0, \tau_n]$ with the process V given by (14.12). One can show that $\hat{V}$ is a continuous semimartingale such that $\mathbb{E}[\sup_{t\in[0,\tau_\infty]} |\hat{V}(t)|^2] < \infty$. Hence, we obtain the convergence

$$\hat{V}(t) = \lim_{n\to\infty} \hat{V}(t \wedge \tau_n) = \lim_{n\to\infty} V(t \wedge \tau_n), \quad 0 \leq t \leq \tau_\infty.$$

Consequently, the process A given by (14.12) can be extended as an a.s. finite process to $[0, \tau_\infty]$. By (14.12) the solution V satisfies $V(t) \geq A(t)D(0)$, $0 \leq t \leq \tau_\infty$.

Notice that we have $R(\tau_\infty)e^{-\hat{L}(\tau_\infty)} = 0$ and we end up with $V(\tau_\infty) = A(\tau_\infty)D(0)$. If $\tau_\infty < T$, i.e. if $R(\tau_\infty) = 0$, then from (14.9) we conclude that $dV(t) = 0$ for $t > \tau_\infty$. Hence, the solution V is defined as constant after τ_∞. We also have

$$\begin{aligned} A(t) &= \sup_{0\leq s\leq t}\left\{\frac{V(s)D(s)}{D(0)}\right\} \\ &= \max\left\{A(\tau_\infty), \sup_{\tau_\infty\leq s\leq t}\left\{\frac{V(s)D(s)}{D(0)}\right\}\right\} \\ &= \max\Big\{A(\tau_\infty), A(\tau_\infty)\sup_{\tau_\infty\leq s\leq t}\big\{D(s)\big\}\Big\} = A(\tau_\infty), \quad t \geq \tau_\infty. \end{aligned}$$

The process V satisfies $V(t) \geq A(t)D(0)$, $0 \leq t \leq T$. This yields $\mathscr{X}(t) \geq \sup_{s\leq t}\{\mathscr{X}(s)\}D(t)$, $0 \leq t \leq T$. The square integrability of $\mathscr{X}$ can be immediately deduced from the square integrability of V. □

We remark that Proposition 14.1.1 gives the dynamics of the investment portfolio under the so-called drawdown constraint. The process $\hat{S}$ can be interpreted as an unconstrained investment portfolio.

We now prove the main result on the time-delayed BSDE.

Theorem 14.1.2 *Let the assumptions of Proposition* 14.1.1 *hold. There exist multiple solutions* $(X^*, \pi^*) \in \mathbb{S}^2(\mathbb{R}) \times \mathscr{A}$ *to the time-delayed BSDE* (14.3) *which are defined by*

$$\begin{aligned} X^*(t) &= \mathscr{X}(t), \quad 0 \le t \le T, \\ \pi^*(t) &= \sup_{0 \le s \le t} \{\mathscr{X}(s)\} D(t) \frac{\sigma^D(t)}{\sigma(t)} \\ &\quad + \frac{\varphi(t)}{\hat{S}(t)} \Big(\mathscr{X}(t) - \sup_{0 \le s \le t} \{\mathscr{X}(s)\} D(t) \Big) \mathbf{1}\{\hat{S}(t) > 0\}, \quad 0 \le t \le T, \end{aligned}$$

where we can choose $x > 0$ *and an* $\mathscr{F}^W$*-predictable process* φ *such that*

$$\mathbb{E}\Big[\int_0^T |\varphi(t)\sigma(t)|^2 \Big] < \infty.$$

Proof First, we show that any solution X to the time-delayed BSDE (14.3) must satisfy $X(t) \ge \sup_{s \le t}\{X(s)\}D(t)$, $0 \le t \le T$, see Theorem 4.1 in Delong (2012c) for details. Such a solution $\mathscr{X} \in \mathbb{S}^2(\mathbb{R})$ is constructed in Proposition 14.1.1. Next, we can notice that the process $\mathscr{X}$ satisfies the terminal condition of the time-delayed BSDE since $\mathscr{X}(T) = V(T) = A(\tau_\infty)D(0) = A(T)D(0) = \xi$. The strategy π^* can be deduced from (14.6) and the dynamics of D and $\hat{S}$. The square integrability of π^* follows from (14.13). Hence, $\pi^* \in \mathscr{A}$. □

The investment strategy derived in Theorem 14.1.2 is an extension of the one-period Option Based Portfolio Insurance strategy which protects the initial capital. Applying the continuous-time strategy from Theorem 14.1.2, we are able to protect the maximal value of the investment process. Under the strategy (X^*, π^*) the assets and the liabilities are matched in the sense that the market-consistent value, or the arbitrage-free value, of the ratchet option contingent on the investment portfolio is equal to the value of the investment portfolio. We remark that the investment strategies π^* yield different final pay-offs $X^{\pi^*}(T)$ depending on φ and risk-return analysis should be applied to choose the strategy for the application.

14.2 Reflected Backward Stochastic Differential Equations

A reflected backward stochastic differential equation is an equation of the form

$$Y(t) = \xi + \int_t^T f\big(s, Y(s), Z(s), U(s, .)\big)ds - \int_t^T Z(s)dW(s)$$

$$-\int_t^T \int_{\mathbb{R}} U(s,z)\tilde{N}(ds,dz) + C(T) - C(t), \quad 0 \le t \le T, \tag{14.14}$$

with the constraints

$$Y(t) \ge R(t), \quad 0 \le t \le T, \quad \int_0^T \big(Y(t) - R(t)\big)dC(t) = 0. \tag{14.15}$$

Given a terminal condition ξ, a generator f and a barrier R, we are interested in finding a quadruple $(Y, Z, U, C) \in \mathbb{S}^2(\mathbb{R}) \times \mathbb{H}^2(\mathbb{R}) \times \mathbb{H}^2_N(\mathbb{R}) \times \mathbb{S}^2_{inc}(\mathbb{R})$ which satisfies (14.14)–(14.15). Since constraints on the solution are introduced, an additional control process has to be added into the dynamics of the BSDE. The process C arises now in (14.14) and its purpose is to force the process Y to stays above the barrier R. The activity of C should be minimal in the sense that the process C acts only when Y touches the lower barrier R. In other words, the process C aims to reflect the process Y on the barrier R. We remark that a solution (Y, Z, U, C) to (14.14)–(14.15) is also called a supersolution to the reflected BSDE (14.14)–(14.15), see Sect. 2.3 in El Karoui et al. (1997b) and Peng (1999).

We prove existence and uniqueness of a solution to the reflected BSDE (14.14)–(14.15) by a fixed point procedure. First, we investigate the case of a generator independent of (Y, Z, U). This case is important in its own right.

Theorem 14.2.1 *Consider the BSDE*

$$Y(t) = \xi + \int_t^T f(s)ds - \int_t^T Z(s)dW(s)$$
$$-\int_t^T \int_{\mathbb{R}} U(s,z)\tilde{N}(ds,dz) + C(T) - C(t), \quad 0 \le t \le T, \tag{14.16}$$

with the constraints

$$Y(t) \ge R(t), \quad 0 \le t \le T, \quad \int_0^T \big(Y(t) - R(t)\big)dC(t) = 0. \tag{14.17}$$

Assume that

(i) *the terminal value* $\xi \in \mathbb{L}^2(\mathbb{R})$,
(ii) *the generator* $f : \Omega \times [0,T] \to \mathbb{R}$ *is predictable,*
(iii) $\mathbb{E}[\int_0^T |f(t)|^2 dt] < \infty$,
(iv) *the barrier* R *is a quasi-left continuous process such that* $R \in \mathbb{S}^2(\mathbb{R})$.

There exists a unique solution $(Y, Z, U, C) \in \mathbb{S}^2(\mathbb{R}) \times \mathbb{H}^2(\mathbb{R}) \times \mathbb{H}^2_N(\mathbb{R}) \times \mathbb{S}^2_{inc}(\mathbb{R})$ *to the reflected BSDE* (14.16)–(14.17) *such that C is continuous. The process Y has the representation*

$$Y(t) = \operatorname*{ess\,sup}_{\tau \in \mathscr{T}_t} \mathbb{E}\Big[\xi \mathbf{1}\{\tau = T\} + R(\tau)\mathbf{1}\{\tau < T\} + \int_t^\tau f(s)ds \,|\, \mathscr{F}_t\Big], \quad 0 \le t \le T, \tag{14.18}$$

where

$$\mathcal{T}_t = \{\tau \text{ is an } \mathcal{F}\text{-stopping time such that } t \leq \tau \leq T\}.$$

Proof Let us consider the process

$$V(t) = \xi \mathbf{1}\{t = T\} + R(t)\mathbf{1}\{t < T\} + \int_0^t f(s)ds, \quad 0 \leq t \leq T. \tag{14.19}$$

The process V is càdlàg, adapted and $\sup_{0\leq t\leq T} |V(t)| \in \mathbb{L}^2(\mathbb{R})$. Hence, V is of Class D, see Sect. III.3 in Protter (2004). We can conclude that there exists the Snell envelope of V, which is the smallest càdlàg supermartingale dominating the process V, see Appendix in Hamadène and Ouknine (2011). The Snell envelope is defined by

$$E^V(t) = \operatorname*{ess\,sup}_{\tau \in \mathcal{T}_t} \mathbb{E}\big[V(\tau)|\mathcal{F}_t\big], \quad 0 \leq t \leq T.$$

The process E^V is also of Class D since

$$\begin{aligned}\mathbb{E}\Big[\sup_{0\leq t\leq T} \big|E^V(t)\big|^2\Big] &\leq \mathbb{E}\Big[\sup_{0\leq t\leq T} \Big|\mathbb{E}\Big[\sup_{0\leq u\leq T} V(u)|\mathcal{F}_t\Big]\Big|^2\Big] \\ &\leq K\mathbb{E}\Big[\sup_{0\leq u\leq T} \big|V(u)\big|^2\Big] < \infty,\end{aligned}$$

where we use an estimate for E^V and the Doob's inequality. Consequently, we can apply the Doob-Meyer theorem, see Theorem III.11 in Protter (2004), and we derive the unique decomposition

$$E^V(t) = M(t) - C(t), \quad 0 \leq t \leq T, \tag{14.20}$$

where M is a càdlàg martingale, and C is a predictable, non-decreasing process such that $C(0) = 0$. Moreover, the process C is square integrable, see Appendix in Hamadène and Ouknine (2011). Hence, from (14.20) we can deduce that the martingale M is square integrable and the predictable representation theorem yields the unique representation

$$M(t) = M(0) + \int_0^t Z(s)dW(s) + \int_0^t \int_{\mathbb{R}} U(s,z)\tilde{N}(ds,dz), \quad 0 \leq t \leq T,$$

where $(Z, U) \in \mathbb{H}^2(\mathbb{R}) \times \mathbb{H}^2_N(\mathbb{R})$. Since the martingale M and the process V are quasi-left continuous and the process V has a positive jump at the terminal time (notice that $\xi = Y(T) \geq R(T)$), the arguments from the proof of Proposition 1.4.a from Hamadène and Ouknine (2003) lead us to the conclusion that the Snell envelope E^V is regular and the process C is continuous.

We define

$$\begin{aligned} Y(t) &= E^V(t) - \int_0^t f(s)ds \\ &= \operatorname*{ess\,sup}_{\tau \in \mathscr{T}_t} \mathbb{E}\left[\xi \mathbf{1}\{\tau = T\} + R(\tau)\mathbf{1}\{\tau < T\} + \int_t^\tau f(s)ds | \mathscr{F}_t\right], \quad 0 \le t \le T. \end{aligned} \tag{14.21}$$

Since $E^V \in \mathbb{S}^2(\mathbb{R})$ and f is square integrable, we have that $Y \in \mathbb{S}^2(\mathbb{R})$. We can observe that

$$Y(t) + \int_0^t f(s)ds = E^V(t) = M(t) - C(t), \quad 0 \le t \le T,$$

and we conclude that the candidate solution $(Y, Z, U, C) \in \mathbb{S}^2(\mathbb{R}) \times \mathbb{H}^2(\mathbb{R}) \times \mathbb{H}^2_N(\mathbb{R}) \times \mathbb{S}^2_{inc}(\mathbb{R})$ satisfies the BSDE (14.16). From representation (14.21) we deduce that our candidate solution also satisfies the first constraint $Y(t) \ge R(t)$, $0 \le t \le T$. We are left with proving the second constraint $\int_0^T (Y(s) - R(s))dC(s) = 0$. Fix $t \in [0, T]$. We define the stopping time

$$\tau_t^* = \inf\{s > t, C(s) > C(t)\} \wedge T.$$

Since E^V is regular, then $E^V(\tau_t^*) = V(\tau_t^*)$ and τ_t^* is the optimal stopping time after t, see Appendix in Hamadène and Ouknine (2011). From (14.19) and (14.21) we obtain

$$\begin{aligned} Y(\tau_t^*) + \int_0^{\tau_t^*} f(s)ds &= E^V(\tau_t^*) \\ &= V(\tau_t^*) = \xi \mathbf{1}\{\tau_t^* = T\} + R(\tau_t^*)\mathbf{1}\{\tau_t^* < T\} + \int_0^{\tau_t^*} f(s)ds, \end{aligned}$$

and we conclude that $Y(\tau_t^*) = R(\tau_t^*)$ if $\tau_t^* < T$. Hence, we have $(Y(s) - R(s))dC(s) = 0$ for $s \in [t, \tau_t^*]$ and, consequently, $\int_0^T (Y(s) - R(s))dC(s) = 0$. The uniqueness of a solution to (14.16)–(14.17) is proved in the first part of the proof of Theorem 14.2.2. □

We remark that the proof of Theorem 14.2.1 relies on quasi-left continuity of the barrier and the jump process related to the random measure (which is our standing assumption in this book). We recall that the assumption of quasi-left continuity is reasonable in actuarial and financial applications, see the end of Sect. 2.1.

Theorem 14.2.1 shows that the process Y which solves the reflected BSDE (14.16)–(14.17) coincides with the optimal value function of the optimal stopping problem (14.18). Since optimal stopping problems are often applied to solve financial and insurance problems, reflected BSDEs are very important for actuarial and financial applications.

Now, we can investigate the reflected BSDE (14.14)–(14.15).

Theorem 14.2.2 *Assume that*

(i) *the terminal value* $\xi \in \mathbb{L}^2(\mathbb{R})$,

(ii) *the generator* $f : \Omega \times [0,T] \times \mathbb{R} \times \mathbb{R} \times L^2_Q(\mathbb{R}) \to \mathbb{R}$ *is predictable and Lipschitz continuous in the sense that*

$$\begin{aligned}
&\left|f(t,y,z,u) - f(t,y',z',u')\right|^2 \\
&\quad \le K\left(\left|y-y'\right|^2 + \left|z-z'\right|^2 + \int_{\mathbb{R}} \left|u(x)-u'(x)\right|^2 Q(t,dx)\eta(t)\right),
\end{aligned}$$

a.e., a.s. $(\omega,t) \in \Omega \times [0,T]$, *for all* $(y,z,u),(y',z',u') \in \mathbb{R} \times \mathbb{R} \times L^2_Q(\mathbb{R})$,

(iii) $\mathbb{E}[\int_0^T |f(t,0,0,0)|^2 dt] < \infty$,

(iv) *the barrier* R *is a quasi-left continuous process such that* $R \in \mathbb{S}^2(\mathbb{R})$.

There exists a unique solution $(Y,Z,U,C) \in \mathbb{S}^2(\mathbb{R}) \times \mathbb{H}^2(\mathbb{R}) \times \mathbb{H}^2_N(\mathbb{R}) \times \mathbb{S}^2_{inc}(\mathbb{R})$ *to the reflected BSDE* (14.14)–(14.15) *such that* C *is continuous.*

Proof 1. The uniqueness of a solution. Assume there are two solutions (Y,Z,U,C), $(Y',Z',U',C') \in \mathbb{S}^2(\mathbb{R}) \times \mathbb{H}^2(\mathbb{R}) \times \mathbb{H}^2_N(\mathbb{R}) \times \mathbb{S}^2_{inc}(\mathbb{R})$ and C and C' are continuous. By the Itô's formula we derive

$$\begin{aligned}
&e^{\rho t}\left|Y(t)-Y'(t)\right|^2 + \rho\int_t^T e^{\rho s}\left|Y(s)-Y'(s)\right|^2 ds + \int_t^T e^{\rho s}\left|Z(s)-Z'(s)\right|^2 ds \\
&\quad + \int_t^T \int_{\mathbb{R}} e^{\rho s}\left|U(s,z)-U'(s,z)\right|^2 Q(s,dz)\eta(s)ds \\
&= -2\int_t^T e^{\rho s}\big(Y(s)-Y'(s)\big)\big(-f\big(s,Y(s),Z(s),U(s)\big) \\
&\quad + f\big(s,Y'(s),Z'(s),U'(s)\big)\big)ds - 2\int_t^T e^{\rho s}\big(Y(s)-Y'(s)\big)\big(-dC(s)+dC'(s)\big) \\
&\quad - 2\int_t^T e^{\rho s}\big(Y(s-)-Y'(s-)\big)\big(Z(s)-Z'(s)\big)dW(s) \\
&\quad - 2\int_t^T \int_{\mathbb{R}} e^{\rho s}\big(Y(s-)-Y'(s-)\big)\big(U(s,z)-U'(s,z)\big)\tilde{N}(ds,dz) \\
&\quad - \int_t^T \int_{\mathbb{R}} e^{\rho s}\left|U(s,z)-U'(s,z)\right|^2 \tilde{N}(ds,dz), \quad 0 \le t \le T. \qquad (14.22)
\end{aligned}$$

We also notice that

$$\begin{aligned}
&\big(Y(s)-Y'(s)\big)\big(dC(s)-dC'(s)\big) \\
&\quad = \big(Y(s)-R(s)\big)dC(s) + \big(Y'(s)-R(s)\big)dC'(s) \\
&\qquad - \big(Y(s)-R(s)\big)dC'(s) - \big(Y'(s)-R(s)\big)dC(s) \\
&\quad = -\big(Y(s)-R(s)\big)dC'(s) - \big(Y'(s)-R(s)\big)dC(s) \le 0. \qquad (14.23)
\end{aligned}$$

Applying inequality (14.23) and following the arguments from the proof of Lemma 3.1.1 which led to (3.5) and (3.7), we obtain the estimate

$$\mathbb{E}\Big[\sup_{s\in[0,T]} e^{\rho s}\big|Y(s)-Y'(s)\big|^2 ds+\int_0^T e^{\rho s}\big|Z(s)-Z'(s)\big|^2 ds$$
$$+\int_0^T\int_{\mathbb{R}} e^{\rho s}\big|U(s,z)-U'(s,z)\big|^2 Q(s,dz)\eta(s)ds\Big]\leq 0.$$

Hence, there exists a unique triple $(Y,Z,U)\in\mathbb{S}^2(\mathbb{R})\times\mathbb{H}^2(\mathbb{R})\times\mathbb{H}^2_N(\mathbb{R})$ which solves (14.14)–(14.15). The uniqueness of the process $C\in\mathbb{S}^2_{inc}(\mathbb{R})$ follows from (14.14).

2. *The existence of a solution.* Let $Y^0(t)=Z^0(t)=U^0(t,z)=0$, $(t,z)\in[0,T]\times\mathbb{R}$ and consider the recursive equation

$$Y^{n+1}(t)=\xi+\int_t^T f\big(s,Y^n(s),Z^n(s),U^n(s)\big)ds-\int_t^T Z^{n+1}(s)dW(s)$$
$$-\int_t^T\int_{\mathbb{R}} U^{n+1}(s,z)\tilde{N}(ds,dz)+C^{n+1}(T)-C^{n+1}(t),\quad 0\leq t\leq T,\tag{14.24}$$

with the constraints

$$Y^{n+1}(t)\geq R(t),\quad 0\leq t\leq T,\qquad \int_0^T\big(Y^{n+1}(t)-R(t)\big)dC^{n+1}(t)=0.\tag{14.25}$$

By Theorem 14.2.1 there exists a sequence of unique solutions $(Y^{n+1},Z^{n+1},U^{n+1},C^{n+1})\in\mathbb{S}^2(\mathbb{R})\times\mathbb{H}^2(\mathbb{R})\times\mathbb{H}^2_N(\mathbb{R})\times\mathbb{S}^2_{inc}(\mathbb{R})$ to the reflected BSDEs (14.24)–(14.25). Property (14.23) and the fixed point arguments from the proof of Theorem 3.1.1 yield the convergence

$$\big\|Y^{n+1}-Y^n\big\|^2_{\mathbb{S}^2}+\big\|Z^{n+1}-Z^n\big\|^2_{\mathbb{H}^2}+\big\|U^{n+1}-U^n\big\|^2_{\mathbb{H}^2_N}\to 0,\quad n\to\infty.\tag{14.26}$$

Since

$$C^{n+1}(t)=Y^{n+1}(0)-Y^{n+1}(t)-\int_0^t f\big(s,Y^n(s),Z^n(s),U^n(s)\big)ds$$
$$+\int_0^t Z^{n+1}(s)dW(s)+\int_0^t\int_{\mathbb{R}} U^{n+1}(s,z)\tilde{N}(ds,dz),\quad 0\leq t\leq T,$$

we obtain

$$\sup_{t\in[0,T]}\big|C^{n+1}(t)-C^n(t)\big|^2$$
$$\leq K\Big(\sup_{t\in[0,T]}\big|Y^{n+1}(t)-Y^n(t)\big|^2$$

$$+\int_0^T \left|f\big(s, Y^n(s), Z^n(s), U^n(s)\big) - f\big(s, Y^{n-1}(s), Z^{n-1}(s), U^{n-1}(s)\big)\right|^2 ds$$

$$+ \sup_{t\in[0,T]}\left|\int_0^t \big(Z^{n+1}(s) - Z^n(s)\big)dW(s)\right|^2$$

$$+ \sup_{t\in[0,T]}\left|\int_0^t \int_{\mathbb{R}} \big(U^{n+1}(s,z) - U^n(s,z)\big)\tilde{N}(ds,dz)\right|^2\bigg).$$

By the Lipschitz property of the generator f and the Burkholder-Davis-Gundy inequality we derive the estimate

$$\|C^{n+1} - C^n\|^2_{\mathbb{S}^2} \leq K\big(\|Y^{n+1} - Y^n\|^2_{\mathbb{S}^2} + \|Z^{n+1} - Z^n\|^2_{\mathbb{H}^2} + \|U^{n+1} - U^n\|^2_{\mathbb{H}^2_N}$$
$$+ \|Y^n - Y^{n-1}\|^2_{\mathbb{S}^2} + \|Z^n - Z^{n-1}\|^2_{\mathbb{H}^2} + \|U^n - U^{n-1}\|^2_{\mathbb{H}^2_N}\big),$$

and the sequence $(C^n)_{n\in\mathbb{N}}$ converges by (14.26). We can conclude that there exists a unique limit $(Y, Z, U, C) \in \mathbb{S}^2(\mathbb{R}) \times \mathbb{H}^2(\mathbb{R}) \times \mathbb{H}^2_N(\mathbb{R}) \times \mathbb{S}^2_{inc}(\mathbb{R})$ of the sequence $(Y^n, Z^n, U^n, C^n)_{n\in\mathbb{N}}$. It is easy to show that the limit (Y, Z, U, C) satisfies the BSDE (14.14), see the proof of Theorem 3.1.1. Moreover, the process C is continuous by the uniform convergence and non-decreasing, and $Y(t) \geq R(t)$, $0 \leq t \leq T$. One can also prove that $\int_0^T (Y(s) - R(s))dC(s) = 0$, see step 6 in the proof of Theorem 1.2 from Hamadène and Ouknine (2003). Hence, the limit (Y, Z, U, C) solves the reflected BSDE (14.14)–(14.15). □

It is worth pointing out the connection in the spirit of a non-linear Feynman-Kac formula between the solution to a reflected FBSDE and the solution to variational inequalities. It is well known that in a Markovian setting the optimal value function of the stopping problem (14.18) can be characterized as a unique solution to variational inequalities, see Theorem 2.2 in Øksendal and Sulem (2004). It turns out that the unique solution to a reflected FBSDE provides a probabilistic representation of the unique viscosity solution to variational inequalities, see El Karoui et al. (1997a) and Crépey (2011) for details.

We give examples of insurance and finance optimal stopping problems which can be studied with reflected BSDEs.

Example 14.3 We consider the financial market (7.1)–(7.2) and an American put option. The buyer of an American put option has the right to sell the stock S for a given price K at the time favorable to him. The goal is to find a replicating strategy and a replicating portfolio for the American put option. Let $\pi \in \mathscr{A}$ be an admissible replicating strategy. Since the American option can be exercised at any time, the replicating portfolio X^π must satisfy the constraint $X^\pi(t) \geq (K - S(t))^+$, $0 \leq t \leq T$. It is reasonable to assume that the replicating portfolio X^π is given by the dynamics

$$dX^\pi(t) = \pi(t)\big(\mu(t)dt + \sigma(t)dW(t)\big) + \big(X^\pi(t) - \pi(t)\big)r(t)dt - dC(t).$$

The process C represents the premium for the right to exercise the option at a favorable time and the value of the replicating portfolio (or the price of the American option) decreases whenever the right to exercise the option should be executed. Since the right to exercise the option should only be executed if the price of the option is equal to $(K - S(t))^+$, we should also impose the constraint $\int_0^T (X^\pi(t) - (K - S(t))^+) dC(t) = 0$. Consequently, the problem of finding a replicating portfolio and a replicating strategy for the American put option is equivalent to the problem of solving the reflected BSDE

$$\begin{aligned} Y(t) &= \big(K - S(T)\big)^+ + \int_t^T \big(-Y(s)r(s) - Z(s)\theta(s)\big)ds \\ &\quad - \int_t^T Z(s)dW(s) + C(T) - C(t), \quad 0 \le t \le T, \\ Y(t) &\ge \big(K - S(t)\big)^+, \quad 0 \le t \le T, \quad \int_0^T \big(Y(t) - \big(K - S(t)\big)^+\big)dC(t) = 0. \end{aligned}$$

By the change of measure and Theorem 14.2.1 the unique replicating portfolio has the representation

$$Y(t) = \sup_{\tau \in \mathscr{T}_t} \mathbb{E}^{\mathbb{Q}}\big[e^{-\int_t^\tau r(s)ds}\big(K - S(\tau)\big)^+ | \mathscr{F}_t\big], \quad 0 \le t \le T, \tag{14.27}$$

and, as expected, the arbitrage-free price of the American option is given by a solution to an optimal stopping problem.

Example 14.4 American options arise in life insurance. The holder of a unit-linked policy is entitled to a terminal guarantee F and, in addition, he has the right to surrender the policy with a benefit G. The value of the embedded surrender option can be treated as the value of an American option, see Milevsky and Salisbury (2002). The value of the policy is defined by

$$\sup_{\tau \in \mathscr{T}_t} \mathbb{E}^{\mathbb{Q}}\big[e^{-\int_t^T r(s)ds}\big(F\mathbf{1}\{\tau = T\} + G(\tau)\mathbf{1}\{\tau < T\}\big) | \mathscr{F}_t\big], \quad 0 \le t \le T. \tag{14.28}$$

By Theorem 14.2.1 the value of the surrender option (14.28) can be characterized as a unique solution to a reflected BSDE. We point out that the optimal stopping problem (14.28) and a reflected BSDE arise if we assume that the policyholder makes an optimal (rational) decision to surrender the policy. Let us recall that in Example 7.7 and Chaps. 9–12 we considered irrational lapse decisions modelled by inaccessible jump times of jump measures.

Example 14.5 A recallable option or an Israeli's option is an example of an American option under which the buyer of the option has the right to sell a stock and the issuer of the option has the right to recall the option. A convertible bond gives the holder of the contract the right to convert a bond into a stock and the issuer of the contract has the right to recall the bond. In both cases we deal with a so-called Dynkin game. The value of the contract is defined as a solution to the optimal stopping problem

$$\sup_{\tau\in\mathcal{T}_t}\inf_{\delta\in\mathcal{T}_t}\mathbb{E}^{\mathbb{Q}}\big[e^{-\int_t^\delta r(s)ds}F^s(\delta)\mathbf{1}\{\delta<\tau\}+e^{-\int_t^\tau r(s)ds}F^b(\tau)\mathbf{1}\{\tau\leq\delta<T\}$$
$$+e^{-\int_t^T r(s)ds}F(T)\mathbf{1}\{\tau=\delta=T\}|\mathcal{F}_t\big],\quad 0\leq t\leq T. \tag{14.29}$$

The value function (14.29), the price of an Israeli's option and the price of a convertible bond, can be characterized as a solution to a doubly reflected BSDE, see Hamadène (2006), Bielecki et al. (2009), Hamadène and Wang (2009), Crépey and Matoussi (2008), Hamadène and Hassani (2006). A doubly reflected BSDE is an equation in which the solution is reflected at a lower and an upper barrier. We remark that the existence of a solution to a doubly reflected BSDE is proved under the so-called Mokobodski condition.

Example 14.6 Doubly reflected BSDEs can also be used to solve starting and stopping problems (two modes switching problems), see Hamadène and Jeanblanc (2007) and Hamadène and Zhang (2010).

14.3 Constrained Backward Stochastic Differential Equations

In the case of a reflected BSDE the process Y is constrained to stay above a barrier. Now, we introduce constraints on all components of the solution to a BSDE.

A constrained backward stochastic differential equation is an equation of the form

$$Y(t)=\xi+\int_t^T f\big(s,Y(s),Z(s),U(s,.)\big)ds-\int_t^T Z(s)dW(s)$$
$$-\int_t^T\int_{\mathbb{R}}U(s,z)\tilde{N}(ds,dz)+C(T)-C(t),\quad 0\leq t\leq T, \tag{14.30}$$

with the constraint

$$\psi\big(t,Y(t),Z(t),U(t,z)\big)\geq 0,\quad \text{a.s., }\vartheta\text{-a.e. }(\omega,t,z)\in\Omega\times[0,T]\times\mathbb{R}. \tag{14.31}$$

Given a terminal condition ξ, a generator f and a constraint function ψ, we are interested in finding a quadruple $(Y,Z,U,C)\in\mathbb{S}^2(\mathbb{R})\times\mathbb{H}^2(\mathbb{R})\times\mathbb{H}^2_N(\mathbb{R})\times\mathbb{S}^2_{inc}(\mathbb{R})$ which satisfies (14.30)–(14.31). A solution (Y,Z,U,C) is also called a supersolution to the constrained BSDE (14.30)–(14.31).

We assume that the random measure N is generated by a compound Poisson process with intensity λ and jump size distribution q. We find a solution to the constrained BSDE (14.30)–(14.31) by a penalization argument. We introduce the sequence of BSDEs

$$\begin{aligned} Y^n(t) = \xi &+ \int_t^T \Big(f\big(s, Y^n(s), Z^n(s), U^n(s,.)\big) \\ &+ n\int_{\mathbb{R}} \psi^-\big(s, Y^n(s), Z^n(s), U^n(s,z)\big)\lambda q(dz)\Big)ds \\ &- \int_t^T Z^n(s)dW(s) - \int_t^T\int_E U^n(s,z)\tilde{N}(ds,dz), \quad 0 \le t \le T, \end{aligned} \tag{14.32}$$

where ψ^- denotes the negative part of ψ. In the sequel we denote

$$C^n(t) = n\int_0^t\int_{\mathbb{R}} \psi^-\big(s, Y^n(s), Z^n(s), U^n(s,z)\big)\lambda q(dz)ds, \quad 0 \le t \le T.$$

We show that the constrained BSDE (14.30)–(14.31) has a minimal solution which can be approximated by the sequence of unique solutions to the penalized BSDEs (14.32). The negative part of the constraint function ψ^- is interpreted as a penalty. We point out that penalization arguments are often applied in the theory of BSDEs. For example, the solution to a reflected BSDE can be found by a penalization argument, see Hamadène and Ouknine (2003).

Theorem 14.3.1 *Let us consider the filtration $\mathscr{F}$ generated by a Brownian motion and a compound Poisson process with intensity λ and jump size distribution q. We investigate the constrained BSDE* (14.30)–(14.31) *and the BSDE* (14.32). *We assume that*

(i) *the generator $f : \Omega \times [0,T] \times \mathbb{R} \times \mathbb{R} \times L_q^2(\mathbb{R}) \to \mathbb{R}$ is $\mathscr{F}$-predictable and satisfies*

$$\big|f(t,y,z,u) - f\big(t,y',z',u\big)\big| \le K\big(|y-y'| + |z-z'|\big),$$

$$f(t,y,z,u) - f\big(t,y,z,u'\big) \le \int_{\mathbb{R}} \delta^{y,z,u,u'}(t,x)\big(u(x) - u'(x)\big)\lambda q(dx),$$

a.s., a.e. $(\omega,t) \in \Omega \times [0,T]$, for all $(y,z,u), (y',z',u), (y,z,u') \in \mathbb{R} \times \mathbb{R} \times L_q^2(\mathbb{R})$, where $\delta^{y,z,u,u'} : \Omega \times [0,T] \times \mathbb{R} \to (-1,\infty)$ is an $\mathscr{F}$-predictable process such that the mapping $t \mapsto \int_{\mathbb{R}} |\delta^{y,z,u,u'}(t,x)|^2 q(dx)$ is uniformly bounded in (y,z,u,u'),

(ii) *the constraint function $\psi : \Omega \times [0,T] \times \mathbb{R} \times \mathbb{R} \times \mathbb{R} \to \mathbb{R}$ is $\mathscr{F}$-predictable, Lipschitz continuous in the sense that*

$$\big|\psi(t,y,z,v) - \psi\big(t,y',z',v'\big)\big| \le K\big(|y-y'| + |z-z'| + |v-v'|\big),$$

a.s., a.e. $(\omega,t) \in \Omega \times [0,T]$, for all $(y,z,v), (y',z',v') \in \mathbb{R} \times \mathbb{R} \times \mathbb{R}$, and the mapping $v \mapsto \psi(t,y,z,v)$ is non-increasing for all $(t,y,z) \in [0,T] \times \mathbb{R} \times \mathbb{R}$,

(iii) $\mathbb{E}[|\xi|^2] < \infty$, $\mathbb{E}[\int_0^T |f(t,0,0,0)|^2 dt] < \infty$ *and* $\mathbb{E}[\int_0^T |\psi(t,0,0,0)|^2 dt] < \infty$,
(iv) *there exists a solution* $(\hat{Y}, \hat{Z}, \hat{U}, \hat{C}) \in \mathbb{S}^2(\mathbb{R}) \times \mathbb{H}^2(\mathbb{R}) \times \mathbb{H}^2_N(\mathbb{R}) \times \mathbb{S}^2_{inc}(\mathbb{R})$ *to the constrained BSDE* (14.30)–(14.31).

The following results hold:

(a) *For each* $n \in \mathbb{N}$ *there exists a unique solution* $(Y^n, Z^n, U^n) \in \mathbb{S}^2(\mathbb{R}) \times \mathbb{H}^2(\mathbb{R}) \times \mathbb{H}^2_N(\mathbb{R})$ *to the BSDE* (14.32).
(b) *There exists a unique minimal solution* $(Y^*, Z^*, U^*, C^*) \in \mathbb{S}^2(\mathbb{R}) \times \mathbb{H}^2(\mathbb{R}) \times \mathbb{H}^2_N(\mathbb{R}) \times \mathbb{S}^2_{inc}(\mathbb{R})$ *to the constrained BSDE* (14.30)–(14.31) *such that* C^* *is* $\mathscr{F}$*-predictable. Moreover, we have the strong convergence*

$$\lim_{n\to\infty}\left\{\mathbb{E}\left[\int_0^T \left|Y^n(t) - Y^*(t)\right|^2 dt\right] + \mathbb{E}\left[\int_0^T \left|Z^n(t) - Z^*(t)\right|^p dt\right]\right.$$
$$\left. + \mathbb{E}\left[\int_0^T \int_{\mathbb{R}} \left|U^n(t,z) - U^*(t,z)\right|^p \lambda q(dz) dt\right]\right\} = 0, \quad 1 \le p < 2,$$

and C^* *is the weak limit of* C^n *in* $\mathbb{H}^2(\mathbb{R})$.

Proof (a) It is easy to notice that the generator of the BSDE (14.32) is Lipschitz continuous in the sense of (A2) from Sect. 3.1. Hence, by Theorem 3.1.1 there exists a unique solution (Y^n, Z^n, U^n, C^n) to (14.32). Moreover, the generator of the BSDE (14.32) satisfies the property

$$\begin{aligned} f(t,y,z,u) &+ n\int_{\mathbb{R}} \psi^-\big(t,y,z,u(x)\big)\lambda q(dx) \\ &- f\big(t,y,z,u'\big) - n\int_{\mathbb{R}} \psi^-\big(t,y,z,u'(x)\big)\lambda q(dx) \\ &\le \int_{\mathbb{R}} \big(\delta^{y,z,u,u'}(t,x) + Kn\mathbf{1}_{\{u(x)\ge u'(x)\}}\big)\big(u(x) - u'(x)\big)\lambda q(dx), \end{aligned}$$

a.s., a.e. $(\omega,t) \in \Omega \times [0,T]$, for all $(y,z,u), (y,z,u') \in \mathbb{R} \times \mathbb{R} \times L^2_q(\mathbb{R})$. Consequently, the assumptions of Theorem 3.2.2 are satisfied and a comparison principle can be applied.

(b) *1. The convergence of* $(Y^n)_{n\in\mathbb{N}}$. The comparison principle yields $Y^n(t) \le Y^{n+1}(t)$, $0 \le t \le T$, for any $n \in \mathbb{N}$. Since $\psi^-(t, \hat{Y}(t), \hat{Z}(t), \hat{U}(t,z)) = 0$ for ϑ-a.e. $(t,z) \in [0,T] \times \mathbb{R}$, we can write

$$\begin{aligned} \hat{Y}(t) = \xi + \int_t^T \Big(& f\big(s, \hat{Y}(s), \hat{Z}(s), \hat{U}(s,.)\big) ds \\ & + n\int_{\mathbb{R}} \psi^-\big(s, \hat{Y}(s), \hat{Z}(s), \hat{U}(s,z)\big)\lambda q(dz)\Big) ds \\ & - \int_t^T \hat{Z}(s) dW(s) - \int_t^T \int_{\mathbb{R}} \hat{U}(s,z)\tilde{N}(ds,dz) + \hat{C}(T) - \hat{C}(t), \quad 0 \le t \le T. \end{aligned}$$

Following the arguments that led to (3.29), we obtain

$$\begin{aligned}
\hat{Y}(t) - Y^n(t) = &-\int_t^T \bar{Z}(s)e^{\int_t^s \Delta_y f(u)du} dW(s) + \int_t^T \bar{Z}(s)e^{\int_t^s \Delta_y f(u)du} \Delta_z f(s)ds \\
&- \int_t^T \int_{\mathbb{R}} \bar{U}(s,z)e^{\int_t^s \Delta_y f(u)du} \tilde{N}(ds,dz) \\
&+ \int_t^T \Big(f\big(s, Y^n(s), Z^n(s), \hat{U}(s)\big) \\
&+ n\int_{\mathbb{R}} \psi^-\big(s, Y^n(s), Z^n(s), \hat{U}(s,z)\big)\lambda q(dz) \\
&- f\big(s, Y^n(s), Z^n(s), U^n(s)\big) \\
&- n\int_{\mathbb{R}} \psi^-\big(s, Y^n(s), Z^n(s), U^n(s,z)\big)\lambda q(dz)\Big) e^{\int_t^s \Delta_y f(u)du} ds \\
&+ \int_t^T e^{\int_t^s \Delta_y f(u)du} d\hat{C}(s), \quad 0 \le t \le T,
\end{aligned}$$

and we conclude that $Y^n(t) \le \hat{Y}(t)$, $0 \le t \le T$, for any $n \in \mathbb{N}$. Since the sequence $(Y^n)_{n\in\mathbb{N}}$ is monotone, lower-bounded by $Y^0 \in \mathbb{S}^2(\mathbb{R})$ and upper-bounded by $\hat{Y} \in \mathbb{S}^2(\mathbb{R})$, the sequence $(Y^n)_{n\in\mathbb{N}}$ has a square integrable limit Y^*. Moreover, the dominated convergence theorem yields the strong convergence $\mathbb{E}[\int_0^T |Y^n(t) - Y^*(t)|^2 dt] \to 0, n \to \infty$.

2. The convergence of $(Z^n, U^n, C^n)_{n\in\mathbb{N}}$. From (14.32) we deduce

$$\begin{aligned}
\mathbb{E}\big[|C^n(T)|^2\big] &\le K\Big(\mathbb{E}[|\xi|^2] + \mathbb{E}\Big[\int_0^T |f(s,0,0,0)|^2 ds\Big] + \mathbb{E}\Big[\sup_{0\le s\le T}|Y^n(s)|^2\Big] \\
&\quad + \mathbb{E}\Big[\int_0^T |Z^n(s)|^2 ds\Big] + \mathbb{E}\Big[\int_0^T\int_{\mathbb{R}} |U^n(s,z)|^2 \lambda q(dz)ds\Big]\Big) \\
&\le K\Big(\mathbb{E}[|\xi|^2] + \mathbb{E}\Big[\int_0^T |f(s,0,0,0)|^2 ds\Big] \\
&\quad + \mathbb{E}\Big[\sup_{0\le s\le T}|\hat{Y}(s)|^2\Big] + \mathbb{E}\Big[\sup_{0\le s\le T}|Y^0(s)|^2\Big] \\
&\quad + \mathbb{E}\Big[\int_0^T |Z^n(s)|^2 ds\Big] + \mathbb{E}\Big[\int_0^T\int_{\mathbb{R}} |U^n(s,z)|^2 \lambda q(dz)ds\Big]\Big) \\
&\le K\Big(1 + \mathbb{E}\Big[\int_0^T |Z^n(s)|^2 ds\Big] + \mathbb{E}\Big[\int_0^T\int_{\mathbb{R}} |U^n(s,z)|^2 \lambda q(dz)ds\Big]\Big),
\end{aligned} \tag{14.33}$$

where we use the growth condition for f, the bound for Y^n and Theorem 2.3.3. From the dynamics of $|Y^n(t)|^2$, see (3.8), we also obtain

$$\begin{aligned}
&|Y^n(0)|^2 + \mathbb{E}\Big[\int_0^T |Z^n(s)|^2 ds\Big] + \mathbb{E}\Big[\int_0^T \int_{\mathbb{R}} |U^n(s,z)|^2 \lambda q(dz) ds\Big] \\
&= \mathbb{E}[|\xi|^2] + 2\mathbb{E}\Big[\int_0^T Y^n(s) f\big(s, Y^n(s), Z^n(s), U^n(s)\big) ds\Big] \\
&\quad + 2\mathbb{E}\Big[\int_0^T Y^n(s) dC^n(s)\Big] \\
&\le K\Big(\mathbb{E}[|\xi|^2] + \frac{1}{\alpha}\mathbb{E}\Big[\int_0^T |f(s,0,0,0)|^2 ds\Big] + \Big(\alpha + \frac{1}{\alpha} + \beta\Big)\Big(\mathbb{E}\Big[\sup_{0\le s\le T} |\hat{Y}(s)|^2\Big] \\
&\quad + \mathbb{E}\Big[\sup_{0\le s\le T} |Y^0(s)|^2\Big]\Big) + \frac{1}{\alpha}\mathbb{E}\Big[\int_0^T |Z^n(s)|^2 ds\Big] \\
&\quad + \frac{1}{\alpha}\mathbb{E}\Big[\int_0^T \int_{\mathbb{R}} |U^n(s,z)|^2 \lambda q(dz) ds\Big] + \frac{1}{\beta}\mathbb{E}\big[|C^n(T)|^2\big]\Big),
\end{aligned} \tag{14.34}$$

where we use the growth condition for f, inequality (3.10) and the bound for Y^n. Choosing α and β sufficiently large and combining (14.33) with (14.34), we can derive the uniform bound

$$\mathbb{E}\Big[\int_0^T |Z^n(s)|^2 ds\Big] + \mathbb{E}\Big[\int_0^T \int_{\mathbb{R}} |U^n(s,z)|^2 \lambda q(dz) ds\Big] + \mathbb{E}\big[|C^n(T)|^2\big] \le K. \tag{14.35}$$

We can conclude that there exists a subsequence of $(Z^n, U^n, C^n)_{n\in\mathbb{N}}$ which converges weakly to (Z^*, U^*, C^*) in $\mathbb{H}^2(\mathbb{R}) \times \mathbb{H}^2_N(\mathbb{R}) \times \mathbb{H}^2(\mathbb{R})$.

3. The existence of a minimal solution to (14.30)–(14.31). One can show that $(Y^*, Z^*, U^*, C^*) \in \mathbb{S}^2(\mathbb{R}) \times \mathbb{H}^2(\mathbb{R}) \times \mathbb{H}^2_N(\mathbb{R}) \times \mathbb{S}^2_{inc}(\mathbb{R})$, the convergence results from item b) hold and (Y^*, Z^*, U^*, C^*) satisfies the equation

$$\begin{aligned}
Y^*(t) = \xi &+ \int_t^T f\big(s, Y^*(s), Z^*(s), U^*(s,.)\big) ds - \int_t^T Z^*(s) dW(s) \\
&- \int_t^T \int_{\mathbb{R}} U^*(s,z) \tilde{N}(ds,dz) + C^*(T) - C^*(t), \quad 0 \le t \le T.
\end{aligned}$$

Moreover, the process C^* is predictable. We omit the details and we refer to Lemma 3.5 in Kharroubi et al. (2010) and Proposition 4.2 in Bouchard and Elie (2008). We now prove that the triple (Y^*, Z^*, U^*) satisfies the constraint (14.31) and that Y^* is a minimal solution to (14.30)–(14.31), i.e. $Y^*(t) \le Y(t)$, $0 \le t \le T$, for any other solution $(Y, Z, U, C) \in \mathbb{S}^2(\mathbb{R}) \times \mathbb{H}^2(\mathbb{R}) \times \mathbb{H}^2_N(\mathbb{R}) \times \mathbb{S}^2_{inc}(\mathbb{R})$ to (14.30)–(14.31). By the Lipschitz property of ψ^- we have

$$\begin{aligned}
&\Big|\mathbb{E}\Big[\int_0^T \int_{\mathbb{R}} \psi^-\big(t, Y^n(t), Z^n(t), U^n(t,z)\big) \lambda q(dz) dt\Big] \\
&\quad - \mathbb{E}\Big[\int_0^T \int_{\mathbb{R}} \psi^-\big(t, Y^*(t), Z^*(t), U^*(t,z)\big) \lambda q(dz) dt\Big]\Big|
\end{aligned}$$

$$\leq K\left(\mathbb{E}\left[\int_0^T \left|Y^n(t) - Y^*(t)\right|dt\right] + \mathbb{E}\left[\int_0^T \left|Z^n(t) - Z^*(t)\right|dt\right] \right.$$
$$\left. + \mathbb{E}\left[\int_0^T \int_{\mathbb{R}} \left|U^n(t,z) - U^*(t,z)\right|\lambda q(dz)dt\right]\right), \tag{14.36}$$

and from the convergence of $(Y^n, Z^n, U^n)_{n\in\mathbb{N}}$ we deduce that the left-hand side of (14.36) converges to zero. Hence, we get

$$\lim_{n\to\infty} \frac{\mathbb{E}[C^n(T)]}{n} = \lim_{n\to\infty} \mathbb{E}\left[\int_0^T \int_{\mathbb{R}} \psi^-\big(s, Y^n(s), Z^n(s), U^n(s,z)\big)\lambda q(dz)ds\right]$$
$$= \mathbb{E}\left[\int_0^T \int_{\mathbb{R}} \psi^-\big(s, Y^*(s), Z^*(s), U^*(s,z)\big)\lambda q(dz)ds\right]. \tag{14.37}$$

By (14.35) the left-hand side of (14.37) converges to zero. Hence, the constraint (14.31) is satisfied for (Y^*, Z^*, U^*). Since the sequence $(Y^n)_{n\in\mathbb{N}}$ is upper-bounded by the solution $\hat{Y}$, the limit Y^* is also bounded by $\hat{Y}$ and, consequently, Y^* is a minimal solution to (14.30)–(14.31) in the sense that $Y^*(t) \leq Y(t)$, $0 \leq t \leq T$, for any solution Y to (14.30)–(14.31).

4. The uniqueness of a minimal solution to (14.30)–(14.31). If Y is a minimal solution, then Y is unique by the definition of a minimal solution. Assume that we have two minimal solutions $(Y, Z, U, C), (Y, Z', U', C') \in \mathbb{S}^2(\mathbb{R}) \times \mathbb{H}^2(\mathbb{R}) \times \mathbb{H}^2_N(\mathbb{R}) \times \mathbb{S}^2_{inc}(\mathbb{R})$ to the BSDE (14.30) and C and C' are predictable. Then, from (14.30) we obtain

$$\int_0^t \int_{\mathbb{R}} \big(U(s,z) - U'(s,z)\big)N(ds,dz)$$
$$= \int_0^t \Big(f\big(s, Y(s), Z(s), U(s)\big) - f\big(s, Y(s), Z'(s), U'(s)\big)\Big)ds$$
$$- \int_0^t \big(Z(s) - Z'(s)\big)dW(s) + \int_0^t \int_{\mathbb{R}} \big(U(s,z) - U'(s,z)\big)\lambda q(dz)ds$$
$$+ C'(t) - C(t), \quad 0 \leq t \leq T. \tag{14.38}$$

Since the process on the right-hand side of (14.38) is predictable, the process on the left-hand side of (14.38) is predictable as well. Recalling Theorem 2.3.2, Definition 3.5.1 and the remark after that definition, we deduce that

$$\int_{\mathbb{R}} \big(U(t,z) - U'(t,z)\big)N\big(\{t\}, dz\big) = 0, \quad 0 \leq t \leq T, \tag{14.39}$$

and, consequently, we get

$$\int_0^t \big(Z(s) - Z'(s)\big)dW(s)$$
$$= \int_0^t \Big(f\big(s, Y(s), Z(s), U(s)\big) - f\big(s, Y(s), Z'(s), U'(s)\big)\Big)ds$$

$$-\int_0^t \int_{\mathbb{R}} \big(U(s,z) - U'(s,z)\big)\lambda q(dz)ds + C'(t) - C(t), \quad 0 \le t \le T. \tag{14.40}$$

From (14.39) we get that $U(t,z) = U'(t,z)$ a.s., ϑ-a.e. $(\omega,t,z) \in \Omega \times [0,T] \times \mathbb{R}$. From (14.40) we have $Z(t) = Z'(t)$ a.s., a.e. $(\omega,t) \in \Omega \times [0,T]$ since a finite variation, continuous martingale is constant. The uniqueness of C follows from (14.30). □

Assumptions under which there exists a solution $(\hat{Y},\hat{Z},\hat{U},\hat{C}) \in \mathbb{S}^2(\mathbb{R}) \times \mathbb{H}^2(\mathbb{R}) \times \mathbb{H}^2_N(\mathbb{R}) \times \mathbb{S}^2_{inc}(\mathbb{R})$ to the constrained BSDE (14.30)–(14.31) can be found in Kharroubi et al. (2010) and Elie and Kharroubi (2013). In the proof of Theorem 14.3.1 we applied the so-called monotonic limit theorem for BSDEs to establish the convergence of the sequence $(Y^n, Z^n, U^n, C^n)_{n\in\mathbb{N}}$, see Peng (1999), Kharroubi et al. (2010) and Elie and Kharroubi (2013).

We now investigate an investment problem which naturally leads to a constrained BSDE. The next example also shows why in finance we should be interested in finding a minimal solution to a BSDE.

Example 14.7 We consider the financial market (7.1)–(7.2). The goal is to find a replicating strategy and a replicating portfolio for a financial claim F. However, admissible replicating strategies $\pi \in \mathscr{A}$ are now constrained to satisfy $\psi(t,\pi(t)) \ge 0$, $0 \le t \le T$, e.g. short-selling of the stock is not allowed. We can deduce that the replicating portfolio X^π is given by the dynamics

$$dX^\pi(t) = \pi(t)\big(\mu(t)dt + \sigma(t)dW(t)\big) + \big(X^\pi(t) - \pi(t)\big)r(t)dt - dC(t).$$

The initial value of the replicating portfolio should be sufficiently high so that the investor can replicate the claim despite the constraint imposed on the strategy. As time passes, the capital invested in the portfolio may become too large, the constraint may not be binding and the investor can withdraw capital from the replicating portfolio. The process C represents the cumulative amount withdrawn from the replicating portfolio (i.e. the profit to the hedger). We can notice that the problem of replicating a financial claim F under a constraint ψ on the replicating strategy is equivalent to the problem of solving the constrained BSDE

$$\begin{aligned} Y(t) &= F + \int_t^T \big(-Y(s)r(s) - Z(s)\theta(s)\big)ds \\ &\quad - \int_t^T Z(s)dW(s) + C(T) - C(t), \quad 0 \le t \le T, \qquad (14.41) \\ \psi\big(t, Z(t)\big) &\ge 0, \quad \text{a.s., a.e. } (\omega,t) \in \Omega \times [0,T]. \end{aligned}$$

Moreover, we are interested in finding the smallest replicating portfolio for the claim (the smallest price process). Hence, we should aim to find a minimal solution (Y^*, Z^*, C^*) to the constrained BSDE (14.41). Such a solution exists by

Theorem 14.3.1. Investment problems under constraints on investment portfolios and investment strategies and constrained BSDEs are investigated in Cvitanić et al. (1998).

Let us now require that the control process U related to the jump component of a BSDE is non-positive. Such a constraint is of great importance for the theory and applications.

Proposition 14.3.1 *Let the random measure N be generated by a compound Poisson process. We assume that the compensator of the random measure is defined on $\mathscr{B}([0,T]) \times \mathscr{B}(E)$ where $E \subset \mathbb{R}$ is a compact set. We investigate the constrained FBSDE*

$$\begin{aligned}
\mathscr{X}(t) &= x + \int_0^t \mu\big(\mathscr{X}(s-)\big)ds + \int_0^t \sigma\big(\mathscr{X}(s-)\big)ds \\
&\quad + \int_0^t \int_E \gamma\big(\mathscr{X}(s-), z\big)N(ds, dz), \quad 0 \le t \le T, \\
Y(t) &= g\big(\mathscr{X}(T)\big) + \int_t^T f\big(s, \mathscr{X}(s-)\big)ds - \int_t^T Z(s)dW(s) \\
&\quad - \int_t^T \int_E \Big(U(s,z) - R\big(\mathscr{X}(s-), z\big)\Big)N(ds, dz) \\
&\quad + C(T) - C(t), \quad 0 \le t \le T, \\
U(t,z) &\le 0, \quad a.s., \vartheta\text{-a.e. } (\omega, t, z) \in \Omega \times [0,T] \times \mathbb{R}.
\end{aligned} \tag{14.42}$$

There exists a unique minimal solution $(Y^*, Z^*, U^*, C^*) \in \mathbb{S}^2(\mathbb{R}) \times \mathbb{H}^2(\mathbb{R}) \times \mathbb{H}^2_N(\mathbb{R}) \times \mathbb{S}^2_{inc}(\mathbb{R})$ to the constrained FBSDE (14.42). Moreover, the process Y^* has the representation

$$\begin{aligned}
Y^*(t) &= \sup_{(\tau_i, \chi_i) \in (t,T] \times E} \mathbb{E}\Big[g\big(\mathscr{X}^{t,x}(T)\big) + \int_t^T f\big(s, \mathscr{X}^{t,x}(s-)\big)ds \\
&\quad + \sum_{t < \tau_i \le T} R\big(\mathscr{X}^{t,x}(\tau_i-), \chi_i\big)\Big], \quad 0 \le t \le T, \\
\mathscr{X}^{t,x}(s) &= x + \int_t^s \mu\big(\mathscr{X}^{t,x}(r-)\big)dr + \int_t^s \sigma\big(\mathscr{X}^{t,x}(r-)\big)dW(r) \\
&\quad + \sum_{t < \tau_i \le s} \gamma\big(\mathscr{X}^{t,x}(\tau_i-), \chi_i\big), \quad t \le s \le T.
\end{aligned} \tag{14.43}$$

For detailed formulation and the proof we refer to Kharroubi et al. (2010).

By Proposition 14.3.1 the minimal solution Y^* to the constrained FBSDE (14.42) coincides with the optimal value function of the impulse control problem (14.43).

Since impulse control processes are common in insurance and finance, constrained BSDEs are also useful for solving actuarial and financial optimization problems.

Example 14.8 Let $\mathscr{X}$ denote a wealth process of a financial institution. The shareholders are allowed to pay out dividends $(\chi_i)_{i\geq 1}$. However, the dividend payment χ yields cost $\alpha+\beta\chi$. The goal is to find an optimal dividend plan. Let g and R denote two utility functions. We are interested in solving the following optimization problem

$$\begin{aligned}
Y^*(t) = & \sup_{(\tau_i,\chi_i)\in(t,T]\times E} \mathbb{E}\Big[e^{-\delta(T-t)}g\big(\mathscr{X}^{t,x}(T)\big) \\
& + \sum_{t<\tau_i\leq T} e^{-\delta(\tau_i-t)}R(\chi_i)\Big], \quad 0\leq t\leq T, \\
\mathscr{X}^{t,x}(s) = & x+\int_t^s \mu\big(\mathscr{X}^{t,x}(r-)\big)dr+\int_t^s \sigma\big(\mathscr{X}^{t,x}(r-)\big)dW(r) \\
& - \sum_{t<\tau_i\leq s}(\alpha+\beta\chi_i), \quad t\leq s\leq T.
\end{aligned}$$

By Proposition 14.3.1 the optimal value function Y^* can be characterized as a minimal solution to a constrained BSDE.

Example 14.9 Let us consider a swing option. The holder of a swing option has the right to sell, whenever he wants over a time period $[0,T]$, an underlying asset S against a fixed strike K. The holder can exercise the right at most n times and the interval between two consecutive exercise date must be at least δ. The price of the swing option can be characterized as a minimal solution to a constrained BSDE, see Bernhart et al. (2010).

Finally, it is worth pointing out the connection in the spirit of a non-linear Feynman-Kac formula between the solution to a constrained FBSDE and the solution to quasi-variational inequalities. It is well known that in a Markovian setting the optimal value function of the impulse control problem (14.43) can be characterized as a unique solution to quasi-variational inequalities, see Theorem 6.2 in Øksendal and Sulem (2004). It turns out that the minimal solution to a constrained FBSDE provides a probabilistic representation of the unique viscosity solution to quasi-variational inequalities, see Kharroubi et al. (2010) for details.

Bibliographical Notes Time-delayed BSDEs were introduced in Delong and Imkeller (2010a) and Delong and Imkeller (2010b). The proof of Proposition 14.1.1 is a slight modification of the proof from Delong (2012c) and it uses the arguments developed by Cvitanić and Karatzas (1994). Reflected BSDEs were introduced by El Karoui et al. (1997a). The proofs of Theorems 14.2.1 and 14.2.2 are taken from Hamadène and Ouknine (2003). For reflected BSDEs with non-Lipschitz generators we refer to Kobylanski et al. (2002), Bahlali et al. (2002), Bahlali et al. (2005). Constrained BSDEs were introduced in Kharroubi et al. (2010) and Elie and Kharroubi (2013). The proof of Theorem 14.3.1 is taken from Elie and Kharroubi (2013). Numerical methods for reflected BSDEs have been well-studied, and we refer to Gobet and Lemor (2006) for the regression-based approach and to Peng and Xu (2011) for a random walk approximation. A numerical method for solving a constrained BSDE was suggested by Kharroubi et al. (2010). We point out that in the literature we also find second order BSDEs which were introduced by Cheridito et al. (2007) and developed by Soner et al. (2012), as well as reflected second order BSDEs which were introduced by Matoussi et al. (2013).

References

Aazizi, S.: Discrete time approximation of decoupled forward-backward SDE's driven by a pure jump Lévy process. Preprint (2011)

Adam, A.: Handbook of Asset and Liability Management. Wiley, New York (2007)

Ankirchner, S., Heine, G.: Cross hedging with stochastic correlation. Finance Stoch. **16**, 17–43 (2012)

Ankirchner, S., Imkeller, P.: Quadratic hedging of weather and catastrophe risk by using short term climate prediction. Preprint (2008)

Ankirchner, S., Imkeller, P., Reis, G.: Classical and variational differentiability of BSDEs with quadratic growth. Electron. J. Probab. **12**, 1418–1453 (2007)

Ankirchner, S., Blanchet-Scalliet, C., Eyraud-Loisel, A.: Credit risk premia and quadratic BSDEs with a single jump. Int. J. Theor. Appl. Finance **13**, 1103–1129 (2010)

Ankirchner, S., Imkeller, P., Reis, G.: Pricing and hedging derivatives based on nontradable underlyings. Math. Finance **20**, 289–312 (2010)

Applebaum, D.: Lévy Processes and Stochastic Calculus. Cambridge University Press, Cambridge (2004)

Arrow, K.J.: Uncertainty and the welfare of medical care. Am. Econ. Rev. **53**, 941–973 (1963)

Artzner, P., Delbaen, F., Eber, J.M., Heath, D.: Coherent measures of risk. Math. Finance **9**, 203–228 (1999)

Bahlali, K., Essaky, E.H., Oknine, Y.: Reflected backward stochastic differential equation with jumps and locally Lipschitz coefficient. Random Oper. Stoch. Equ. **10**, 335–350 (2002)

Bahlali, K., Essaky, E.H., Oknine, Y.: Reflected backward stochastic differential equation with locally monotone coefficient. Stoch. Anal. Appl. **22**, 939–970 (2005)

Barles, G., Buckdahn, R., Pardoux, E.: Backward stochastic differential equations and integral-partial differential equations. Stoch. Stoch. Rep. **60**, 57–83 (1997)

Barrieu, P., El Karoui, N.: Optimal derivatives design under dynamic risk measures. In: Yin, G., Zhang, Q. (eds.) Proceedings of an AMS-IMS-SIAN. Mathematics of Finance, pp. 13–26 (2004)

Barrieu, P., El Karoui, N.: Pricing, hedging and optimally designing derivatives via minimization of risk measures. In: Carmona, R. (ed.) Indifference Pricing-Theory and Applications, pp. 77–141. Princeton University Press, Princeton (2005)

Barrieu, P., El Karoui, N.: Monotone stability of quadratic semimartingales with applications to unbounded general quadratic BSDEs. Ann. Probab. **41**, 1831–1863 (2013)

Bayraktar, E., Young, V.: Hedging life insurance with pure endowments. Insur. Math. Econ. **40**, 435–444 (2007)

Bayraktar, E., Young, V.: Pricing options in incomplete equity markets via the instantaneous Sharpe ratio. Ann. Finance **4**, 399–429 (2008)

Ł. Delong, *Backward Stochastic Differential Equations with Jumps and Their Actuarial and Financial Applications*, EAA Series, DOI 10.1007/978-1-4471-5331-3,

Bayraktar, E., Milevsky, M., Promislow, S., Young, V.: Valuation of mortality risk via the instantaneous Sharpe ratio: applications to life annuities. J. Econ. Dyn. Control **33**, 676–691 (2009)

Becherer, D.: Bounded solutions to BSDE's with jumps for utility optimization and indifference hedging. Ann. Appl. Probab. **16**, 2027–2054 (2006)

Becherer, D.: From bounds on optimal growth towards a theory of good-deal hedging. In: Albecher, H., Runggaldier, W., Schachermayer, W. (eds.) Advanced Financial Modelling, pp. 27–52. de Gruyter, Berlin (2009)

Bender, C., Denk, R.: A forward scheme for backwards SDEs. Stoch. Process. Appl. **117**, 1793–1812 (2007)

Bender, C., Kohlmann, M.: Optimal superhedging under nonconvex constraints—a BSDE-approach. Int. J. Theor. Appl. Finance **11**, 1–18 (2008)

Bening, V.E., Korolev, V.Y.: Generalised Poisson Models and Their Applications in Insurance and Finance. VSP, Utrecht (2002)

Bernhart, M., Pham, H., Tankov, P., Warin, X.: Swing options valuation: a BSDE with constrained jumps approach. In: Carmona, R., Del Moral, P., Hu, P., Oudjane, N. (eds.) Numerical Methods in Finance, Chapter 12. Springer, Berlin (2010)

Bielecki, T., Jeanblanc, M., Rutkowski, R.: Hedging of defaultable claims. In: Carmona, R. (ed.) Indifference Pricing, Paris-Princeton Lectures on Mathematical Finance, pp. 1–132. Springer, Berlin (2004)

Bielecki, T., Jeanblanc, M., Rutkowski, R.: Pricing and trading credit default swaps in a hazard process model. Ann. Appl. Probab. **18**, 2495–2529 (2008)

Bielecki, T., Crépey, S., Jeanblanc, M., Rutkowski, R.: Defaultable game options in a hazard process model. J. Appl. Math. Stoch. Anal. **2009**, Article ID 695798 (2009)

Bismut, J.M.: Conjugate convex functions in optimal stochastic control. J. Math. Anal. Appl. **44**, 384–404 (1973)

Björk, T., Slinko, I.: Towards a general theory of good deal bounds. Rev. Finance **10**, 221–260 (2006)

Blake, D., De Waegenaere, A., MacMinn, R., Nijman, T.: Longevity risk and capital markets: the 2008–2009 update. Insur. Math. Econ. **46**, 135–138 (2010)

Blanchet-Scalliet, C., Jeanblanc, M.: Hazard rate for credit risk and hedging defaultable contingent claims. Finance Stoch. **8**, 145–159 (2004)

Blanchet-Scalliet, C., Eyraud-Loisel, A., Royer-Carenzi, M.: Hedging of defaultable claims using BSDE with uncertain time horizon. Preprint (2008)

Boekel, P., Delft, L., Hoshino, T., Ino, R., Reynolds, C., Verheugen, H.: Replicating portfolios. Milliman Research. http://publications.milliman.com/research/life-rr/pdfs/replicating-portfolios-rr.pdf (2009)

Borch, K.: Equilibrium in a reinsurance market. Econometrica **30**, 424–444 (1962)

Bouchard, B., Elie, R.: Discrete time approximation of decoupled forward backward SDE with jumps. Stoch. Process. Appl. **118**, 53–75 (2008)

Bouchard, B., Touzi, N.: Discrete time approximation and Monte-Carlo simulation of backward stochastic differential equations. Stoch. Process. Appl. **111**, 175–206 (2004)

Bouchard, B., Warin, X.: Monte Carlo valuation of American options—new algorithm to improve on existing methods. In: Carmona, R., Del Moral, P., Hu, P., Oudjane, N. (eds.) Numerical Methods in Finance, Chapter 7. Springer, Berlin (2010)

Bouchard, B., Ekeland, I., Touzi, N.: On the Malliavin approach to Monte Carlo approximation of conditional expectations. Finance Stoch. **8**, 45–71 (2004)

Brémaud, P.: Point Processes and Queues. Springer, New York (1981)

Briand, P., Delyon, B., Mémin, J.: On the robustness of backward stochastic differential equations. Stoch. Process. Appl. **97**, 229–253 (2002)

Briand, P., Delyon, B., Hu, Y., Pardoux, E., Stoica, L.: $\mathbb{L}^p$ solutions of backward stochastic differential equations. Stoch. Process. Appl. **108**, 109–129 (2003)

Buckdahn, R., Engelbert, H.J., Răşcanu, A.: On weak solutions of backward stochastic differential equations. Theory Probab. Appl. **49**, 70–108 (2004)

Bühlmann, H.: Mathematical Methods of Risk Theory. Springer, Berlin (1970)

Carmona, R.: Indifference Pricing: Theory and Applications. Princeton University Press, Princeton (2008)
Chen, Z., Epstein, L.: Ambiguity, risk and asset returns in continuous time. Econometrica **70**, 1403–1443 (2002)
Chen, Z., Kulperger, R.: Minimax pricing and Choquet expectations. Insur. Math. Econ. **38**, 518–528 (2006)
Chen, Z., Chen, T., Davison, M.: Choquet expectation and Peng's g-expectations. Ann. Probab. **33**, 1179–1199 (2005)
Chen, Z., Kulperger, R., Wei, G.: A comonotonic theorem for BSDE. Stoch. Process. Appl. **115**, 41–54 (2005)
Cheridito, P., Hu, Y.: Optimal consumption and investment in incomplete markets with general constraints. Stoch. Dyn. **11**, 283–301 (2011)
Cheridito, P., Soner, H.M., Touzi, N., Victoir, N.: Second order BSDE's and fully nonlinear PDE's. Commun. Pure Appl. Math. **60**, 1081–1110 (2007)
Choquet, G.: Theory of capacities. Ann. Inst. Fourier **5**, 131–195 (1953)
Cochrane, J., Saá-Requejo, J.: Beyond arbitrage: good-deal asset price bounds in incomplete markets. J. Polit. Econ. **1008**, 79–119 (2000)
Cohen, S.N.: Representing filtration consistent non-linear expectations as g-expectations in general probability spaces. Preprint (2011)
Cont, R., Tankov, P.: Financial Modelling with Jump Processes. Chapman and Hall/CRC Press, London (2004)
Coquet, F., Hu, Y., Mémin, J., Peng, S.: A general converse comparison theorem for backward stochastic differential equations. C. R. Acad. Sci., Sér. 1 Math. **333**, 577–581 (2001)
Coquet, F., Hu, Y., Mémin, J., Peng, S.: Filtration-consistent non-linear expectations and related g-expectations. Probab. Theory Relat. Fields **123**, 1–27 (2002)
Crépey, S.: About the pricing equation in finance. In: Carmona, R.A., Çinlar, E., Ekeland, I., Jouini, E., Scheinkman, J.A., Touzi, N. (eds.) Paris-Princeton Lectures in Mathematical Finance 2010, pp. 63–203. Springer, Berlin (2011)
Crépey, S., Matoussi, A.: Reflected and doubly reflected BSDEs with jumps: a priori estimates and comparison. Ann. Appl. Probab. **18**, 2041–2069 (2008)
Cvitanić, J., Karatzas, I.: On portfolio optimization under drawdown constraints. IMA Vol. Math. Appl. **65**, 35–46 (1994)
Cvitanić, J., Ma, J.: Hedging options for a large investor and forward-backward SDE's. Ann. Appl. Probab. **6**, 370–398 (1996)
Cvitanić, J., Zhang, J.: Contract Theory in Continuous-Time Models. Springer, Berlin (2012)
Cvitanić, J., Karatzas, I., Soner, M.: Backward stochastic differential equations with constraints on the gain-process. Ann. Probab. **26**, 1522–1551 (1998)
Dahl, M., Møller, T.: Valuation and hedging of life insurance liabilities with systematic mortality risk. Insur. Math. Econ. **39**, 193–217 (2006)
Dahl, M., Melchior, M., Møller, T.: On systematic mortality risk and risk-minimization with survival swaps. Scand. Actuar. J. **2**, 114–146 (2008)
Dassios, A., Jang, J.: Pricing of catastrophe reinsurance and derivatives using the Cox process with shot noise intensity. Finance Stoch. **7**, 73–95 (2003)
De Scheemaekere, X.: A converse comparison theorem for backward stochastic differential equations with jumps. Stat. Probab. Lett. **81**, 298–301 (2011)
Delbaen, F.: The structure of m-stable sets and in particular of the set of risk neutral measures. In: Émery, M., Yor, M. (eds.) Séminaire de Probabilités XXXIX, pp. 215–258. Springer, Berlin (2006)
Delbaen, F., Schachermayer, W.: A general version of the fundamental theorem of asset pricing. Math. Ann. **300**, 463–520 (1994)
Delbaen, F., Hu, Y., Richou, A.: On the uniqueness of solutions to quadratic BSDEs with convex generators and unbounded terminal conditions. Ann. Inst. Henri Poincaré B, Probab. Stat. **47**, 559–574 (2011)

Delong, Ł.: An optimal investment strategy for a stream of liabilities generated by a step process in a financial market driven by a Lévy process. Insur. Math. Econ. **47**, 278–293 (2010)

Delong, Ł.: No-good-deal, local mean-variance and ambiguity risk pricing and hedging for an insurance payment process. ASTIN Bull. **42**, 203–232 (2012a)

Delong, Ł.: Exponential utility maximization, indifference pricing and hedging for a payment process. Appl. Math. **39**, 211–229 (2012b)

Delong, Ł.: Applications of time-delayed backward stochastic differential equations to pricing, hedging and portfolio management. Appl. Math. **39**, 463–488 (2012c)

Delong, Ł., Imkeller, P.: Backward stochastic differential equations with time delayed generators—results and counterexamples. Ann. Appl. Probab. **20**, 1512–1536 (2010a)

Delong, Ł., Imkeller, P.: On Malliavin's differentiability of BSDEs with time-delayed generators driven by a Brownian motion and a Poisson random measure. Stoch. Process. Appl. **120**, 1748–1775 (2010b)

Denuit, M., Dhaene, J., Goovaerts, M., Kass, R.: Modern Actuarial Risk Theory. Kluwer Academic, Boston (2001)

Detemple, J., Rindisbacher, M.: Dynamic asset liability management with tolerance for limited shortfalls. Insur. Math. Econ. **43**, 281–294 (2008)

Di Nunno, G., Øksendal, B., Proske, F.: Malliavin Calculus for Lévy processes with Applications to Finance. Springer, Berlin (2009)

Dos Reis, G.: Some Advances on Quadratic BSDE: Theory–Numerics–Applications. LAP Lambert Academic Publishing, Saarbrücken (2011)

Douglas, J., Ma, J., Protter, P.: Numerical methods for forward-backward stochastic differential equations. Ann. Appl. Probab. **6**, 940–968 (1996)

Dudley, R.M.: Winer functionals as Itô integrals. Ann. Probab. **5**, 140–141 (1977)

Duffie, D., Epstein, L.: Stochastic differential utility. Econometrica **60**, 353–394 (1997)

Dupire, B.: Pricing and hedging with smiles. In: Dempster, M.A., Pliska, S.R. (eds.) Mathematics of Derivative Securities, pp. 103–111. Cambridge University Press, Cambridge (1997)

El Karoui, N., Hamadène, S.: BSDEs and risk-sensitive control, zero-sum and nonzero-sum game problems of stochastic functional differential equations. Stoch. Process. Appl. **107**, 145–169 (2003)

El Karoui, N., Mazliak, L.: Backward Stochastic Differential Equations. Longman, Harlow (1997)

El Karoui, N., Quenez, M.C.: Dynamic programming and pricing of contingent claims in an incomplete market. SIAM J. Control Optim. **33**, 29–66 (1995)

El Karoui, N., Ravanelli, C.: Cash sub-additive risk measures and interest rate ambiguity. Math. Finance **19**, 561–590 (2009)

El Karoui, N., Kapoudjan, C., Pardoux, E., Peng, S., Quenez, M.C.: Reflected solutions of backward SDEs and related obstacle problems for PDEs. Ann. Probab. **25**, 702–737 (1997a)

El Karoui, N., Peng, S., Quenez, M.C.: Backward stochastic differential equations in finance. Math. Finance **7**, 1–71 (1997b)

Elie, R., Kharroubi, I.: Adding constraints to BSDEs with jumps: an alternative to multidimensional reflection. ESAIM Probab. Stat. (2013, in press)

European Commission: Fifth quantitative impact study: call for advice and technical specifications. http://ec.europa.eu/internal_market/insurance/solvency/index_en.htm (2010)

Fan, S.J., Jiang, L.: Uniqueness result for the BSDE whose generator is monotonic in y and uniformly continuous in z. C. R. Acad. Sci., Ser. 1 Math. **348**, 89–92 (2010)

Filipovic, D.: Term-Structure Models. Springer, Berlin (2009)

Föllmer, H., Schied, A.: Convex measures of risk and trading constraints. Finance Stoch. **6**, 429–447 (2002)

Fouque, J.P., Papanicolaou, G., Sircar, K.R.: Derivatives in Financial Markets with Stochastic Volatility. Cambridge University Press, Cambridge (2000)

Gihman, I.I., Skorohod, A.V.: The Theory of Stochastic Processes III. Springer, Berlin (1979)

Gobet, E., Lemor, J.P.: Numerical simulation of BSDEs using empirical regression methods: theory and practice. Preprint (2006)

Gobet, E., Lemor, J.P., Warin, X.: A regression-based Monte Carlo method to solve backward stochastic differential equations. Ann. Appl. Probab. **15**, 2172–2202 (2005)

Gobet, E., Lemor, J.P., Warin, X.: Rate of convergence of an empirical regression method for solving generalized backward stochastic differential equations. Bernoulli **5**, 889–916 (2006)

Goovaerts, M.J., Kaas, R., Dhaene, J., Tang, Q.: A unified approach to generate risk measures. ASTIN Bull. **33**, 173–191 (2003)

Gundlach, M., Lehrbass, F.: CreditRisk+ in the Banking Industry. Springer, Berlin (2004)

Hamadène, S.: Mixed zero-sum differential game and American game options. SIAM J. Control Optim. **45**, 496–518 (2006)

Hamadène, S., Hassani, M.: BSDEs with two reflecting barriers driven by a Brownian motion and an independent Poisson noise and related Dynkin game. Electron. J. Probab. **11**, 121–145 (2006)

Hamadène, S., Jeanblanc, M.: On the starting and stopping problem: application in reversible investments. Math. Oper. Res. **32**, 182–192 (2007)

Hamadène, S., Lepeltier, J.P.: Zero-sum stochastic differential games and backward equations. Syst. Control Lett. **24**, 259–263 (1995)

Hamadène, S., Ouknine, Y.: Reflected backward stochastic differential equations with jumps and random obstacle. Electron. J. Probab. **8**, 1–20 (2003)

Hamadène, S., Ouknine, Y.: Reflected backward stochastic differential equations with general jumps. Preprint (2011)

Hamadène, S., Wang, H.: BSDEs with two reflecting obstacles driven by a Brownian motion Poisson measure and related mixed zero-sum games. Stoch. Process. Appl. **119**, 2881–2912 (2009)

Hamadène, S., Zhang, J.: Switching problem and related system of reflected backward SDEs. Stoch. Process. Appl. **120**, 403–426 (2010)

He, S., Wang, J., Yan, J.: Semimartingale Theory and Stochastic Calculus. CRC Press, Boca Raton (1992)

Hodges, S.D., Neuberger, A.: Optimal replication of contingent claims under transaction costs. Rev. Futures Mark. **8**, 222–239 (1989)

Hu, Y., Imkeller, P., Müller, M.: Utility maximization in incomplete markets. Ann. Appl. Probab. **15**, 1691–1712 (2005)

Imkeller, P., Dos Reis, G., Zhang, J.: Results on numerics for FBSDE with drivers of quadratic growth. In: Chiarella, C., Novikov, A. (eds.) Contemporary Quantitative Finance, pp. 159–182. Springer, Heidelberg (2010)

Jacod, J., Shiryaev, A.N.: Limit Theorems for Stochastic Processes. Springer, Berlin (2003)

Jeanblanc, M., Le Cam, Y.: Progressive enlargement of filtrations with initial times. Stoch. Process. Appl. **8**, 2523–2543 (2009)

Jeanblanc, M., Rutkowski, M.: Default risk and hazard process. In: Geman, H., Madan, D., Pliska, S.R., Vorst, T. (eds.) Mathematical Finance—Bachelier Congress 2002, pp. 281–313. Springer, Berlin (2000)

Jeanblanc, M., Mania, M., Santacroce, M., Schweizer, M.: Mean-variance hedging via stochastic control and BSDEs for general semimartingales. Ann. Appl. Probab. **22**, 2388–2428 (2012)

Jiang, L.: Convexity, translation invariance and subadditivity for g-expectations and related risk measures. Ann. Appl. Probab. **18**, 245–258 (2008)

Jiao, Y., Kharroubi, I., Pham, H.: Optimal investment under multiple default risk: a BSDE decomposition approach. Ann. Appl. Probab. **23**, 455–491 (2013)

Karatzas, I., Shreve, S.E.: Brownian Motion and Stochastic Calculus. Springer, New York (1988)

Kazamaki, N.: Continuous Exponential Martingales and BMO. Springer, Berlin (1994)

Kharroubi, I., Lim, T.: Progressive enlargement of filtrations and backward SDEs with jumps. J. Theor. Probab. (2012, in press)

Kharroubi, I., Ma, J., Pham, H., Zhang, J.: Backward SDEs with constrained jumps and quasi-variational inequalities. Ann. Probab. **38**, 794–840 (2010)

Kharroubi, I., Lim, T., Ngoupeyou, A.: Mean-variance hedging on uncertain time horizon in a market with a jump. Preprint (2012)

Kobylanski, M.: Backward stochastic differential equations and partial differential equations with quadratic growth. Ann. Probab. **28**, 558–602 (2000)

Kobylanski, M., Lepeltier, J.P., Quenez, M.C., Torres, S.: Reflected BSDE with superlinear quadratic coefficient. Probab. Math. Stat. **22**, 51–83 (2002)

Kohlmann, M., Tang, S.: Global adapted solutions of one-dimensional backward stochastic Riccati equations, with applications to the mean-variance hedging. Stoch. Process. Appl. **97**, 255–288 (2002)

Kohlmann, M., Tang, S.: Multidimensional backward stochastic Riccati equations and applications. SIAM J. Control Optim. **41**, 1696–1721 (2003)

Kohlmann, M., Xiong, D., Ye, Z.: Mean-variance hedging in a general jump model. Appl. Math. Finance **27**, 29–57 (2010)

Kunita, H.: Stochastic differential equations based on Lévy processes and stochastic flows of diffeomorphisms. In: Rao, M.M. (ed.) Real and Stochastic Analysis, pp. 305–375. Birkhäuser, Basel (2004)

Laeven, R.J.A., Stadje, M.: Robust portfolio choice and indifference valuation. Preprint (2012)

Lazrak, A., Quenez, M.C.: A generalized stochastic differential utility. Math. Oper. Res. **28**, 154–180 (2003)

Leitner, J.: Pricing and hedging with globally and instantaneously vanishing risk. Stat. Decis. **25**, 311–332 (2007)

Lejay, A., Mordecki, E., Torres, S.: Numerical approximations of backward stochastic differential equations with jumps. Preprint (2010)

Lepeltier, J.P., Martin, J.S.: Backward stochastic differential equations with continuous coefficients. Stat. Probab. Lett. **34**, 425–430 (1997)

Lepeltier, J.P., Martin, J.S.: Existence for BSDE with superlinear-quadratic coefficients. Stoch. Stoch. Rep. **63**, 227–240 (1998)

Liang, Z., Yuen, K.C., Guo, J.: Optimal proportional reinsurance and investment in a stock market with Ornstein–Uhlenbeck process. Insur. Math. Econ. **49**, 207–215 (2011)

Lim, A.: Quadratic hedging and mean-variance portfolio selection with random parameters in an incomplete market. Math. Oper. Res. **29**, 132–161 (2004)

Lim, A.: Mean-variance hedging when there are jumps. SIAM J. Control Optim. **44**, 1893–1922 (2005)

Lim, T., Quenez, M.C.: Exponential utility maximization in an incomplete market with defaults. Electron. J. Probab. **16**, 1434–1464 (2011)

Liu, Y., Ma, J.: Optimal reinsurance/investment problems for general insurance models. Ann. Appl. Probab. **19**, 1495–1528 (2009)

Lo, A.W.: The statistics of Sharpe ratios. Financ. Anal. J. **58**, 36–52 (2002)

Ludkovski, M., Young, V.R.: Indifference pricing of pure endowments and life annuities under stochastic hazard and interest rates. Insur. Math. Econ. **42**, 14–30 (2008)

Ma, J., Yong, J.: Solvability of forward-backward SDEs and the nodal set of Hamilton–Jacobi–Bellman equations. Chin. Ann. Math., Ser. B **16**, 279–298 (1995)

Ma, J., Yong, J.: Forward-Backward Stochastic Differential Equations and their Applications. Springer, Berlin (2000)

Ma, J., Protter, P., Yong, J.: Solving forward-backward stochastic differential equations explicitly—a four step scheme. Probab. Theory Relat. Fields **98**, 339–359 (1994)

Ma, J., Yong, J., Zhao, Y.: Four step scheme for general Markovian forward-backward SDEs. J. Syst. Sci. Complex. **23**, 546–571 (2010)

Ma, J., Wu, Z., Zhang, D., Zhang, J.: On wellposedness of forward-backward SDEs—a unified approach. Preprint (2011)

Mania, M., Schweizer, M.: Dynamic exponential utility indifference valuation. Ann. Appl. Probab. **15**, 2113–2143 (2005)

Markowitz, H.: Portfolio selection. J. Finance **7**, 77–91 (1952)

Matoussi, A., Possamaï, D., Zhou, C.: Reflected second-order backward stochastic differential equations. Ann. Appl. Probab. (2013, in press)

McNeil, A.J., Frey, R., Embrechts, P.: Quantitative Risk Management. Princeton University Press, Princeton (2005)
Merton, R.: Lifetime portfolio selection under uncertainty: the continuous time case. Rev. Econ. Stat. **1**, 247–257 (1969)
Mikosch, T.: Non-life Insurance Mathematics. Springer, Berlin (2009)
Milevsky, M., Salisbury, T.: The real option to lapse and the valuation of death-protected investments. Preprint (2002)
Møller, T.: Risk minimizing hedging strategies for insurance payment processes. Finance Stoch. **5**, 419–446 (2001)
Møller, T., Steffensen, M.: Market Valuation Methods in Life and Pension Insurance. Cambridge University Press, Cambridge (2007)
Morlais, M.: Utility maximization in a jump market model. Stochastics **81**, 1–27 (2009)
Nguyen, H., Pham, U., Tran, H.: On some claims related to Choquet integral risk measures. Ann. Oper. Res. **195**, 5–31 (2012)
Nualart, D.: The Malliavin Calculus and Related Topics. Springer, Berlin (1995)
Øksendal, B., Hu, Y.: Partial information linear quadratic control for jump diffusions. SIAM J. Control Optim. **47**, 1744–1761 (2008)
Øksendal, B., Sulem, A.: Applied Stochastic Control of Jump Diffusions. Springer, Berlin (2004)
Øksendal, B., Sulem, A.: Portfolio optimization under model uncertainty and BSDE games. Quant. Finance **11**, 1665–1674 (2011)
Øksendal, B., Sulem, A.: Forward-backward SDE games and stochastic control under model with uncertainty. J. Optim. Theory Appl. (2012, in print)
Pardoux, E., Peng, S.: Adapted solution of a backward stochastic differential equation. Syst. Control Lett. **14**, 55–61 (1990)
Pelsser, A.: Pricing in incomplete markets. Preprint (2011)
Peng, S.: Backward SDE and related g-expectations. In: El Karoui, N., Mazliak, L. (eds.) Backward Stochastic Differential Equations, Pitman Research Notes, pp. 141–161. Pitman, London (1997)
Peng, S.: Monotonic limit theorem of BSDE and non-linear decomposition of Doob-Meyer's type. Probab. Theory Relat. Fields **113**, 473–499 (1999)
Peng, S.: Modelling derivatives pricing mechanisms with their generating functions. Preprint (2006)
Peng, S., Xu, M.: Numerical algorithms for backward stochastic differential equations with 1-d Brownian motion: convergence and simulations. Math. Model. Anal. **45**, 335–360 (2011)
Perera, R.S.: Optimal consumption, investment and insurance with insurable risk for an investor in a Lévy market. Insur. Math. Econ. **46**, 479–484 (2010)
Petrou, E.: Malliavin calculus in Lévy spaces and applications to finance. Electron. J. Probab. **13**, 852–879 (2008)
Pham, H.: Continuous-Time Stochastic Control and Optimization with Financial Applications. Springer, Berlin (2009)
Protter, P.: Stochastic Integration and Differential Equations. Springer, Berlin (2004)
Quenez, M.C., Sulem, A.: BSDEs with jumps, optimization and applications to dynamic risk measures. Preprint (2012)
Rolski, T., Schmidili, H., Schmidt, V., Teugels, J.: Stochastic Processes for Insurance and Finance. Wiley, New York (1999)
Rong, S.: On solutions of backward stochastic differential equations with jumps and applications. Stoch. Process. Appl. **66**, 209–236 (1997)
Rong, S.: Theory of Stochastic Differential Equations with Jumps and Applications. Springer, Berlin (2005)
Rosazza Gianin, E.: Risk measures via g-expectations. Insur. Math. Econ. **39**, 19–34 (2006)
Royer, M.: Backward stochastic differential equations with jumps and related non-linear expectations. Stoch. Process. Appl. **116**, 1358–1376 (2006)
Russo, V., Giacometti, R., Ortobelli, S., Rachev, S., Fabozzi, F.: Calibrating affine stochastic mortality models using term assurance premiums. Insur. Math. Econ. **49**, 53–60 (2011)

Schied, A.: Optimal investments for robust utility functionals in complete market models. Math. Oper. Res. **30**, 750–764 (2005)
Schied, A.: Risk Measures and Robust Optimization Problems. Lecture Notes (2006). http://people.orie.cornell.edu/~schied/PueblaNotes8.pdf
Schmidli, H.: Stochastic Control in Insurance. Springer, Berlin (2007)
Schrager, D.: Affine stochastic mortality. Insur. Math. Econ. **38**, 81–97 (2006)
Schweizer, M.: Option hedging for semimartingales. Stoch. Process. Appl. **37**, 339–363 (1991)
Schweizer, M.: Local risk minimization for multidimensional assets and payment streams. Banach Cent. Publ. **83**, 213–229 (2008)
Schweizer, M.: Mean-variance hedging. In: Cont, R. (ed.) Encyclopedia of Quantitative Finance, pp. 1177–1181. Wiley, New York (2010)
Shreve, S.E.: Stochastic Calculus for Finance II: Continuous-Time Models. Springer, Berlin (2004)
Skiadas, C.: Robust control and recursive utility. Finance Stoch. **7**, 475–489 (2003)
Solé, J.L., Utzet, F., Vives, J.: Canonical Lévy process and Malliavin calculus. Stoch. Process. Appl. **117**, 165–187 (2007)
Soner, H.M., Touzi, N., Zhang, J.: Wellposedness of second order backward SDEs. Probab. Theory Relat. Fields **153**, 149–190 (2012)
Stadje, M.: Extending dynamic convex risk measures from discrete time to continuous time: a convergence approach. Insur. Math. Econ. **47**, 391–404 (2010)
Tevzadze, R.: Solvability of backward stochastic differential equations with quadratic growth. Stoch. Process. Appl. **118**, 503–515 (2008)
Vandaele, N., Vanmaele, M.: A locally risk-minimizing hedging strategies for unit-linked life insurance contracts in a Lévy process financial market. Insur. Math. Econ. **42**, 1128–1137 (2008)
Wang, S.: A class of distortion operators for pricing financial and insurance risks. J. Risk Insur. **1**, 15–36 (2000)
Wang, N.: Optimal investment for an insurer with exponential utility preference. Insur. Math. Econ. **40**, 77–84 (2007)
Wills, S., Sherris, M.: Securitization, structuring and pricing of longevity risk. Insur. Math. Econ. **46**, 173–185 (2010)
Xie, S.: Continuous-time mean-variance portfolio selection with liability and regime switching. Insur. Math. Econ. **45**, 148–155 (2009)
Xie, S., Li, Z., Wang, S.: Continuous-time portfolio selection with liability: mean-variance model and stochastic LQ approach. Insur. Math. Econ. **42**, 943–953 (2008)
Yin, J., Mao, X.: The adapted solution and comparison theorem for backward stochastic differential equations with Poisson jumps and applications. J. Math. Anal. Appl. **346**, 345–358 (2008)
Young, V.: Pricing life insurance under stochastic mortality via the instantaneous Sharpe ratio. Insur. Math. Econ. **42**, 691–703 (2008)
Young, V.R., Zariphopoulou, T.: Pricing dynamic risks using the principle of equivalent utility. Scand. Actuar. J. **4**, 246–279 (2002)
Zenios, S.A., Ziemba, W.T.: Handbook of Asset and Liability Management. North-Holland, Amsterdam (2006)
Zhang, J.: A numerical scheme for BSDEs. Ann. Appl. Probab. **14**, 459–488 (2004)

Index

Ł. Delong, *Backward Stochastic Differential Equations with Jumps and Their Actuarial and Financial Applications*, EAA Series, DOI 10.1007/978-1-4471-5331-3,